THE FIRST WAR OF PHYSICS

THE FIRST WAR
OF PHYSICS

THE SECRET HISTORY OF THE ATOM BOMB
1939-1949

JIM BAGGOTT

PEGASUS BOOKS
NEW YORK

THE FIRST WAR OF PHYSICS

Pegasus Books LLC
80 Broad Street
5th Floor
New York, NY 10004

First Pegasus Books cloth edition 2010

Library of Congress Cataloging-in-Publication Data is available.

ISBN: 978-1-60598-084-3

10 9 8 7 6 5 4 3 2 1

Printed in the United States of America
Distributed by W. W. Norton & Company, Inc.
www.pegasusbooks.us

For Jini,
My soulmate

CONTENTS

LIST OF ILLUSTRATIONS

About the author

Jim Baggott is an award-winning science writer. A former academic scientist, he now works as an independent business consultant but maintains a broad interest in science, philosophy and history, and continues to write on these subjects in his spare time. His previous books have been widely acclaimed and include *A Beginner's Guide to Reality* (Penguin, 2005), *Beyond Measure: Modern Physics, Philosophy and the Meaning of Quantum Theory* (Oxford University Press, 2004) and *Perfect Symmetry: The Accidental Discovery of Buckminsterfullerene* (Oxford University Press, 1994).

PREFACE

From the very moment of their conception, atomic weapons have been synonymous with fear. Fear that Hitler's Nazi Germany might be first to build an atomic bomb drove Anglo-American efforts during the Second World War. Fear that America might threaten a first nuclear strike drove the Soviet Union's own bomb programme during the Cold War that followed.

I was born in 1957, and grew up in the shadow of the fear inspired by Cold War rhetoric and the concept of mutual assured destruction. I was just five years old when, during the Cuban missile crisis in October 1962, the US Strategic Air Command carried aloft thermonuclear weapons with a total explosive potential of more than half a million Hiroshimas. As Air Force chief of staff Curtis LeMay urged President John F. Kennedy to hit the Soviet Union with everything in the US nuclear arsenal, the world held its breath.

Just how did this happen? How did this dreadful instrument of fear come to be created? Of course, the bomb was created by some of the world's greatest physicists, many of them Nobel laureates – physicists who, only a few years before, had been leading a series of revolutions in theoretical science and shaking the very foundations of our understanding of physical reality. But how did these men become such profoundly important *military* resources in a war that was to redefine the very meaning of barbarity, a war that was to recalibrate what it means to be inhuman? How did these other-worldly 'eggheads' find themselves centre-stage in such a drama of heroic endeavour, sabotage, espionage, counter-espionage, assassination and terrible destruction that it now seems barely credible as fiction? How did they come, in the words of J. Robert Oppenheimer, to know sin?

These were men such as Niels Bohr, Albert Einstein, Enrico Fermi, Richard Feynman, Otto Frisch, Klaus Fuchs, Werner Heisenberg, Yuli Khariton, Igor Kurchatov, Oppenheimer, Edward Teller, and many, many more. Diverted from their academic preoccupations by the biggest military conflict in human history, they became deeply embroiled in the biggest of human dramas. They found themselves drawn inexorably into a project to build the world's most awful weapon of war, a weapon judged to be 'practically irresistible' at a time when the world was threatened by the darkest evil.

Confronted with bare historical facts, our questions tumble out. The devastation that could be wrought with atomic weapons was obvious to the physicists right from the start, so why did they persist in developing these weapons, without hesitation? Why, despite having a clear lead in nuclear physics at the beginning of the Second World War, did German physicists fail to develop an atomic bomb? Did the Allies really plot to have Heisenberg kidnapped or assassinated? Why, when it became clear that there was no threat of a Nazi weapon, did the Allies use the atomic bomb, without warning, against Japan? To what extent did the Soviet atomic programme rely on intelligence gathered by spies such as Klaus Fuchs, Theodore Hall, David Greenglass and the Rosenbergs? Could the Soviets have developed the bomb without them? What was the full extent of Soviet penetration of the Manhattan Project?

Were the physicists merely instruments in a political game-plan to establish supremacy in a post-war world? Or did they knowingly inspire the arms race? What, if any, lessons can be drawn from this history to inform our perspective on nuclear energy and the proliferation of weapons technology today?

This book is my attempt to answer these and many other questions through a popular, accessible account of the race to build the first atomic bombs, centred on the individual stories of the physicists directly involved. The book spans ten historic years, beginning with the discovery of nuclear fission in early 1939 and closing shortly after 'Joe-1', the first Soviet atomic bomb test in August 1949.

Now, parts of this story have already been told, and told very well. But there are important strands of the story that have emerged only within the last decade or so, relating specifically to aspects of the German and Soviet atomic programmes and penetration of the Manhattan Project by Soviet spies. These new materials allow a single-volume popular history of the

Anglo-American, German and Soviet programmes to be assembled for the first time.

The book is organised in four parts. Part I covers the mobilisation of nuclear physicists around the world following the outbreak of war in September 1939 and early work on atom bomb and reactor physics. Part II recounts the early frustrations and progress in weapon design, the development of bomb and reactor materials in Germany, Britain and America, the spectacular sabotage of the heavy water plant at Vemork by Norwegian commandos, and the establishment of the Soviet espionage operation codenamed ENORMOZ.

In Part III the book addresses the direct involvement of Allied scientists in the hunt for their German counterparts in war-torn Europe following the D-Day landings, the successful Trinity test at Alamogordo in New Mexico, the bombing of Hiroshima and Nagasaki, and the reactions of the captured German scientists on hearing of the Allied success.

Finally, Part IV describes the origins of the Cold War, the acceleration of the Soviet atomic programme, proliferation of weapons technology, the Venona project, the unmasking of Soviet spies, and the first successful Soviet test in August 1949. The book concludes with an extended epilogue which attempts to tie up many of the loose ends, describing the American and Soviet H-bomb programmes and the Cuban missile crisis which brought the world to the very edge of disaster.

For me, this book represents the end of a long journey. I suppose I can trace its beginning all the way back to my first classes in quantum mechanics, as an undergraduate student in Manchester, England, during the cold, damp winter of 1975–76. I probably didn't fully realise it back then, but I was completely captivated. To anyone tutored in the language and the logic of classical physics, quantum mechanics is at once mathematically challenging, maddeningly bizarre and breathtakingly beautiful. I have spent a lifetime trying to understand it.

The physicists who forged this outrageous new theory left their fingerprints all over it, in the form of new laws, physical constants, principles and approximations. It is impossible to study quantum mechanics without tripping over their names. To study quantum mechanics, therefore, is to study the physicists who made it. Many of these same physicists also played crucial roles in the development of the world's first atomic bombs, and this juxtaposition has always fascinated me. To understand them is to understand the various roles they played in this development: the things that drove them, and the things that scared them.

I have drawn extensively on the published works of noted scholars and sources of historical documents that can be found online. I owe particular debts of gratitude to Kai Bird and Martin Sherwin, authors of *American Prometheus: the Triumph and Tragedy of J. Robert Oppenheimer*; Margaret Gowing, author of *Britain and Atomic Energy* and *Independence and Deterrence: Britain and Atomic Energy, 1945–1952*; David Holloway, author of *Stalin and the Bomb*; Richard Rhodes, author of *The Making of the Atomic Bomb* and *Dark Sun*; and Mark Walker, author of *German National Socialism and the Quest for Nuclear Power, 1939–1949* and *Nazi Science: Myth, Truth and the German Atomic Bomb*. I have happily climbed on the shoulders of their scholarship.

I owe a debt of thanks to Jeremy Bernstein, John Fricker, Martin Sherwin, Peter Tallack, Jon Turney and Mark Walker, who read and provided many comments on the first draft manuscript. I am, of course, more than happy to accept full responsibility for all the errors that remain. My thanks must also go to Simon Flynn, my editor at Icon, for being indulgent when the project ran on longer than anticipated, and for being ready to accept an 'epilogue' that surely does violence to the meaning of this term.

LETTER FROM BERLIN

Christmas 1938–September 1939

It was something of a family tradition. Every year Otto Frisch would celebrate Christmas in Berlin with his aunt, Lise Meitner. But not this year. Not this Christmas.

Frisch had left Germany five years earlier, in October 1933. He was a young, personable and inventive physicist at the University of Hamburg. He was also an Austrian Jew, and he had fallen victim to the Law for the Reestablishment of the Career Civil Service, introduced in April that year by the new National Socialist government, the first of 400 such decrees. It provided a legal basis on which the Nazis could forbid Jews from holding positions in the civil service, including academic positions in German universities.

Like most academic physicists, he had until that time paid little attention to politics. His only short-lived political affiliation had been to a student organisation in Vienna, for which he had served on the entertainments committee, arranging dances. When he had taken part in political discussions as a student he had found them stiffly formal, and rather ridiculous. He had dismissed Adolf Hitler's appointment as Chancellor of Germany earlier in 1933 with a shrug of the shoulders. He figured that Hitler could surely be no worse than his predecessors.

1

He was quickly proved wrong. Within a few short months came the stark realisation that he was about to lose his job. He had no alternative but to join the swelling ranks of displaced Jewish physicists, together accounting for a quarter of all the physicists in Germany, including many Nobel prize-winners. He had to rely on the small but close-knit international community to which he belonged, which now scrambled to find grants and positions for those displaced by Hitler's anti-Semitism. He had moved from Hamburg first to Birkbeck College in London before being whisked off a year later to Copenhagen by the celebrated Danish Nobel laureate Niels Bohr.

Meitner had received reassurances from Carl Bosch, a director of the chemicals giant I.G. Farben and a principal sponsor of the prestigious Kaiser Wilhelm Institute for Chemistry, where she worked. She remained stubbornly optimistic for the future, and continued to work in Berlin for another five years before she too succumbed to the inevitable. Frisch's 'short, dark and bossy' aunt was a pioneer of the study of radioactive substances. She had stoutly fought prejudice against women in science for most of her professional life.

Though rather shy and withdrawn, she had earned great respect. In 1919 she had become head of physics at the Kaiser Wilhelm Institute for Chemistry, where she worked with distinguished German chemist Otto Hahn, her collaborator of some 30 years. She had become a professor (a *Privatdozent*) at the University of Berlin in 1926, at the age of 48, the first female professor of physics in Germany. Albert Einstein had once called her a 'German Madame Curie'.

But Meitner was not German. Like Frisch, she was Austrian. She was the third of eight children born into a Jewish family in Vienna. Her Austrian nationality had for five years spared her from the worst excesses of Nazi persecution, although her position and status had gradually and inexorably been eroded.

When German forces marched into a welcoming Austria in the Anschluss of 12 March 1938, Meitner became a German Jew. That she considered herself a Protestant Christian who had withdrawn from the Jewish community and been baptised at the age of 30 held no sway with

German racial laws. Against the prejudices to which she was now fully exposed, there could be no victory, however small. The very next day she was denounced by a Nazi colleague and declared a danger to the Institute.

Hahn had tried to defend her, but lost his nerve. While he was not perceived by his colleagues as a Nazi or even a strong supporter of the National Socialists, he had never really questioned the legitimacy of the new regime. He had told a newspaper reporter in Toronto in 1933 that he believed that those whom the Nazis had jailed during their first months in power had been Communists who also just happened to be Jews. Like many middle-class Germans, Hahn was now learning that there was a significant price to be paid for his passive acceptance. He told Meitner on 20 March that she could no longer work at the Institute. She was stunned. Hahn had, in effect, thrown her out.

Meitner's position had grown more perilous by the day. Jewish academics had previously been allowed to leave Germany with their families and their possessions. But new laws threatened to close all escape routes. In May 1938 she had applied to join Bohr in Copenhagen, where her nephew was working, but the Danish embassy had declared her Austrian passport invalid and refused to grant her a visa. In June she was refused a German passport. Heinrich Himmler, Reichsführer-SS, had declared it undesirable that well-known Jewish scientists be allowed to leave the country. She was at considerable risk of becoming trapped in Nazi Germany.

Bohr appealed to the scientific community on her behalf. Weeks of anxious waiting went by. Eventually, on 13 July 1938, Meitner left Germany for neighbouring Holland. She was taken to the railway station by fellow Austrian Paul Rosbaud, editor of the German scientific journal *Die Naturwissenschaften* and no friend of the Nazis. Hahn had helped her to pack, and had given her a valuable diamond ring to provide for her in an emergency.

Meitner crossed the border into Holland without incident. From Groningen she made her way first to Copenhagen, then to Stockholm, where a job awaited her as a guest researcher in Swedish physicist Manne Siegbahn's research group. Professor Meitner from the prestigious University of Berlin and Kaiser Wilhelm Institute for Chemistry became an ordinary researcher, with a salary that barely covered her living expenses. That she was able to continue her collaboration with Hahn by letter at least offered some small consolation.

She met with Hahn briefly in November 1938 at Bohr's Institute for Theoretical Physics in Copenhagen, but it was not an auspicious meeting. Hahn arrived bearing news that Frisch's father – Meitner's brother-in-law – had been arrested in Vienna. She subsequently discovered that he had been taken to Dachau concentration camp in Bavaria.

As Christmas 1938 approached Frisch grew determined to honour what family traditions remained. Meitner was lonely and depressed, her poor relationship with Siegbahn preventing her from deriving any real pleasure from her work in Stockholm. She was in need of her faithful nephew. Frisch took a break from his work with Bohr in Copenhagen and joined his aunt in the small seaside village of Kungälv – King's River – near Gothenburg, where she had been invited to spend Christmas with some Swedish friends.

For Frisch, this was to prove 'the most momentous visit of my whole life'.

Secrets of the atom

The early decades of the twentieth century had witnessed a remarkable transformation in our understanding of the constitution of physical matter. The once indestructible and indivisible atoms of ancient Greek philosophy had given way to a new model of atoms with discrete internal structures. Atoms had become tiny positively-charged nuclei surrounded by mysterious negatively-charged electron 'wave-particles'.

Attention turned inevitably to the atomic nucleus. With the discovery of the neutron in 1932, a picture emerged of atomic nuclei made up of positively-charged protons and electrically neutral neutrons. In this model, the number of protons in the nucleus determines the nature of the chemical element. Different elements, such as hydrogen, oxygen, sulphur, iron, uranium, and so on, all have different numbers of protons in their atomic nuclei. Atoms containing nuclei with the same numbers of protons but different numbers of neutrons are called *isotopes*. They are chemically identical, and differ only in their relative atomic weight and stability.

New discoveries came thick and fast. It was found that by using powerful magnets and electric fields it was possible to accelerate charged particles such as protons to the kinds of high speeds, and hence high energies, required to smash nuclei apart. At the University of California in Berkeley, a district of San Francisco, Ernest Lawrence invented the *cyclotron*, a new type of particle accelerator, which was used to obtain evidence for artificially-induced nuclear reactions.[1]

But the discovery of the neutron had not only given physicists deeper insight into the structure of the nucleus, it had also provided them with another weapon with which to penetrate its secrets. As an electrically neutral sub-atomic particle, the neutron could be fired into a positively-charged nucleus without being diverted by the force of electrostatic repulsion.

Italian physicist Enrico Fermi and his research team in Rome began a systematic study of the effects of bombarding nuclei with neutrons, starting with the lightest-known elements and working their way through the entire periodic table. When in 1934 they fired neutrons at the heaviest known atomic nuclei – those of uranium – the Italian physicists presumed they had created even heavier elements that did not occur in nature, called *transuranic* elements. This discovery made headline news and was greeted as a triumph for Italian science.

It was a discovery that had caught Hahn's attention in Berlin, and he and Meitner had set about repeating Fermi's experiments and conducting their own, much more detailed, chemical investigations.

All the scientists involved in this work assumed that while neutron bombardment would transform elements, it would do so only in small,

---

[1] Particle accelerators had existed for some years prior to Lawrence's development of the cyclotron in 1929. However, these earlier machines were *linear* accelerators, accelerating particles along straight lines by passing them through a series of plates to which a carefully controlled alternating high voltage was applied. Lawrence's cyclotron was designed to accelerate particles along a *circular* path created by an electromagnet, promising greater efficiencies and higher particle energies.

incremental amounts. Absorbing a neutron was expected to yield products that differed from the original target nucleus by no more than a couple of protons or neutrons. In other words, the products were expected to be found no more than one or two places higher or lower in the periodic table of the elements.

Hahn and his assistant Fritz Strassman carefully repeated work on the neutron bombardment of uranium. The German chemists initially believed that they were producing the highly radioactive element radium, somewhat lighter than uranium. They could find no evidence for transuranic elements.

The most stable, common isotope of uranium contains 92 protons and 146 neutrons, giving a total of 238 altogether (written U-238). Radium nuclei contain 88 protons and can exist as a variety of radioactive isotopes with varying numbers of neutrons, the most common isotope containing 138 neutrons and 226 'nucleons' overall. Transforming uranium into radium – as Hahn and Strassman were suggesting – appeared to require too big a leap, moving the target nucleus four places down the periodic table. This was much bigger than the small, incremental changes that were expected from neutron bombardment.

Late in 1938 Hahn had written to his sorely missed collaborator about these results, but Meitner urged caution. What Hahn and Strassman had found was quite simply unprecedented, and could not be explained by current theories of atomic nuclei.

Mass into energy

Frisch emerged from his hotel room after his first night in Kungälv to find his aunt already at breakfast. It was Christmas Eve 1938. This was an opportunity to set aside his gloom about developments in Germany and fears for the safety of his father. He wanted to talk physics, and he planned to tell Meitner all about a new experiment he was working on. However, he found that she was completely preoccupied. She was clutching a further letter from Hahn, dated 19 December, which contained news of further results from Berlin that were, if anything, even more bizarre.

Hahn and Strassman had repeated their experiments and concluded that the atoms they had created were not those of radium at all. They were, in fact, barium atoms. The most common isotope of barium has just 56 protons and 82 neutrons, totalling 138. This was a remarkable conclusion. The target uranium nucleus had moved not one, two or even four places down the periodic table: it had moved down an astonishing 36 places. *Bombarding uranium with neutrons had caused the uranium nucleus to split virtually in half.*

'I don't believe it,' Frisch declared, 'there's some mistake.' But Meitner claimed that Hahn was just too good a scientist to have made a fundamental error.

Meitner had written back to Hahn a few days before, declaring that these results were 'startling', but going on to say: 'but in nuclear physics we have experienced so many surprises, that one cannot unconditionally say: it is impossible.'

Frisch and Meitner's animated discussion continued after breakfast. They set out from their hotel across the flood plain of the river, crossing the frozen river itself before entering open woods, Frisch on skis and Meitner on foot, all the time debating. How could a single neutron cause the uranium nucleus to fall apart so spectacularly? Of course, nobody really knew how a uranium nucleus would behave in such a reaction. All they could do was reach for analogies with other physical phenomena that were better understood, and hope for the best.

One of these analogies had been proposed ten years before by the Ukrainian-born physicist George Gamow and had been adapted by Bohr to describe nuclear reactions. Meitner now recalled it. In this model, the force binding an atomic nucleus together is imagined to act in much the same way as surface tension binds a drop of liquid. In this 'liquid-drop' model of the nucleus, a balance is maintained between the surface tension which holds it together and the force of repulsion between its positively-charged protons which threatens to tear it apart.

Both forces increase as the size of the nucleus increases, but the force of repulsion increases more rapidly, eventually overwhelming the surface tension when the number of protons reaches about 100.[2] Perhaps in a uranium nucleus, with its 92 protons, this balance is very delicate. Maybe adding a single neutron is enough to cause the nucleus to distort, elongating and forming a narrow waist before splitting to form two smaller 'drops'.

Frisch and Meitner sat on a tree trunk scrambling for pieces of paper on which to draw diagrams and scribble calculations. They quickly deduced that the sizeable positive charge of the uranium nucleus was indeed large enough to offset the surface tension. This suggested an image of 'a very wobbly, unstable drop, ready to divide itself at the slightest provocation, such as the impact of a single neutron'.

But the nuclear fragments created by such a split would each carry away a sizeable amount of energy. Meitner estimated this to be about 200 million electron volts.[3] The fragments would be propelled away from each other by the mutual repulsion of their positive charges. Energy had to be conserved in this process. This was an unquestionable, fundamental law of physics that could not be broken. If they could not account for this energy then their idea would be worthless.

So, where could this energy have come from? Meitner recalled her first meeting with Albert Einstein, in 1909. She had heard him lecture on his theory of relativity, and had watched intently as he had derived his famous formula, $E = mc^2$. The very idea that mass could be converted to energy had left a deep impression on her. She also remembered that the nuclear masses of the fragments created by splitting a uranium nucleus would not quite add up to the mass of the original nucleus. These masses differed by

---

[2] This is the reason why the periodic table does not go on forever. There is an essential limit on the size of an atomic nucleus imposed by the cumulative force of repulsion between its positively-charged constituents.

[3] An electron volt is the amount of energy a single negatively-charged electron gains when accelerated through a one-volt electric field. A 100W light bulb burns energy at the rate of about 600 billion billion electron volts per second. So, 200 million electron volts might sound like small beer, but remember this is energy released by a *single nucleus*. A kilogram of uranium contains billions upon billions of nuclei. In fact, if every nucleus in a kilogram of uranium released 200 million electron volts of energy, this would be equivalent to the energy released by about *22,000 tons* of TNT.

about one-fifth of the mass of a single proton, mass that had gone 'missing' in the nuclear reaction.

The sums checked out. It all fitted together. A neutron causes the uranium nucleus to split in two, converting a tiny amount of mass into energy along the way.[4]

Nuclear fission

Frisch arrived back at Bohr's institute in Copenhagen on 3 January 1939, and rushed to tell Bohr what he and Meitner had discovered. On hearing of their proposal Bohr immediately recognised its basic truth, declaring: 'Oh what idiots we have all been! Oh but this is wonderful. This is just as it must be.' He urged Frisch to publish what they had discovered as soon as possible and promised to keep the news to himself until Frisch and Meitner had established priority.

A new physical process demanded a new name. Frisch saw parallels between the contortions of the uranium nucleus, the 'wobbly, unstable drop', and biological cell division. On advice from a biologist, Frisch borrowed the term *fission* to describe the fragmentation of uranium nuclei in the paper he was hastily drafting with Meitner. Despite Bohr's reservations about the name, it stuck.

Bohr turned his attention to preparations for a trip to Princeton University in the United States. He was intending to continue his debate with Einstein on the interpretation of quantum theory, a debate that had begun in 1927 and which proved to be one of the most important scientific debates of the twentieth century, if not the entire history of science. At issue was the role of uncertainty and probability in the behaviour of fundamental sub-atomic particles, to which Einstein had stubbornly refused to yield

---

[4] To a certain extent, Frisch and Meitner's discovery had been anticipated four years earlier by German chemist Ida Noddack. She had made the suggestion that far from creating transuranic elements, Fermi and his colleagues had actually split the uranium nucleus into several much smaller atomic fragments. At the time nobody took Noddack's proposal seriously. Hahn himself dismissed the idea as absurd.

his insistence that 'God does not play dice'. At stake was the ability of the human mind to comprehend the very nature of physical reality itself.

Bohr was joined on the trip by his son Erik and Léon Rosenfeld, a young protégé whom Bohr would frequently use as a 'sounding board', bouncing ideas off him as a way to sharpen his own thinking. They left for Gothenburg on 7 January, where they embarked on the MS *Drottningholm*, bound for New York. But the subject of discussion in Bohr's stateroom – in which he had arranged to have a blackboard fitted – was not the interpretation of quantum theory, as he had originally intended. It was nuclear fission.

As Bohr crossed the Atlantic, Frisch was busy back in Copenhagen. Nuclear fission in uranium had been discovered through the careful identification of the chemical substances that had resulted from it. The scientists knew what substances they had started with, and they knew what they had finished with, and fission was proposed to account for the journey from start to finish. This was a bit like starting with the opening scenes of Shakespeare's *Hamlet*, finishing with a stage strewn with corpses, and hypothesising about what had happened in between.

Frisch's Czech colleague George Placzek was sceptical. If Frisch and Meitner were right, then surely the fission reaction would be expected to produce a tell-tale burst of energy that should be *physically* detectable. Frisch simply hadn't thought of this. Within a matter of a few days he had retreated to his laboratory, devised and carried out a simple experiment and had found what he was looking for. He had found the tell-tale signature of fission.

But there was yet more to be discovered. Lighter elements in the periodic table tend to have equal numbers of protons and neutrons in their nuclei. But as the number of protons in the nuclei increases, so does the force of repulsion between them. Heavier nuclei therefore require an excess of neutrons over protons to create enough 'surface tension' to remain stable. If the uranium nucleus was splitting up to form lighter elements, then perhaps these were being formed with more neutrons in their nuclei than they could comfortably accommodate.

It was Frisch's Danish colleague Christian Møller who suggested that if the newly-formed fission fragments spit out one or two additional neutrons in their turn (subsequently called *secondary neutrons*), perhaps these could go on to break up more uranium nuclei, releasing more energy and creating more neutrons, and so on and on. The result would be a cascade, a *chain reaction* that could liberate nuclear energy on a large scale. Control the chain reaction and you would have a nuclear 'reactor'.

An uncontrolled chain reaction would be a bomb of unprecedented destructive power.

Verification

The Bohrs and Rosenfeld disembarked at the Hudson River dock on 16 January 1939. They were met by a young Princeton University physicist called John Wheeler, who had worked with Bohr in Copenhagen in 1934 and 1935, and was looking forward to spending a few months with his former colleague.

Wheeler was joined at the dockside by Enrico Fermi and his wife Laura. On 10 December 1938 Enrico had collected a Nobel prize for his work on neutron bombardment, and the Fermis had 'got lost' on their way back from Stockholm. In truth, Fermi had sought to protect Laura from Mussolini's Fascist state, which had introduced its own anti-Semitic laws a few months previously.[5] He had accepted the offer of a professorship at Columbia University and had arrived in New York on 2 January.

Bohr, ever mindful of the importance of establishing priority for his colleagues when a significant discovery had been made, said nothing to either Wheeler or Fermi about nuclear fission. But as Bohr left to spend a day catching up with Fermi, Rosenfeld – unaware of Bohr's concerns – happily spilled the beans to Wheeler. The news spread quickly through

---

[5] The *Manifesto della Razza* had been published on 14 July 1938. It declared that the Italian population is of Aryan race and that Jews did not, therefore, belong. The first anti-Semitic laws were passed in September. Fermi and the children were Catholics, but Laura was Jewish.

the community of physicists in America, which by now included many European émigrés.

Hahn had contacted Paul Rosbaud about their experimental results on 22 December 1938, and Rosbaud had helped rush these into print. Hahn and Strassman's paper on the neutron bombardment of uranium was published in *Die Naturwissenschaften* on 6 January 1939. Although Meitner had been part of the team that had worked on these problems and had continued to collaborate from her exile in Sweden, it was now politically unacceptable for Hahn to name her as a co-author. Meitner, in her turn, had remained reticent about her and Frisch's interpretation of Hahn and Strassman's results.

Frisch and Meitner's paper on nuclear fission in uranium was published in the British scientific journal *Nature* on 11 February. Frisch's paper reporting the results of his simple experiment to verify fission was published in *Nature* a week later, on 18 February. In an effort to ensure that Frisch and Meitner were properly credited with the discovery, Bohr himself published a short paper on the subject in *Nature* on 25 February.

Thanks to Rosenfeld's indiscretion, and a subsequent official report on fission by Bohr at a conference at George Washington University, by the time these papers appeared, experiments repeating and verifying the results had already been carried out in America.

On the West Coast, 27-year-old physicist Luis Alvarez had found out about fission from an article buried in the *San Francisco Chronicle*. He immediately abandoned the barber's chair in which he was sitting, cutting off the barber mid-snip, and ran to Berkeley's Radiation Laboratory to spread the news.[6] His first encounter was with Philip Abelson, one of his own graduate students, who was within a week or so of making the discovery of fission for himself. Abelson quickly verified the results.

Alvarez also informed another young Berkeley professor, regarded by many as the American West Coast's *wunderkind* of theoretical physics.

---

[6] The Radiation Laboratory (or Rad Lab) had been established on the University of California campus at Berkeley by Lawrence in August 1931 specifically to study high-energy physics using the cyclotron.

The young professor 'instantly pronounced the reaction impossible and proceeded to prove mathematically to everyone in the room that someone must have made a mistake'. He became convinced within minutes of being shown the experimental evidence, however, and within a few days a crude design for an atomic bomb had appeared on the blackboard in his office. His name was J. Robert Oppenheimer.

Nuclear fission was now not only an established scientific fact, it was fast becoming a new scientific discipline.

Uranium-235

By March 1939 research groups in America and in France had shown that, on average, between two and four secondary neutrons are released in each fission of a uranium nucleus – more than adequate to support a self-sustaining chain reaction. Growing concern about the possibility of an atomic bomb was, however, quickly dispelled.[7]

Bohr hadn't lost any time. There was clearly much work to be done in the newly-emerging field of fission physics, and Princeton was just as good a place to do this work as Copenhagen. He asked Wheeler if he would like to collaborate, and together they started work on a more detailed theory of the fission process. They were aided by some new experimental results from an apparatus that had been hastily set up in the attic of Princeton's Palmer Laboratory. These results were initially quite puzzling.

The Princeton apparatus had been designed to discover how the rate of nuclear fission in uranium changes as the energy of the bombarding neutrons is varied.[8] It was found that the highest fission rates are obtained at the highest energies, with the rate falling as the energy of the neutrons

---

[7] Strictly speaking, as all the action takes place in the uranium nucleus, this should be referred to as a 'nuclear bomb', just as we refer to a 'nuclear reactor'. However, in this book I will happily stick with the commonly-accepted term 'atomic bomb', or A-bomb.

[8] In practice, physicists are concerned to measure something called the cross-section of the reaction, reported in units of square centimetres. The cross-section can be thought to represent the size of a hypothetical 'window' through which the reaction occurs. The larger the window, the more likely the reaction. The more likely the reaction, the faster it will occur. This simple picture gets a bit fuzzy, however, when we start to consider the wave

falls, largely as expected. But then it was found that at low neutron energies the rate of fission increases once more.

Placzek, having challenged Frisch in Copenhagen to go look for the tell-tale signature of fission, now found himself sitting at breakfast with Bohr and Rosenfeld in Princeton. 'What kind of crazy thing is this big [rate] for both fast and slow?' he exclaimed.

Bohr figured out the answer while walking from breakfast back to his office. The high rate of fission with slow neutrons must, he reasoned, be due to the rare isotope U-235, which makes up only a tiny proportion of naturally-occurring uranium. Bohr and Wheeler now worked out the details. There were two factors at play.

The balance between the force of repulsion of the protons in the uranium nucleus and the surface tension holding the nucleus together is much more delicate in U-235 than in U-238. The three extra neutrons in U-238 help to stabilise the nucleus, thereby increasing the barrier to fission. Faster, higher-energy neutrons are therefore required to get over the hurdle in U-238.

The second factor concerns the nature of the compound nucleus itself. Atomic nuclei exhibit a general preference for even numbers of protons and neutrons, behaviour that can be traced back to the quantum nature of their sub-atomic constituents. Adding an extra neutron to U-235 makes U-236, containing 92 protons and 144 neutrons, both even numbers. Adding an extra neutron to U-238 makes U-239, which contains an odd number of neutrons. This means that U-235 'accommodates' an additional neutron, and so reacts more readily with it, than does U-238.

These two factors combined are enough to account for the significant difference in behaviour of the two uranium isotopes. To fission a U-238 nucleus requires fast neutrons. The much more vulnerable U-235 nucleus can be split with slow neutrons. This meant that in a bomb consisting of a mixture of U-235 and U-238 which relied on slow-neutron fission of

---

nature of sub-atomic particles. So, to avoid complications, I will restrict discussion always to the *rate* (or speed) of the nuclear reaction.

U-235, the result would be a slow chain reaction. An atomic bomb based on a slow chain reaction would fizzle out long before it could explode.

The immediate prospect for a bomb diminished, though it did not disappear entirely. Of course, Bohr declared in discussions with his colleagues in April 1939, it *might* be possible to manufacture a bomb based on pure U-235. But this was a minor isotope, present in naturally-occurring uranium to only one part in 140, a miserly 0.7 per cent. Isotopes are chemically identical and so cannot be separated by chemical means. Physical separation would be required, relying on the tiny difference in the masses of the isotopes. Such physical separation on the scale required to build an atomic bomb – a scale presumed at this stage to be measured in tons – appeared profoundly impractical.

'Yes, it would be possible to make a bomb,' Bohr declared, 'but it would take the entire efforts of a nation to do it.'

Hungarian conspiracy

There was no denying the novelty of atomic energy and the unprecedented scale of energy release promised by nuclear fission. Under any circumstances, the discovery of fission would have provoked interest among not just scientists, but also governments, the military establishment and business enterprises. But these were not any circumstances. Only a few short months had passed since Frisch and Meitner's discovery, made as they scribbled their calculations sitting on a tree trunk in a wood near Kungälv, yet this was time enough to found a whole new physics. And this new physics was being elucidated just as preparations for war in Europe were taking shape.

French physicists Frédéric Joliot-Curie, Hans von Halban and Lew Kowarski[9] summarised the evidence for nuclear chain reactions in uranium in *Nature* on 19 March. Their subsequent report specifying the number of secondary neutrons released in each nuclear fission event appeared in the

---

[9] Hans von Halban was French, of Austrian-Jewish descent. Lew Kowarski was a naturalised Frenchman, of Russian-Polish descent.

same journal on 22 April.[10] These publications provoked a number of breathless accounts in the popular press of an impending 'super-bomb'. They also caused a flurry of activity in government ministries.

At a hastily-convened conference held on 29 April 1939, Abraham Esau, president of the Reich Bureau of Standards and head of the physics section of the Reich Research Council, recommended the establishment of a uranium research project under his leadership. He gathered a number of leading German nuclear physicists together, referring to them as the *Uranverein* (Uranium Club). He charged the Uranverein with the task of investigating the potential for atomic energy, urging that they at once secure all stocks of uranium in Germany and ban all future exports.

A few days earlier, on 24 April, young Hamburg chemist Paul Harteck and his assistant Wilhelm Groth had written to the German War Office, urging the military to take note of the new developments in nuclear physics. They wrote:

We take the liberty of calling your attention to the newest development in nuclear physics, which, in our opinion, will probably make it possible to produce an explosive many orders of magnitude more powerful than the conventional ones ... That country which first makes use of it has an unsurpassable advantage over the others.

The letter was passed to the German Army Weapons Bureau, which subsequently initiated its own, rival, uranium research project under the leadership of physicist Kurt Diebner.

In Britain, Member of Parliament Winston Churchill grew increasingly alarmed by the summer's press accounts of a possible 'super bomb'. His concern was not that such weapons might be built first by Nazi scientists; it was rather that Hitler might seek to use the threat of a new secret weapon in his bargaining with Prime Minister Neville Chamberlain.

---

[10] This report suggested that an average of 3.5 secondary neutrons is produced for each fission of a uranium nucleus. This figure was subsequently revised to 2.5.

Churchill sought advice from his trusted scientific adviser and friend, Oxford physicist Frederick Lindemann, known to Churchill fondly as 'the Prof'. Lindemann had visited him often in the period 1931–34, during his 'wilderness years' at Chartwell, the Churchill family home. Drawing on Lindemann's advice, he wrote to Sir Kingsley Wood, Secretary of State for Air, on 5 August 1939 to advise him that such weapons would not be available for several years.

Nuclear physics research in the Soviet Union was conducted primarily (though not exclusively) at the Leningrad Physicotechnical Institute, commonly known as Fiztekh, which had been established in the early 1930s. The energetic and expressive physicist Igor Kurchatov headed the Institute's nuclear department. Although the Soviet Union had many highly talented scientists, it had become increasingly difficult to maintain research programmes not directly supportive of the country's efforts to achieve rapid industrialisation, and nuclear physics in particular was not seen to have any practical application. Even worse, as Stalin's regime had become more isolated following the Great Purge of 1937–38 (in which 100 Soviet physicists are believed to have been among the seven to eight million citizens arrested), virtually all contact with Western nuclear scientists had ceased.

Soviet physicists had little choice but to follow developments through the pages of Western science journals. Once read, the new discoveries relating to fission were quickly repeated and extended in the Soviet Union. Kurchatov directed his colleagues Georgei Flerov and Lev Rusinov to measure the number of secondary neutrons formed by the fission of uranium. By 10 April they had confirmed that between two and four secondary neutrons are produced, thereby independently confirming the possibility of nuclear chain reactions. By June, Flerov and Rusinov had also indirectly confirmed Bohr's suggestion that U-235 is primarily responsible for fission in uranium. At this stage the physicists did not see fit to bring these discoveries to the attention of the Soviet government or military authorities, or alert them to the potential for an atomic threat.

When Hitler had been appointed Chancellor in January 1933, Hungarian physicist Leo Szilard had packed his belongings into two suitcases, and made ready to leave Berlin at a moment's notice, 'when things got too bad'. That moment arrived just a few months later. Szilard moved first to Vienna, then to London where he helped Lindemann to establish a fund to bring exiled scientists to Britain. He spent several years in London and Oxford anticipating the development of nuclear fission, chain reactions and atomic bombs,[11] before emigrating to the US in early 1938. On hearing of the discovery of fission in uranium in January 1939, he borrowed $2,000 from a successful American inventor and persuaded the chairman of the physics department at Columbia University – where Fermi had recently arrived – to provide laboratory facilities.

Working with Fermi he independently verified the production of secondary neutrons and the possibility of a nuclear chain reaction, thus realising his worst fears. He had been thinking about the potential for releasing atomic energy for many years, and now confronted a real fear that Nazi Germany might be the first to build an atomic weapon. He sought commitments from his scientific colleagues to refrain from publishing their results openly in the scientific literature. Joliot-Curie refused.

Szilard voiced his growing concerns to fellow Hungarian émigré physicists Eugene Wigner and Edward Teller. Wigner had arrived in Princeton from Berlin in October 1930 to take up a temporary lectureship, which had become a permanent position in 1935. He had spent a couple of years at the University of Wisconsin before returning to Princeton in June 1938. Teller had left Göttingen in Germany first for Copenhagen, then University College London, then George Washington University in Washington, DC, where he had arrived with his new wife Mici in August 1935.

All three members of this 'Hungarian conspiracy' had had direct personal experience of the Nazi regime and understood precisely what it could be capable of. The news from Europe suggested that German expansionism might easily engulf Belgium, whose colony in Africa was a rich source of uranium ore. Wigner suggested that they alert the Belgian government to the danger.

---

[11] Szilard filed a patent application based on the idea of a nuclear chain reaction in March 1934.

Szilard remembered that his former colleague Einstein knew Elizabeth, Queen of the Belgians personally, and could perhaps approach her on their behalf. Shortly afterwards, on 16 July, Szilard, Wigner and Einstein met at Einstein's holiday home on Long Island. This was the first that Einstein had heard of the possibility of nuclear chain reactions, and he enthusiastically agreed to help. Einstein dictated a letter in German, and Wigner wrote it down.

After having secured Einstein's agreement, Szilard then found an alternative way to sound the alarm. He had been put in touch with Alexander Sachs, an economic adviser to US President Franklin Roosevelt. Sachs listened carefully to Szilard's concerns, before concluding that this was surely a matter for Roosevelt himself. Sachs promised to provide a statement on the subject directly to the President.

Teller and Szilard held a further meeting with Einstein on 30 July, Teller later remarking that he had 'entered history as Szilard's chauffeur' (Szilard had never learned to drive and did not own a car). Einstein agreed to the change of plan, and the three physicists then worked on a draft letter. The end result, dated 2 August 1939, was communicated to Sachs on 15 August but was not delivered verbally to Roosevelt until October.

The letter warned of 'extremely powerful bombs of a new type'. It warned that Germany had banned the sale of uranium from mines in recently annexed Czechoslovakia and that American work on uranium was now being repeated in Berlin.

Declaration of war

At 4:40am on 1 September 1939 the German Luftwaffe attacked and destroyed the Polish town of Weilun, killing 1,200 people, mostly civilians. This was merely the first in a series of preludes to a full-scale German invasion.

Allied governments declared war on Germany on 3 September.

German Army Ordnance hastily consolidated the country's two uranium research projects and issued call-up papers to selected nuclear scientists. On 16 September the scientists attended a secret conference to establish the consolidated project and discuss some of the scientific problems they were likely to face. Among them were Diebner, Harteck and Hahn.

The paper by Bohr and Wheeler on the theory of nuclear fission and the importance of U-235 had recently been published in the American journal *Physical Review*. It had been widely and eagerly read by the German scientists. One of these – Erich Bagge – suggested that the Uranverein co-opt his professor at the University of Leipzig to investigate further the theory of nuclear chain reactions in uranium. This was Nobel laureate Werner Heisenberg, the country's leading theoretical physicist, famous for his discovery of the uncertainty principle. Heisenberg received notification of his call-up from Bagge himself on 25 September.

The first war of physics had begun.

PART I

MOBILISATION

Chapter 1

THE URANVEREIN

September 1939–July 1940

W erner Heisenberg loved his country. He was a patriot and, by his own standards, a 'good' German. Slightly built, blond, with a warm and welcoming smile, he might have seemed to some the very essence of Aryan manhood. As an impressionable young student in his late teens he had dreamt of a romanticised Third Reich with fellow members of the New German Pathfinders, a youth movement composed of upper-middle-class adolescent males. This was a Reich that was to be forged through a return to the spirit of community and noble leadership characteristic of the medieval crusader knights. It demanded a complete rejection of the corruption and hypocrisy of modern German society and extolled moral purity, honour and chivalry. The movement was firmly apolitical.

The older Heisenberg might have been able to persuade himself that a German victory in the war that had just been unleashed would be ultimately good for Europe, but it was painfully obvious that Hitler's National Socialism was a gross corruption of his youthful ideals. He had managed to convince himself that Hitler's regime would surely be transitory, giving way in the fullness of time to a more moderate and honourable form of government.

In the meantime, many of Heisenberg's Jewish colleagues had fled the country, fearing for their lives and the lives of their families. Heisenberg himself preferred the inner exile of political reticence and conformity to the prospect of physical exile that had been afforded by the offers of academic positions he had received from abroad. In reaching this conclusion he was guided by Max Planck, the great grandfather of the quantum, now president of the Kaiser Wilhelm Society. Planck had counselled that emigration would be an empty gesture, and that Heisenberg could perform a higher service by offering support to the next generation of German physicists, needed by the country long after the Nazis had gone.

It was a morally ambiguous position. Physics and physicists had to be defended without offence to Nazi ideology, a task requiring painstakingly careful steps along a very fine line. It was a path that was to involve considerable personal danger and many shameful compromises.

Heisenberg himself was intimately aware of the dangers. He had been publicly denounced two years before for his association with the kind of physics that Nazi purists had branded 'Jewish', largely because of its departure from classical preconceptions and because of the prevalence of Jews in its discovery and development. The archetypal Jewish physicist was Einstein, and Einstein's theories of relativity had come to epitomise Jewish physics.

At that time Heisenberg had been waiting for news about his appointment to a professorial chair at the University of Munich. This was a position vacated by Arnold Sommerfeld, Heisenberg's former doctoral adviser, who had retired a few years previously. His appointment had seemed certain. Then came an article by Nazi physicist Johannes Stark in the SS newspaper, *Das Schwarze Korps*, on 15 July 1937. 'How secure the "White Jews" feel in their position,' Stark wrote, 'is proven by the actions of the Professor for Theoretical Physics in Leipzig, Prof. Werner Heisenberg, who ... declared Einstein's theory of relativity to be "the obvious basis for further research ..."' Stark went on to accuse Heisenberg of anti-regime views, of being a 'Jew lover' and a 'Jewish pawn'.

In itself the attack proved enough to deny Heisenberg the Munich chair. But he was now faced with a dark choice. Silence in the face of such accusations would imply complicity, placing both himself and his new (and now pregnant) wife Elisabeth in a danger from which physical exile from Germany and German science would be the only escape. The Nazi dogs would hound him out of the country he loved. The alternative was to defend what he saw to be his 'honour', declare his patriotism and, by inference, his loyalty to the Nazi cause. 'Now I actually see no other possibility but to ask for my dismissal [from his professorship in Leipzig] if the defence of my honour is refused here', he wrote to Sommerfeld.

Towards the end of July, Heisenberg wrote directly to Heinrich Himmler, asking that Himmler either approve or disapprove of Stark's attack on him. Approval of Stark's denouncement by Himmler would lead Heisenberg to resign his position. Disapproval would lead him to demand that his honour be restored and that he be protected from any future such attacks.

This was not a letter that could be trusted to the usual channels, as these would work too slowly, if at all. Instead Heisenberg's mother offered to pass the letter to Himmler via Himmler's mother, whom she knew personally. They met in either late July or early August 1937, Heisenberg's mother appealing to Mrs Himmler's maternal instincts: '... we mothers know nothing about politics – neither your son's nor mine,' she confided, 'But we know that we have to care for our boys. That is why I have come to you.'

Himmler probably received the letter later that August, and launched a preliminary internal investigation. This evolved into a more intensive SS investigation that lasted more than eight months. Heisenberg would have come to know real fear during this time. The Gestapo bugged his home and placed spies in his physics classes. An apparent preference for the company of young men and the apparently unseemly haste with which the 35-year-old Werner had married twenty-year-old Elisabeth Schumacher evolved into dark hints of homosexuality, a crime punishable by immediate imprisonment in a concentration camp. Such allegations were frequently used by the SS to extract confessions for lesser crimes.

What, one wonders, might have been said to Heisenberg during his interrogations in the notorious cellars of SS headquarters in Prinz Albert Strasse in Berlin, where a sign hanging on the wall reminded all exposed to such questioning to 'Breathe calmly and deeply'? Heisenberg was not physically harmed, but would return home from each interrogation exhausted and deeply disturbed.

Several of the SS investigation team had studied physics and Heisenberg had actually acted as doctoral thesis examiner in Leipzig for one of them. The investigation concluded positively, clearing Heisenberg of all the charges levelled by Stark. The application of some further, gentle diplomatic pressure on Himmler finally led to a compromise, and a conclusion of the affair, a year after Stark's accusations had first appeared in print. Himmler expressed his disapproval of the attack, his belief that '… Heisenberg is decent, and we could not afford to lose or silence this man, who is relatively young and can educate a new generation'. He instructed Reinhard Heydrich, head of the Nazi intelligence service – the SD – that Heisenberg should be protected from any future attack.

Such protection was dearly bought. The compromise meant that relativity theory could continue to be taught to the next generation of German physicists, but it had to be divorced from Einstein's name. Indeed, the argument went, the foundations of relativity theory had, surely, been laid by good Aryan physicists. The Jew Einstein had merely profiteered from their ideas. Compared to the evils visited upon Jews by the Nazis in the time since they had come to power, denial of the role they had played in the development of modern physics was, perhaps, a bargain that was not so difficult for Heisenberg to accept. But the Faustian nature of the bargain was now crystal clear.

Preventing catastrophe

Heisenberg's American colleagues, and those European physicists who had found sanctuary in America, could not understand his decision to stay in Germany.

He visited America in the summer of 1939, probably judging this to be the last opportunity to do so for some time to come. He lectured in Chicago and at Purdue University in Indiana, before moving on to Ann Arbor to attend a summer school organised by Dutch physicist Samuel Goudsmit, then on the faculty of the University of Michigan.

It was in America that Fermi caught up with him, and together they discussed the prospect for a new kind of super-weapon based on nuclear chain reactions. Heisenberg shared the commonly-held view that this was a remote, long-term possibility. Fermi insisted that, should war break out, nuclear physicists of all nations would surely be expected to devote all their energies to building these new weapons. Heisenberg conceded the point but played down the potential for success: 'I believe that the war will be over long before the first atom bomb is built', he said.

In Ann Arbor Heisenberg faced a friendlier, though no less intense, interrogation. What was Heisenberg going to do? Why was he staying on in Nazi Germany? How could he continue to do physics under the auspices of such an evil regime? Why was he in such a hurry to get back? Goudsmit pursued him relentlessly. Laura Fermi remarked that anyone must be crazy to stay in Germany. Exasperated, Heisenberg responded in kind: 'People must learn to prevent catastrophes,' he argued, 'not to run away from them.'

Before returning to Germany Heisenberg stopped over in New York, where once again he received the offer of an academic position at Columbia University, an offer that had first been made during his darkest hours in 1937. Once again he turned the offer down.

Perhaps Heisenberg had not received the kind of reception in America that he had anticipated. His insensitivity to the effect on his friends and colleagues of some of his more casual remarks – emphasising that he needed to return to his German army reserve unit for machine-gun practice, for example – would certainly not have helped. He boarded the SS *Europa* in early August. It was virtually empty. On the journey back to Germany he would have had plenty of time to ponder what the future held.

Warfare for physics

As each day had passed since the outbreak of war on 1 September 1939, Heisenberg had anticipated the arrival of his call-up papers, just as he had nervously, but eagerly, anticipated a call to arms with his reserve infantry brigade during the Sudeten crisis a year before, a crisis averted when

Czechoslovakia's allies traded appeasement of Hitler's aggressive expansionism for 'peace in our time'. When Erich Bagge returned to Leipzig on 25 September and advised him that he was to report not to the infantry but to the next meeting of the Uranverein, he was both greatly relieved and excited. He had been presented with an opportunity to contribute to Germany's war effort by doing what he loved most: research.

In his own mind Heisenberg had already dismissed the prospect of an atomic super-weapon as a remote one, but the German military authorities were nevertheless willing to engage the services of nuclear physicists and provide research funds and facilities to explore the possibilities. Here was an opportunity to contribute to the war effort and at the same time carry out fundamental research. 'We must make use of physics for warfare', was the official slogan of the Nazi government. Heisenberg thought to turn this on its head: 'We must make use of warfare for physics', he wrote years later of his reaction to the news.

Many scientists down the centuries have fallen prey to such impeccable, but arrogant, logic. When the ends are deemed to be improbably achievable or irrelevant, the means become the most important consideration. But these same scientists have tended to fail spectacularly to see all possible ends. So, Werner Heisenberg, Nobel laureate, discoverer of quantum mechanics and the uncertainty principle, and one of the most talented theoretical physicists of his time, accepted the challenge to work on atomic weapons for Hitler's Nazi Germany. He accepted eagerly and without hesitation. A darker and potentially much more dangerous Faustian bargain had now been struck.

The Uranverein was to meet again in Berlin the very next day, 26 September. Heisenberg journeyed to Berlin that night.

Reactors and bombs

The second meeting of the Uranverein, whose very existence was now classified as a military secret, was held in the research offices of German Army Ordnance in Berlin. The research branch of Army Ordnance was headed by Erich Schumann,[1] who had wrested control of the Uranverein from Esau at the Reich Research Council, part of the Ministry of Education. Schumann appointed Diebner to direct the project, supported by Bagge. Diebner had studied physics in Innsbruck and in Halle before joining the German Bureau of Standards and the Army Weapons Bureau in 1934. Bagge had studied in Munich and Berlin and gained his doctorate under Heisenberg at Leipzig in 1938. Both were loyal Nazis.

Heisenberg now joined Diebner, Bagge, Harteck, Hahn and other Uranverein physicists, including Carl Friedrich von Weizsäcker, one of Heisenberg's former students and a close friend. Weizsäcker had studied in Berlin and Copenhagen before gaining his doctorate under Heisenberg in Leipzig in 1933. He was a talented young theoretical physicist and philosopher, the son of Ernst von Weizsäcker, Secretary of State under Foreign Minister Joachim von Ribbentrop. Just 24 days prior to the Uranverein meeting, on the second day of the war, his younger brother Heinrich had been killed fighting with the Ninth Infantry Regiment near Danzig.

Diebner and Bagge had drafted an outline of the research programme a few days prior to the meeting, and had allocated tasks to each of the scientists involved. There was still considerable uncertainty regarding the physical principles of a fission chain reaction in uranium and there were few hard measurements available, but there was enough understanding to make a start.

Bohr and Wheeler had argued that U-235 is responsible for fission in uranium, and that fission can be triggered by bombardment of U-235 with slow neutrons. Fission in the much more abundant isotope U-238 requires much faster, higher-energy neutrons. However, there are certain characteristic neutron energies, called 'resonant' energies, at which a U-238 nucleus will capture a neutron to form the unstable isotope U-239 without undergoing fission. At these relatively high energies, neutrons would therefore be removed from any chain reaction, the U-238 nuclei acting as a 'sink', preventing the neutrons from going on to fission more U-235 nuclei.

---

[1] Grandson of the composer Robert Schumann.

Obtaining a self-sustaining chain reaction in a nuclear reactor based on naturally-occurring uranium was therefore a simple matter of population statistics. Secondary neutrons produced by fission of U-235 would be formed with a range of energies, or speeds. If, on average, one or more neutrons survived long enough to encounter other U-235 nuclei, then there was a chance that these would cause fission, sustaining the chain reaction. If, on the other hand, the secondary neutrons were captured by the more abundant U-238, leaving, on average, less than one neutron to find a U-235 nucleus, then the chain reaction would be unsustainable and would quickly fizzle out.

The solution was obvious. To give the secondary neutrons as much chance as possible of finding and fissioning more U-235 nuclei, it would be necessary to incorporate a *moderator* in the reactor design. This would be a material containing light atoms capable of slowing the neutrons down without absorbing them. By slowing the neutrons down to an energy below the threshold of the U-238 resonance, they would be prevented from being absorbed in their turn by the U-238 nuclei. Suitable candidates for a moderator included so-called 'heavy' water, in which the hydrogen atoms of ordinary water are replaced by heavier deuterium isotopes,[2] or pure carbon in some readily available form, such as graphite. Harteck had already done some preliminary work on a reactor design consisting of alternating layers of uranium and heavy water.

It was also fairly clear at this early stage that a compact reactor or a bomb could not be built without separating U-235 from U-238 or, at the very least, greatly enriching the proportion of U-235 present in a mixture. There were few options available and, as Bohr had observed months earlier to his colleagues in Princeton, the prospects for large-scale separation of U-235 were dim. A thermal diffusion method, based on a process devised in 1938 by German chemists Klaus Clusius and Gerhard Dickel, appeared to be the best bet. This process relies on the tiny differences in the diffusion properties of gaseous forms of the isotopes when exposed to a temperature

---

[2] A hydrogen nucleus consists of a single proton. The heavier deuterium nucleus consists of one proton and one neutron.

differential. To get uranium into a gaseous form would require working with uranium hexafluoride, a highly unpleasant substance that corrodes just about anything it comes into contact with.

The Uranverein physicists were faced with two hurdles. They needed to make some basic measurements to assess the suitability of various materials for use as a moderator and so figure out the optimal configuration for a nuclear reactor. They also needed to work out how to separate U-235 on a large scale.

Bagge was assigned the task of determining the suitability of heavy water as a moderator. Harteck was asked to continue with some preliminary work on isotope separation using thermal diffusion methods and to examine the effect of different reactor configurations on the production of secondary neutrons. Heisenberg was asked to assess the feasibility of achieving a self-sustaining chain reaction in uranium based on the known physical properties of the materials likely to be required.

Schumann announced that the War Office had requisitioned the Kaiser Wilhelm Institute for Physics in Berlin to house the uranium project and those Uranverein physicists based in other cities were now asked to relocate. Almost all of them refused, preferring instead to remain where they were and if necessary travel to Berlin once or twice a week. Although they were all keen to make their contribution to the effort, from their perspective this was simply another research project to be added to their existing projects and teaching commitments. There was as yet no sense of the kind of urgency that would warrant a major disruption to their academic schedules.

Heavy water

Heisenberg immersed himself in the scientific literature and in December 1939 produced the first part of a detailed report to the German War Office entitled *The Possibility of Technical Energy Production from Uranium Fission*. This paper was to chart the future course of the German nuclear programme.

From the outset Heisenberg focused his attentions on the physics of a nuclear reactor, or 'uranium burner'. He saw no need to differentiate between this physics and the physics of a uranium bomb, perceiving them as extreme ends of a continuous spectrum. At one end of this spectrum would be a reactor formed from naturally-occurring uranium and a suitable moderator. At the other end would be an explosive formed from uranium greatly enriched in U-235, to the point of being 'almost pure'.

Heisenberg estimated that a reactor capable of achieving a self-sustaining chain reaction would require over a ton of uranium and around a ton of heavy water combined in a spherical configuration. Such a reactor should settle down to stable operation at a temperature of around 800° Celsius. Adopting the layer configuration advocated by Harteck could be expected to reduce the size of the reactor somewhat. Heisenberg concluded his report with the observation that enriching the proportion of U-235 would help to reduce the size of the reactor further, and that enrichment was 'the only method of producing explosives several orders of magnitude more powerful than the strongest explosives yet known'. At this stage Heisenberg expressed no preference for either heavy water or graphite as a moderator.

The War Office issued a contract for the production and delivery of quantities of refined uranium oxide to the Berlin-based Auer company, which had access to uranium from the Joachimsthal mines in Czechoslovakia. The director of Auer's Radiological Laboratory was Russian chemist Nikolaus Riehl, who had studied nuclear chemistry and physics under Hahn and Meitner. Riehl immediately established production facilities at Oranienburg, about twenty miles north of Berlin, and the first ton of uranium oxide was delivered in early 1940.

Sourcing suitable quantities of heavy water was more problematic. The only facility producing heavy water in commercial quantities was a fertiliser plant owned by the Norwegian company Norsk Hydro, which produced it as a by-product. The plant, which had begun production in 1934, was perched high in the fjords at Vemork near the town of Rjukan in the remote Telemark region of Norway, about 150 miles west of Oslo.

Graphite was very much the preferred candidate for the choice of moderator because of its ready availability in large quantities, in pure form. But initial data from German chemist Walther Bothe's team in Heidelberg – supported by theoretical predictions from Weizsäcker and his group in Berlin – suggested that graphite might absorb neutrons too readily and so prove to be unsuitable.

In Heisenberg's second report to the War Office, delivered in February 1940, he was already leaning towards use of heavy water as a moderator. This was a much less attractive option because of the problems of isolating enough of this substance to meet the needs of the research project. Diebner asked if it was necessary for Germany to construct its own production facility. Heisenberg suggested that they first acquire a few litres of heavy water with which to check its suitability, and Diebner promised to procure ten litres from the Norsk Hydro plant.

However, the Norwegians were not very co-operative. Norsk Hydro was approached by a representative of the German chemicals giant I.G. Farben, which owned stock in the Norwegian company, with an offer to buy up all the available heavy water. At this stage the Vemork plant was producing about ten litres a year, more than enough to meet the esoteric needs of the research laboratories which were its principal customers. When asked why so much heavy water was needed, the I.G. Farben representative would not say. The Norwegians gave their regrets: they would not comply with the German request.

When Norsk Hydro was subsequently approached by Jacques Allier the response was very different. Allier was a representative of the Banque de Paris et des Pays Bas, which had a controlling interest in the company. He was also a lieutenant in the Deuxième Bureau, the French military intelligence agency. Joliot-Curie in Paris had also identified heavy water as a potential moderator and had advised the French Ministry of Armaments of its importance in nuclear research.

Arriving in Oslo under an assumed name and carrying a credit note for FF36 million, Allier had intended to negotiate the purchase of all the available heavy water. But when it became clear what purpose the heavy water served, the Norsk Hydro managing director Axel Aubert pledged the entire stock to the French government at no cost: 'Say that our company will accept not one centime for the product you are taking, if it will aid France's victory.' The heavy water was removed from Vemork and smuggled first by air to Edinburgh, then by rail and ferry to Paris.

The fall of France

The situation changed dramatically on 9 April 1940, when German forces invaded Denmark and Norway in Operation Weserübung. The Danish government quickly capitulated under threat from the Luftwaffe, and signed a non-aggression pact to secure a measure of political independence. Niels Bohr, long aware of the impending catastrophe, had now become trapped in Copenhagen.

The German forces met more resistance in Norway. King Haakon VII, together with other members of the Norwegian royal family and key government ministers, were eventually able to escape to Britain with the nation's gold reserves. They formed a government in exile, leaving the Nazi sympathiser Vidkun Quisling to declare himself premier in a coup d'etat, broadcast on radio.

Fighting had been fierce around Rjukan, and this was the last town to yield in southern Norway. German troops entered the town on 3 May. This time there would be no negotiations. The Germans now learned that all existing stocks of heavy water had been smuggled to France, but there appeared little to prevent production from being accelerated to meet the needs of the German nuclear project. An increase in output to 1.5 tons a year was promised.

On 10 May German forces invaded France and the Low Countries. German armoured divisions scythed through the Ardennes forest, cutting off Allied units that had taken up positions in Belgium, including the British Expeditionary Force, consisting of ten infantry divisions despatched to the Franco-Belgian border following the German invasion of Poland. The Luftwaffe quickly gained superiority in the air over Belgium and Holland. Following the carpet bombing of Rotterdam, the Dutch army surrendered on 14 May. The encircled British Expeditionary Force, and many French soldiers, were evacuated from Dunkirk on 26 May, in a rout that was heralded as little short of a miracle. Belgium capitulated on 28 May.

With the north of the country secure, German forces launched south into France on 5 June. Italy declared war on Britain and France on 10 June, Paris fell on 14 June and the French government fled to Bordeaux. French resistance quickly collapsed, and the government signed an armistice with Germany on 22 June. This was signed at Compiègne in the same railway car, and in the same forest, as the armistice of 1918.

The Soviet Union had signed a non-aggression pact with the Nazis in August 1939, and had invaded Finland in November that year. With the fall of France, only Britain, Greece, the Commonwealth and the exiled forces of European Allies stood between Germany and the conquest of all of Europe.

Union Minière in Belgium had thus far fulfilled orders received from Germany for about a ton of refined uranium compounds a month. Now under German occupation it received an order from the Auer company for 60 tons.

Uranverein physicists hastened to Joliot-Curie's laboratory in occupied Paris towards the end of June. Bothe was first to visit, followed by Schumann and Diebner. All but Joliot-Curie himself had fled. With his co-operation, Diebner assimilated the results of the work of the French nuclear physicists and arranged for the completion of the assembly of the cyclotron that they had begun.

Joliot-Curie could not hide the fact that he had accepted deliveries of uranium ore from Belgium and heavy water from the Vemork plant in Norway. When the Uranverein physicists demanded to know where these materials were, he simply stated that the uranium ore had disappeared 'south' along with the French government (it had in fact gone to Algeria) and that the heavy water had been loaded onto a ship known to have been sunk (it had actually gone to Britain, along with Joliot-Curie's colleagues Halban and Kowarski).

Element 93

In his second report to the German War Office Heisenberg had been reticent on the subject of a bomb. His reasons are unclear. It may be that, although Harteck had begun construction in Hamburg of a large-scale Clusius–Dickel apparatus to separate U-235, and had reasons to be optimistic, separation on the scale required for a bomb still appeared incredibly daunting.

The key question was one of scale: precisely how much U-235 would actually be required? There is no evidence in the historical record for this period (to spring 1940) of any formal calculation to determine the quantity of U-235 that would be required for a bomb. If Heisenberg or any other Uranverein physicist had made such a calculation at this time, it did not survive. It is possible that no such calculation had been carried out. For whatever reason, the possibility of a bomb based on 'almost pure' U-235 was not pursued.

A second route to an explosive device was potentially available in the form of an unstable reactor based on uranium enriched with U-235, a reactor on the edge of a runaway chain reaction. Calculations by one of Heisenberg's Uranverein co-workers suggested that such a 'reactor-bomb' would need to contain 70 per cent more U-235 than U-238. It was, of course, very difficult to imagine how such a reactor-bomb might be delivered to its target. And enrichment on the scale required for a reactor-bomb still appeared beyond the bounds of possibility in any timeframe likely to affect the course of the war.

But then a completely unanticipated third avenue appeared. Heisenberg's close friend and Uranverein colleague Weizsäcker would pass time on Berlin's underground railway reading papers on nuclear fission which were still being published in American scientific journals, oblivious to the suspicious glances of his fellow commuters.

Hahn's group in Berlin had found that U-239, formed from U-238 by the capture of a neutron, is unstable and undergoes radioactive decay within about 23 minutes. It was believed that emission of a beta particle[3] from U-239, which turns a neutron into a positively-charged proton, would transmute the uranium nucleus, characterised by its 92 protons, into a new element with 93 protons. Hahn thought this element might be chemically similar to the element rhenium and had called it eka-rhenium, or *eka re*. Weizsäcker suspected that this new element might be fissionable, just like U-235.

On the surface this proposal seems innocent enough. But it is far from innocent. Unlike U-235, element 93 does not occur in nature and is *chemically* distinct from uranium. Weizsäcker realised that it would be possible to separate element 93 from uranium by chemical means. In essence, he was suggesting that if element 93 could be produced in a uranium reactor in significant quantities, it could be separated relatively easily and used to make a fission bomb.

That element 93 could indeed be produced by bombarding U-238 with neutrons was demonstrated by American physicists Edwin McMillan and Philip Abelson at the Radiation Laboratory in Berkeley. But they also noted that this element was relatively unstable, decaying within a matter of days. Astonishingly, they published their results in the open scientific literature in June 1940. Here was concrete proof of the practical feasibility of using a uranium reactor to produce fissionable material for a bomb. In July 1940 Weizsäcker wrote a paper for the Army Weapons Research Bureau in which he enthusiastically recommended that this possibility be pursued.

It was now clear to the Uranverein that the construction of a bomb depended on first solving the problems related to the construction of a reactor. There remained the question of the most appropriate moderator. Some initial measurements of the rate of absorption of neutrons by graphite were reported in a confidential paper first issued by the Heidelberg group in June 1940. The results were rather inconclusive. It seemed that the rate was too high for graphite to be used successfully as a moderator,

---

[3] A beta particle is a fast-moving electron ejected directly from a neutron inside the nucleus during beta radioactive decay. During this process the neutron is transformed into a proton.

although it was conceded that part of the problem lay in the homogeneity and purity of the graphite used. At this stage Bothe was reasonably confident that further tests with purer samples would demonstrate the potential of graphite as a moderator.

The Virus House

The War Office takeover of the Kaiser Wilhelm Institute for Physics created great difficulties for its director, the esteemed Dutch physicist Pieter Debye. The German authorities presented Debye with an ultimatum: take German nationality and continue as director or take a temporary leave of absence. Debye refused to cede his Dutch nationality. He left Germany in January 1940 and embarked on a lecture tour of America. He never returned.

Debye's departure left the directorship open. Schumann favoured Diebner, but Diebner's appointment was resisted by the Kaiser Wilhelm Foundation. Weizsäcker and fellow Uranverein physicist Karl Wirtz, who expressed concern that they now 'had Nazis in the institute', conspired to bring Heisenberg to Berlin. Diebner was appointed as interim director, and Heisenberg agreed to travel once a week from Leipzig.

Heisenberg now had considerable influence over the work of the theoretical group, the reactor experiments that were being established in Berlin, and the reactor experiments that he himself was setting up in collaboration with his colleague Robert Döpel in Leipzig. Heisenberg was not director of the uranium research project, but he was running a substantial part of the show.

The German atomic programme was not a coherent research effort driven relentlessly by the demands of war. Rather, it was a loose association of rival research teams that would sometimes squabble over supplies of uranium and heavy water.

But for those able to read them, the signs were ominous.

The Uranverein physicists now had access to thousands of tons of refined uranium. They were building their first cyclotron in Joliot-Curie's captured Paris laboratory. They had the promise of substantial quantities of heavy water. Separation of U-235 was proving to be as difficult as had been anticipated, but some of the greatest minds in chemical and physical science were being applied to the search for a solution.

In July 1940 work was begun on a new building to house an experimental nuclear reactor at the Kaiser Wilhelm Institute for Biology and Virus Research, next door to the Institute of Physics in Berlin. To limit unwanted attention, the building was called the *Virus House*.

Chapter 2

ELEMENT 94

September 1939–September 1940

M uch to Leo Szilard's frustration, Einstein's letter to Roosevelt was slow to have any kind of impact. The letter had been drafted in early August 1939 but as the days and weeks passed he heard nothing from Sachs. In the meantime, war in Europe had begun.

When Szilard and Wigner visited Sachs towards the end of September, they discovered to their dismay that Sachs still had the letter in his possession. He had tried repeatedly to gain an audience with Roosevelt to discuss the matter, but had not so far managed to get past Roosevelt's secretary.

Sachs finally gained access to Roosevelt in the Oval Office on 11 October. He prepared the ground with a parable about Napoleon, and this prompted Roosevelt to ask for a carafe of Napoleon brandy and a couple of glasses. As Sachs sipped brandy with Roosevelt he tried to present a verbal summary of the content of Einstein's letter. But Roosevelt appeared distracted and inattentive, and asked if Sachs could return the next day. Fearing he had blown his chance, he returned next morning with some trepidation. But this time Roosevelt was ready and willing to listen.

Using his own 800-word précis of Einstein's letter, Sachs chose to emphasise the peaceful uses of nuclear power, mentioning last of all the threat of 'bombs of hitherto unenvisaged potency and scope'. He concluded with the observation that we 'can only hope that [man] will not use [sub-atomic energy] exclusively in blowing up his next door neighbour'.

Roosevelt got the message. 'Alex,' he said, 'what you are after is to see that the Nazis don't blow us up.' He called for immediate action, and responded to Einstein's letter a week later.

It was quickly agreed that the administration would establish an Advisory Committee on Uranium to be headed by Lyman J. Briggs, director of the US National Bureau of Standards. The committee consisted of nuclear physicists and ordnance experts from the US Army and Navy. To Szilard and his fellow Hungarian conspirators, it looked as though something was finally going to happen.

The first meeting of the Advisory Committee took place on 21 October in Washington. Szilard and Wigner held a pre-meeting with Sachs at the Carlton Hotel to discuss tactics, before joining Teller and other members of the Advisory Committee at the Bureau of Standards offices in the Department of Commerce. Einstein had been invited to attend, but declined.

Szilard explained the scientific background and the importance of putting the theory of nuclear chain reactions to the test in large-scale reactor experiments, which he proposed should be constructed from uranium oxide and graphite. These were experiments he had been trying, but failing, to set up with Fermi at Columbia University since July. The ordnance experts were openly sceptical of the physicists' claims. The destructive potential of an atomic bomb was simply way beyond their reckoning. It takes two wars, Lieutenant Colonel Keith Adamson declared, before one can know if a new weapon is any good or not.

The physicists themselves were relatively ill-prepared. When asked directly how much was needed from Treasury funds to start work on Szilard's proposed experiments, they were at a loss for a reasoned answer. Teller leapt forward with a request. He asked for just $6,000. 'My friends blamed me,' he later said, 'because the great enterprise of nuclear energy was to start with such a pittance: they haven't forgiven me yet.'

After the meeting, Szilard – who would shortly estimate that they needed at least $33,000 for the graphite alone – nearly murdered Teller for the modesty of his impromptu request.

Despite the paltry nature of the sum that had been mentioned, Adamson bridled. 'Gentlemen,' he berated them, 'armaments are not what decides war and makes history. Wars are won through the morale of the civilian population.' Wigner, normally polite and formal in his dealings with colleagues, became angry and spoke up for the first time in the meeting. 'If that is true,' he declared in his high-pitched voice, 'then perhaps we should cut the Army budget thirty per cent and spread that wonderful morale through the civilian population.'

Adamson visibly flushed, and muttered that the physicists would get their money.

Szilard drafted a blueprint for the American uranium research project and mailed this to Briggs five days after the committee's first meeting. In it he suggested which experiments should be conducted and identified the American laboratories that should be involved. He also urged that all future research reports be subject to the strictest secrecy and withheld from publication in the open scientific literature.

But the Advisory Committee lacked resolve. It reported back to Roosevelt on 1 November a commitment to explore controlled chain reactions in uranium as a potential power source for submarines which, if the reaction turned out to be explosive, could be further explored as a source of highly destructive bombs. It agreed to supply four tons of purified graphite to support Fermi and Szilard's experiments, to be followed by 50 tons of uranium oxide, if this could be subsequently justified.

Briggs was respected but obsessed with secrecy and dogged by poor health. He was unable to imbue the committee or its sponsors with any real sense of urgency. The war in Europe was, after all, a long way away. What's more, he was reluctant to commit large sums of money to the project. The money that had been promised at the 21 October meeting was not quickly forthcoming.

Szilard might have been elated by the fact that the importance of uranium fission had now been recognised, but this gave way to more frustration as the first months of 1940 unfolded. He was still without formal employment, and uncertain how long his loose affiliation with Columbia University could be maintained. He was not in a position to repay the $2,000 he had borrowed to carry out experiments to verify the production of secondary neutrons and was obliged to go back to his sponsor to declare this a bad debt.

He heard nothing from Briggs.

Zeal for secrecy

News that a secret German research project on nuclear fission had begun at the Kaiser Wilhelm Institute in Berlin reached America in January 1940 through Pieter Debye, recently expelled from his position at the Institute and now on extended 'leave of absence'. Debye played down the significance of the project. The Uranverein physicists were very well aware of the German army's objectives but considered success 'improbable', he claimed. In the meantime the German physicists had a splendid opportunity to carry out fundamental research at the army's expense. On the whole, Debye was inclined to consider the situation a good joke on the German army.

Debye visited Fermi at Columbia University shortly after arriving in America. Fermi too, it seemed, was unconcerned by Debye's news. The Uranverein physicists were working at laboratories all over Germany, he observed, and would not be able to make any kind of concerted effort towards a bomb.

But the news had precisely the opposite effect on Szilard. He had spent the previous weeks working on a couple of theoretical papers on self-sustaining nuclear chain reactions,[1] work which no doubt convinced him that a nuclear explosive of some kind was now inevitable. The existence of a German fission project greatly alarmed him. He discussed the matter with Einstein at Princeton, and together they decided to draft another letter, this time to Sachs.

---

[1] The longer of these papers cites H.G. Wells's *The World Set Free*, which was published in 1913 and which first introduced the idea of 'atomic bombs'.

In this letter they emphasised that interest in uranium had intensified in Germany since the outbreak of war, that nuclear research had been taken over by the German government and was being conducted in great secrecy. The implications were reasonably clear: whether they liked it or not, they were now locked in a race with the Nazis to build an atomic bomb. The letter also contained a threat: unless there was a change of policy, Szilard would publish his latest research on nuclear chain reactions in the open literature.

The letter was sent to Sachs on 7 March 1940. A week later, Sachs wrote of these new developments to Roosevelt, who called for a further meeting of the Advisory Committee. Progress was still painfully slow: the meeting was not scheduled until 27 April. Einstein was again invited, but again declined. At least the further letter to Roosevelt prompted the release of the $6,000 that had been promised.

By the time the meeting was held, Alfred Nier at the University of Minnesota and John Dunning at Columbia had gathered experimental evidence confirming that U-235 is indeed responsible for slow-neutron fission in uranium, vindicating Bohr and Wheeler's original hypothesis. They had used tiny quantities of U-235 and U-238 obtained from uranium compounds of chlorine and bromine. They went on to conclude that a fission chain-reaction would not be possible without separation of U-235.

The opinion of the Advisory Committee was split. Briggs expressed doubts that a chain reaction would be possible in natural uranium. Sachs urged that they should nevertheless move ahead with experiments on the uranium–graphite reactor that Szilard had proposed. All agreed that they should wait for the results of measurements on neutron absorption by graphite.

The funds were transferred to Columbia University and used to purchase a quantity of purified graphite. Szilard had been careful to specify high levels of purity. At lunch with representatives of the National Carbon Company, Szilard had probed for details about likely impurities in commercially-available graphite. He specifically mentioned potential contaminants that would absorb neutrons and render meaningless any attempts to measure neutron absorption by graphite itself. Half-jokingly, he said: 'You wouldn't put boron into your graphite, or would you?'

His visitors looked at each other in embarrassed silence. One of the principal uses of graphite is in the manufacture of electrodes for electric arcs, and boron is typically a component in the manufacturing process. Any graphite they supplied would therefore likely be contaminated. They agreed to supply a quantity of graphite manufactured using different methods, without the use of boron.

Four tons of graphite duly arrived at the Columbia laboratory in the form of carefully-wrapped bricks. The simple process of unwrapping and stacking the bricks in a neat pile was enough to give the researchers the appearance of coal miners. The results of the neutron absorption measurements were, however, strongly positive: graphite could indeed be used satisfactorily as a moderator. The idea of a nuclear reactor in the form of a uranium–graphite 'pile' took a critically important step towards becoming a reality.[2]

Szilard urged Fermi not to publish the results of these experiments. The relationship between the two had to this point been quite tense, but now it reached breaking point. They were two quite different personalities. Szilard was a loner, always ready to challenge conventional wisdom and norms of behaviour, sometimes outrageously. Fermi was an out-and-out scientist, much more collaborative and polite, caring little for the world outside the domain of science. Szilard's experiences had led him to be extremely wary of the world outside science, and he fervently believed that scientists had a duty to behave responsibly in matters likely to have a significant impact on this world. 'Fermi and I had disagreed from the very start of our collaboration about every issue that involved not science but principles of action in the face of the approaching war', he later wrote.

---

[2] The term 'pile' crept into common usage over the course of several years' experimental effort on the first nuclear reactor. Readers suspecting some deep scientific significance in this term should know that Fermi used it because the first reactor was literally a pile of graphite and uranium blocks.

Szilard could also be intensely irritating, and Fermi now lost his temper. He thought Szilard's zeal for secrecy absurd, but eventually relented under pressure. The results were not published.

Super-cyclotron

Ernest Lawrence was a visionary. The inventor of the cyclotron was a rather atypical physicist. A blond, blue-eyed Midwesterner of Norwegian parentage, he carried the values of his Lutheran upbringing into adulthood, and into his science. His preference for smart suits and his magisterial manner lent him an appearance that was more businessman than scientist. And, in truth, managing the kind of scientific enterprise that he was keen to establish at Berkeley's Radiation Laboratory – or 'Rad Lab' – demanded a much more businesslike approach. His teenage experiences as a kitchenware salesman had given him the necessary selling skills, and had taught him the rudiments of fund-raising.

Lawrence had invented the cyclotron in 1929. Use a magnet to confine a stream of protons to move in a circle while accelerating them to higher and higher speeds using an alternating electric field and, Lawrence had figured, you had a machine for penetrating the secrets of the atomic nucleus. He built a small demonstration model for just $25. It was four inches in diameter and covered in red sealing wax. Although it didn't quite deliver the proton energies that Lawrence claimed it should, it was enough to impress his scientific colleagues and prove the principle. However, the machine's scientific name, the *magnetic resonance accelerator*, was too abstract and clumsy. *Cyclotron* sounded much more futuristic, and therefore much more appealing to potential sponsors.

He was already thinking on a larger scale, and there quickly followed a succession of such machines. A cyclotron containing a magnet with an eleven-inch diameter pole face delivered proton energies of over a million electron volts. This was followed by a 27-inch machine, which then quickly became a 37-inch cyclotron. When news of the discovery of nuclear fission in uranium reached Berkeley in January 1939, Lawrence was planning a 60-inch cyclotron that would deliver proton energies of the order of 20 million electron volts. It would need a magnet weighing 200 tons.

The 60-inch machine was barely operational at the Rad Lab's Crocker Laboratory before Lawrence was busy designing the next one. This was to be a gargantuan 120-inch super-cyclotron with a magnet weighing 2,000 tons. Lawrence estimated that this would deliver proton energies of 100 million electron volts, on the threshold of nuclear-scale energies. Lawrence approached the Rockefeller Foundation with requests for support. His pitch was greatly strengthened when, in the middle of a game at the Berkeley Tennis Club on 9 November, he was informed that he had just won the 1939 Nobel prize for physics.

Suitably emboldened, as Christmas approached Lawrence escalated the scale of the super-cyclotron even further, to include a magnet with 184-inch pole faces (the largest diameter of commercially-available steel plate), weighing 5,000 tons. It would cost an estimated $1.5 million to build.

The outbreak of war in Europe in September had an immediate personal impact on Lawrence – after several days of anxious waiting he heard that his brother John had survived the sinking of the *Athenia* by a German submarine on 2 September. But life at the Rad Lab continued pretty much as normal. There were interesting experiments to be performed on uranium using the 60-inch cyclotron, but this was work that would have been carried out irrespective of the war. There was no sense yet that the Rad Lab was in any way involved in 'war work'.

A photograph from around this time shows the Rad Lab faculty, gathered in three rows beneath the magnet of the 60-inch cyclotron. Lawrence is sitting in the centre of the front row. Oppenheimer is standing in the centre of the back row. At the extreme right of the first and second rows are two Rad Lab physicists who were now busy at work on the uranium problem – Edwin McMillan and Philip Abelson.

McMillan, a native Californian, had worked on Lawrence's cyclotrons for many years, and when the discovery of fission had been announced he had devised some simple experiments to confirm the phenomenon. He had now become intrigued by some of the discovery's more subtle aspects. Bombarding uranium with neutrons produced a radioactive substance which decayed in a characteristic time of about 23 minutes. Like Hahn, Strassman and Meitner, McMillan surmised that this substance was U-239, formed by the resonant capture of a neutron by the predominant isotope, U-238. But there was another radioactive substance produced, which had a characteristic decay time of about two days.

He believed this second substance to be a new element, formed by emission of a beta particle from U-239, in the process turning a neutron into a proton. Just as Weizsäcker had done, sitting on the Berlin underground railway, so McMillan had reasoned that this was element 93, perhaps the first in a series of transuranic elements. And, just as Hahn had done, McMillan further surmised that element 93 might behave somewhat like the element rhenium.

With the help of a Berkeley research associate, Emilio Segrè, who had worked previously with Fermi in Rome, McMillan tried to obtain evidence of rhenium-like chemical properties. But they could find nothing of the sort. It seemed that, after all, the transuranics would continue to remain elusive. Segrè published the results in *Physical Review*, as an 'unsuccessful search for transuranic elements'.

McMillan had now refined the measurement of the decay time of this second mysterious substance to 2.3 days and became determined to identify precisely what it was. In the spring of 1940 he used the 60-inch cyclotron to investigate it further, and was joined in the quest by Abelson, who had by this time moved to the Carnegie Institution in Washington but had returned to Berkeley in April for a working vacation. Abelson had studied chemistry as well as physics and turned his attention to the chemical identification of the mysterious substance.

It turned out to have properties not so very different from uranium itself. Bohr had in fact already suggested some time before that the transuranics – if they existed – might behave chemically more like uranium. Further work demonstrated unambiguously that the substance with the 2.3-day decay time was formed directly from U-239, with its characteristic 23-minute decay time. There was only one conclusion: the second substance was element 93.

McMillan had already devised a name for the new substance – *neptunium* – though he chose to withhold it for the time being. Just as element 93 is one step further along the periodic table from uranium, so Neptune is one planet further along in the solar system from Uranus. Unaware of any reasons for secrecy, on 27 May McMillan and Abelson submitted a paper describing the results of their work to the American journal *Physical Review*. The paper was published on 15 June, and read with great interest by Weizsäcker when the journal reached him in Berlin in July.[3]

Of course, this work raised a further question. If element 93 was radioactive, with a characteristic decay time of 2.3 days, what was it decaying into? McMillan had his suspicions. He thought that element 93 might decay through a further emission of a beta particle, turning another neutron into a proton and so forming element 94. He immediately began work to find evidence for it.

Wild enough speculation

Szilard was probably unaware of McMillan and Abelson's paper until it was published. The physicists had not thought to send it to him to seek his advice on the safety or otherwise of its publication in the open literature. But, by pure coincidence, on the same day that McMillan and Abelson submitted their paper to *Physical Review*, Szilard received a manuscript from Princeton theoretical physicist Louis Turner on precisely the same subject.

In January 1940 Turner had surveyed the literature on uranium fission and published a review in the journal *Reviews of Modern Physics*. This work had set him thinking. While all the attention had so far been diverted towards U-235, Turner now nagged away at the idea of producing atomic energy from the stable, and much more abundant, isotope U-238. The resonant capture of neutrons by U-238 was considered something of a

---

[3] British physicists strongly protested at the publication of the McMillan–Abelson paper and Lawrence received a formal dressing-down by an attaché from the British embassy for giving away secrets to the Germans.

nuisance, to be avoided in a reactor through the use of a suitable moderator. Now Turner followed much the same logic as Weizsäcker, McMillan and Abelson. Neutron capture by U-238 would create an unstable U-239 isotope, which would decay to form element 93. But Turner did not stop there. He had figured from theoretical principles that element 93 would be relatively unstable and would decay quite quickly, creating element 94.

Element 94 opened up an altogether different kind of prospect. It would consist of 94 protons and 145 neutrons, making a total of 239. In this sense it paralleled the pattern in U-235, with 92 protons and 143 neutrons. Some simple calculations suggested that this new element would be even more fissionable than U-235. It would be produced from the abundant isotope U-238 and, because it was a new element with its own distinct chemical properties, it could be chemically separated from its uranium parent. Turner anticipated that element 94 could represent a new source of fissionable material for nuclear chain reactions.

Turner had drafted a paper for submission to *Physical Review* and wanted Szilard's opinion on whether or not it was safe to publish. 'It seems as if it was wild enough speculation so that it could do no possible harm, but that is for someone else to say', he told Szilard.

Speculation it might have been, but Szilard was a master of this game. He was stunned by the implications. 'With this remark of Turner,' he later said, 'a whole landscape of the future of atomic energy rose before our eyes.' Szilard suspected that achieving self-sustaining chain reactions – and bombs – might be a lot easier with element 94 than with uranium itself.

He recommended that Turner delay publication of his paper 'indefinitely'.

Undoubtedly a Fascist

Despite these revelations, the Advisory Committee on Uranium still moved at a snail's pace. Briggs, it seemed, was an innately cautious man. He could move at only one speed – full ahead slow.

Things were about to change, however. Vannevar Bush had vacated his vice presidency of the Massachusetts Institute of Technology (MIT) to accept the presidency of Washington's Carnegie Institution in the summer of 1939. Bush had trained as an electrical engineer and had gone on to become a highly pragmatic scientific administrator. During the First World War he had worked on the development of a magnetic device capable of detecting submarines. The device worked well enough, but was never put into operation. This experience had taught him all he needed to know about the importance of proper liaison between military and civilian research in the development of weapons in a time of war.

From his position as president of the Carnegie Institution, Bush lobbied for the establishment of a national organisation for just this kind of liaison. On 12 June 1940 he presented his arguments to Roosevelt, summarised in four short paragraphs in the middle of a sheet of paper. The groundwork had been done for him by Harry Hopkins, a Roosevelt aide, and the National Defense Research Council (NDRC) came into being. Its purpose was to direct all scientific research for military purposes.

One of its first actions was to take over the Advisory Committee on Uranium. The need for censorship was immediately agreed – all research papers on uranium fission would henceforth be subject to strict secrecy. Briggs remained as chairman of the committee, reporting to James Bryant Conant, president of Harvard, who had joined the NDRC at Bush's invitation. The dependence on sceptical military advisers for funding was now greatly reduced.

Not that this made a great deal of difference, however. Bush and Conant were very aware of the potential threat of a German atomic bomb, but instead of lobbying to secure greatly increased funding for the American nuclear programme they preferred rather to focus research efforts on proving that a bomb was impossible. After all, if it could not be done then there would never be a threat from a German weapon. In a report to the NDRC dated 1 July 1940, Briggs summarised the progress to date and requested $40,000 for further critical research on the nuclear properties of the materials involved and $100,000 for experiments on a large-scale uranium–graphite pile. Briggs got the $40,000.

Szilard was left to wait a while longer.

The creation of the NDRC produced one unlooked-for side-effect. This was an American organisation involved in secret military research projects – only US citizens could be members. Fermi, Szilard, Teller and Wigner were now suddenly excluded from the proceedings. This was obviously absurd, and Sachs argued strenuously that the entire work of the Advisory Committee had depended on the efforts of émigré scientists who were now to be barred from future involvement.

Military security checks were duly carried out. The security report on Fermi labelled him 'undoubtedly a Fascist' (he was not) and recommended that he should not be employed on secret work. The report on Szilard suggested that he was 'very pro-German' and had 'remarked on many occasions that he thinks the Germans will win the war'. The report recommended that Szilard, too, be barred from employment on secret work. Both reports quoted 'highly reliable sources'. The irony was lost on them. The only secrets worth protecting were in the minds of the very scientists the authorities wanted to exclude.

The reports were sent to J. Edgar Hoover in August 1940 with a request for further FBI security checks. The FBI merely repeated what the military authorities had already claimed. It appeared not to matter. Sachs' arguments won the day. All four émigré physicists were allowed to continue to make their contributions to the project, but now as advisers to the NDRC rather than full members.

Despite its now greatly raised profile, the work still proceeded slowly. In fairness, the results obtained thus far painted a rather confused picture. U-235 was clearly responsible for slow-neutron fission in uranium but separation of this isotope from U-238 was going to be an incredibly difficult feat. The early signs pointing to the feasibility of a uranium reactor were both encouraging and discouraging. Graphite in a suitably pure form would serve as a suitable moderator, but it was still not yet known if a self-sustaining chain reaction would develop in a uranium reactor without considerable enrichment of the minor U-235 isotope. Nier and Dunning's conclusions in this regard had not been very promising. If a working reactor could be built, resonant absorption of neutrons by U-238 in such a reactor might produce element 94, which could be more easily separated from uranium and which might prove to be fissionable in its own right.

To cap it all, Teller had carried out some calculations which suggested that a uranium bomb would require a mass of more than 30 tons. Even if it could be made to explode, it was difficult to see how such a bomb could be delivered to its target.

Bush remained sceptical of the science. It was difficult to see all this as anything other than a wild goose chase.

Thousands of times more powerful

The reckless, naked aggression that had been unleashed in Europe by Nazi Germany shaped the attitudes of all who observed it from across the Atlantic, but its effect on European émigrés was particularly profound. In the spring of 1940 Teller had been playing out an internal moral debate. He was at once uneasy with the prospect of working on weapons of such potentially massive destructive power, but at the same time understood enough about German military and technical superiority to develop real fear of a Nazi victory. 'At that time,' he later said, 'I believed that Hitler would conquer the world unless a miracle happened.'

Teller had not seen fit to involve himself in politics or pay any attention to the pronouncements of politicians. He had initially thought not to accept an invitation to attend a Pan American Science Congress in Washington, which was to be addressed by Roosevelt, but Hitler's rape of Europe in May 1940 caused him to change his mind. In the event, Roosevelt's speech helped determine his moral position and gave him a resolve that was to remain unshakeable for the rest of his life.

Of course Teller was well aware of Einstein's letter to Roosevelt and its consequences in terms of America's stuttering nuclear programme. He had never met Roosevelt but, sitting in the audience listening to his speech, Teller was overcome by the eerie sensation that the president was talking to him directly. After pointing out how small the world had now become, Roosevelt cautioned that America could not depend on its 'mystic immunity' from a European war that threatened the very kind of civilisation so

valued by Americans. He then turned to the role of the scientists themselves:

> You who are scientists may have been told that you are in part respon-
> sible for the debacle of today ... but I assure you that it is not the scien-
> tists of the world who are responsible ... What has come about has been
> caused solely by those who would use, and are using, the progress that
> you have made along lines of peace in an entirely different cause.

For Teller this was both rallying-cry and moral absolution. He had been
fortunate to have escaped the tyranny that was now overrunning Europe
and threatening to engulf the whole world. 'I had the obligation to do
whatever I could to protect freedom', he said.

His mind was now firmly made up.

The news from Europe became ever more depressing. After unleashing
his blitzkrieg on continental Europe, Hitler had anticipated negotiating a
peace with Britain before turning his attention to Russia, his notional ally.
Churchill had become prime minister of a new coalition government in
early May. Unlike his predecessor Neville Chamberlain, who after the fall
of France wanted to sue for peace, Churchill was not minded to negotiate.
Hitler was left with no choice but to subdue Britain first, and this meant
gaining air superiority over the south-east of England and the English
Channel.

The Luftwaffe launched air assaults from their newly-acquired bases
in northern France, harrying British convoys crossing the Channel, their
purpose being not only to sink ships but also to lure British fighters out
over the sea. In August Hermann Göring, head of the Luftwaffe, ordered
assaults on coastal airfields and radar stations, and then inland airfields
and aircraft production centres. The Battle of Britain had begun.

On 7 September Göring launched a series of massive air raids against
London, partly in reprisal for a British bombing raid on Berlin and as a
prelude to Operation Sealion, the full-scale invasion of Britain. Göring
dispatched nearly 400 bombers and more than 600 fighters in two waves

against London's East End. When 200 German bombers returned later that night in a further wave of attacks, London was still burning.

As news of the bombing reached Szilard, he remarked quietly: 'Before this war is over there will be bombs thousands of times more powerful than those in the blitz.'

Chapter 3

CRITICAL MASS

September 1939–November 1940

In January 1939 Otto Frisch at last received some good news. He learned that his father, though still imprisoned in Dachau concentration camp, had been granted a Swedish visa. Shortly afterwards he was released to rejoin Frisch's mother in Vienna. Both then made their way to Stockholm and safety.

However, even this good news could not relieve the sense of dark foreboding that had begun to overwhelm him. He became increasingly depressed by the prospect of impending war, and he saw no value in continuing his research in Copenhagen. His sense of vulnerability grew. When British physicist Patrick Blackett and Australian Mark Oliphant came to visit Bohr's laboratory, Frisch asked them for help.

Oliphant was a native of Adelaide whose initial leanings towards medicine and dentistry had been diverted towards physics while at university. A speech by New Zealander Ernest Rutherford had further directed the impressionable student towards nuclear physics. He joined Rutherford's research group at the Cavendish Laboratory in Cambridge in 1927, where he witnessed at first hand many of the remarkable discoveries in nuclear physics of the early 1930s. In 1934 he had published a paper with

Rutherford on nuclear fusion reactions involving deuterium, or heavy hydrogen.[1] German chemist Paul Harteck was a co-author.

In 1937 Oliphant had been appointed to a professorship at the University of Birmingham in England and was now head of the physics department. He was very sympathetic to Frisch's appeals and subsequently wrote to suggest that Frisch visit Birmingham in the summer of 1939 to see what could be done. Oliphant's confidence and calmness had impressed the now desperate Frisch, and he did not need a second invitation. He packed two small suitcases and made his way to England, 'just like any tourist'.

Oliphant found Frisch a job as an assistant lecturer. It was a very informal position. Oliphant would deliver a lecture to a group of students and at the end would hand over to Frisch those who had struggled to comprehend the subject matter. Frisch would sit with a few dozen students who would fire questions at him. The discussions were very lively and much to Frisch's taste.

At Birmingham Frisch joined fellow émigré Rudolf Peierls.[2] Peierls had been born in Berlin, of assimilated Jewish parents, and had studied physics in Berlin, Munich and Leipzig, where he gained his doctorate under Heisenberg in 1928. He moved to Zurich in Switzerland before taking a Rockefeller Scholarship in 1932 to study first with Fermi in Rome, then in Cambridge, England, to work with the theoretician Ralph Fowler. He was in Cambridge when Hitler came to power in 1933, and it became obvious shortly thereafter that he would not be able to return to Germany. At the end of his scholarship he moved to Manchester to work with Lawrence Bragg, then back to Cambridge for a couple of years before successfully applying for a professorship in mathematics at Birmingham University in 1937.

As war commenced in September 1939 the laboratory facilities at Birmingham were largely given over to essential – and secret – war-related research. Much of this activity centred on the development of the cavity magnetron, used to generate intense microwave radiation for ground and

---

[1] Nuclear fusion involves the joining together of two light nuclei to form a heavier nucleus, accompanied by the release of energy.

[2] Pronounced 'piles'.

airborne radar systems in what C.P. Snow later called 'the most valuable English scientific innovation in the Hitler war'.

As enemy aliens, Frisch and Peierls were not meant to know anything about this work, but the secrecy surrounding it was all a bit of a charade. Occasionally, Oliphant would ask Peierls hypothetical questions beginning, 'If you were faced with the problem ...' Years later Frisch wrote: 'Oliphant knew that Peierls knew, and I think that Peierls knew that Oliphant knew that he knew. But neither of them let on.'

Frisch's teaching commitments were relatively light and, with time on his hands, he turned his attention back to problems related to nuclear fission. He used what spare laboratory space he could find to carry out some small-scale experiments. Bohr and Wheeler had argued that the fission discovered in uranium was primarily due to the less stable U-235 isotope. Frisch decided to try to gain some experimental evidence for this, by making measurements on samples that were slightly enriched in the minor isotope. He set up an apparatus to separate a small quantity of U-235 based on the thermal diffusion technique developed by Clusius and Dickel. Progress was slow.

In the meantime he had received a request from the British Chemical Society for a review on recent progress in nuclear science that would be of interest to chemists. He wrote the article sitting in his bedsit, his typewriter on his knees, wearing a greatcoat, huddled over a gas fire in an attempt to stay warm as the winter temperature dipped, sometimes to minus 18 Celsius. On cold nights the tumbler of water at his bedside would freeze solid.

In the section on fission he repeated the consensus view that while it might one day be possible to create a self-sustaining chain reaction, its dependence on slow neutrons meant that it would build too slowly to make an effective bomb. 'The result would be no worse than setting fire to a similar quantity of old-fashioned gunpowder', he concluded. Frisch did not believe an atomic bomb was possible.

But the task of writing the review article had set him thinking. The problem that had been identified by Bohr and Wheeler related to slow neutrons. Because of the tendency for U-238 to capture fast neutrons with certain characteristic 'resonant' energies, or speeds, slow neutrons are necessary to achieve a chain reaction in naturally-occurring uranium. Slow neutrons mean a slow build-up of energy. The energy released in a slow-neutron reaction would heat up the uranium, possibly melting or even evaporating it long before it could explode. As the uranium heated up, more and more neutrons would escape the surface and eventually the chain reaction would grind to a halt.

The Uranverein physicists had come to precisely the same conclusion. But, Frisch now wondered, what would happen if *fast* neutrons were used? U-235 was expected to be fissioned by fast as well as slow neutrons. Fast secondary neutrons generated by fission of U-235 would be of no use in mixtures containing large amounts of U-238, as a potentially high proportion would be removed through resonance capture by U-238. However, there would be no such constraint if pure or nearly pure U-235 were used. Frisch had set up his small Clusius–Dickel apparatus to separate U-235 without much difficulty. This was clearly not a technique that could be expected to produce quantities of pure U-235 measured in tons, but might it be possible that a much smaller amount of U-235 would be required to support a fast-neutron chain reaction?

A fast-neutron chain reaction in pure U-235. Insofar as the atomic bomb ever had a 'secret', Frisch had just found it.

Frisch shared his thoughts with Peierls, who had in early June 1939 refined a mathematical formula for calculating the critical mass of material required to support a nuclear chain reaction, a formula that had originally been developed by French theoretician Francis Perrin. For mixtures of isotopes with a high proportion of U-238, Peierls had used his revised formula to calculate a critical mass of the order of tons, totally unsuitable for a weapon.

Now Frisch was demanding a rather different calculation, based on fast rather than slow neutrons in pure U-235. The problem was that nobody knew the rate at which U-235 would be fissioned by fast neutrons as nobody had ever separated enough of the isotope to measure it.

They had no choice but to speculate. It was clear from the work of Bohr and Wheeler that U-235 nuclei could be fissioned rather easily by slow neutrons, so it made sense to assume that fast neutrons would be just as effective, perhaps even to the point that fission would occur every time a U-235 nucleus was hit by a fast neutron. As Peierls later put it: 'The work of Bohr and Wheeler seemed to suggest that every neutron that hits a [U-]235 nucleus should produce fission.' This assumption greatly simplified the calculation. The rate they needed to estimate was just the rate at which fast neutrons would hit the U-235 nuclei.

They plugged the numbers into Peierls' formula, and were profoundly shocked by the result. This was no longer a matter of tons. They had estimated a critical mass of only a *few pounds*. For a substance as dense as uranium this was a critical mass about the size of a golf ball.[3] Frisch estimated that this much U-235 could be separated in a matter of weeks using about 100,000 Clusius–Dickel tubes like the one he had assembled in the laboratory in Birmingham.

'At that point we stared at each other and realised that an atomic bomb might after all be possible.'

Fast-neutron fission

In Liverpool, Polish-born physicist Joseph Rotblat had reached much the same conclusion. He had read about the discovery of nuclear fission and had conducted his own experiments at the University of Warsaw to verify the production of secondary neutrons. He had quickly realised the potential for a bomb, and grew greatly concerned about what the Nazis would do with such a weapon: 'I had no doubt that the Nazis would not hesitate to use any device, however inhumane, if it gave their doctrine world domination.'

He had very little modern equipment with which to perform nuclear experiments at the University of Warsaw. He was aware that James

---

[3] Note that the critical mass is the mass of fissionable material in which production of neutrons is balanced by their capture without further fission or escape to the surrounding environment. A mass of U-235 equal to the critical mass is therefore not explosive. To create an explosion, it is necessary to assemble a mass that is in excess of the critical mass. This is sometimes referred to as a super-critical mass.

Chadwick, winner of the 1935 Nobel prize in physics for his discovery of the neutron and Britain's leading experimental nuclear physicist, was building Britain's first cyclotron – based on Lawrence's design – in the basement of his laboratory at the University of Liverpool. Rotblat dreamed of one day building a cyclotron in Warsaw. He had approached Chadwick in the spring of 1939 with a request to join his group for a short period to observe the latter stages of construction of the cyclotron. Chadwick had agreed and, with a small stipend from Warsaw, Rotblat had leapt at the opportunity to make what was to be his first trip outside his native Poland. He left his new wife, Tola, behind. Just for now, or so he thought.

Although he battled with the English language, the drab slums of Liverpool and the rather primitive conditions of the laboratory, he settled in and was quick to impress Chadwick with his skills as an experimentalist. So much so that in August 1939 Chadwick offered him an Oliver Lodge Fellowship, the department's most prestigious award. This was the first time the Fellowship had been awarded to a foreign scientist. The stipend associated with it was enough for Rotblat to bring his wife to England.

Rotblat returned to Poland towards the end of August, but Tola was recovering from appendicitis and could not travel. The news blackout in Poland meant that neither of them realised how dangerous the situation had become. Rotblat left for England only a few days before the German invasion began, on what was one of the last trains to leave Poland. His young wife was now trapped. Despite repeated attempts to get her out, he never saw her again. She died in Nazi-occupied Poland. Rotblat never remarried.

The brutal invasion of his native country led him in late November to suggest to Chadwick that they work together on developing an atomic bomb. He feared that those physicists who had stayed on in Nazi Germany might already be working on such a bomb, which Hitler could then use to conquer the world. 'It was a terrible time for me, perhaps the worst dilemma a scientist could experience', he said later. 'Working on a weapon of mass destruction was against all my ideas – all my ideas of what science should do – but those ideas were in danger of being eradicated if Hitler acquired the bomb.'

Chadwick himself had been caught on the hop by the outbreak of war.[4] He had been on a trout-fishing holiday in a remote part of northern Sweden with his wife and daughters. When news of the German invasion of Poland reached them, they had immediately headed for Stockholm only to discover that all flights to London had been cancelled. They flew to Holland instead, encountering the author H.G. Wells in their Amsterdam hotel before finally making their way across the North Sea on a tramp steamer.

Rotblat had independently come to the conclusion that slow neutron fission of U-235 would not be sufficient to support an explosive release of nuclear energy, but that fast-neutron fission might. Late one evening in November 1939 he approached Chadwick, who just grunted. But Chadwick's own early scepticism about the prospects for a bomb had given way to interest, and Rotblat's logic might have simply served to confirm Chadwick's own thoughts on the subject. The Liverpool cyclotron had come on stream a few months before. A few days later, Chadwick sat down with Rotblat and together they agreed what experiments needed to be done.

Frisch–Peierls memorandum

Frisch and Peierls discussed their results with Oliphant, who was immediately convinced by their arguments. He recommended they write it all up in a short memorandum. They produced two short typewritten notes, both dated March 1940. The first was largely concerned with the physical principles and practical feasibility of a super-bomb based on U-235. The second, titled *Memorandum on the Properties of a Radioactive 'Super-bomb'*, was to prove remarkably prescient. It argued that development of an atomic weapon would be 'practically irresistible'; that it could not be used without 'killing large numbers of civilians'; that it was 'quite conceivable that Germany is, in fact, developing this weapon', though Frisch and Peierls went on to admit that it was possible that 'nobody in Germany has yet

---

[4] For Chadwick this was history repeating itself. He had been trapped in Germany in August 1914 when, following Germany's invasion of Belgium, Britain declared war.

realised that separation of uranium isotopes would make the construction of a super-bomb possible'. The memorandum hints at the threat to come:

> If one works on the assumption that Germany is, or will be, in the possession of this weapon, it must be realized that no shelters are available that would be effective and that could be used on a large scale. The most effective reply would be a counter-threat with a similar bomb.

Frisch and Peierls had already realised that the only defence would be deterrence.

Oliphant sent the memorandum to Henry Tizard, an Oxford chemist and chairman of the Aeronautical Research Committee. Although its preoccupation was radar, this was one of the most important committees concerned with the application of science in wartime. Tizard recommended that a small advisory committee be set up, which was eventually to consist of Oliphant, George Thomson, professor of physics at Imperial College London,[5] and Patrick Blackett, professor of physics at Manchester University. Thomson was appointed as chairman of the committee. Blackett was involved with other war projects and did not join the committee immediately. John Cockcroft, who with Ernest Walton had been the first physicist to 'split the atom', also joined. With the outbreak of war, Cockcroft had taken the position of Assistant Director of Scientific Research in the Ministry of Supply, working primarily on radar. As enemy aliens, Frisch and Peierls were excluded.

The pace was forced somewhat at the beginning of April by a visit to London by Jacques Allier, who reported to Thomson, Oliphant and Cockcroft on the work of the French nuclear scientists in Paris and the German physicists' interest in heavy water. The British committee met for the first time on 10 April 1940 at the Royal Society in London. Denmark and Norway had been invaded by German forces just the day before.

---

[5] George P. Thomson was the son of J.J. Thomson, who had discovered the electron in 1897. In a nice twist of history, J.J. Thomson won the 1906 Nobel prize for physics for showing that the electron is a particle, whereas his son won the 1937 Nobel prize for showing that it is a wave.

However, when the committee considered the Frisch–Peierls memorandum, it was met with general scepticism. Some small-scale experiments on separating U-235 using uranium hexafluoride might be justified, the committee concluded, but this was not likely to become a project of real military significance. Tizard thought that the French had become 'unnecessarily excited' by the existence of the German nuclear programme. The comparison with the first meeting of Briggs' Advisory Committee on Uranium seemed complete.

But there the similarities ended. In a letter dated 16 April, Thomson invited Chadwick to join the committee, which met for the second time eight days later. When he heard of the details of the Frisch–Peierls proposals, Chadwick was embarrassed. He admitted that he had reached similar conclusions, but had not felt justified in raising the possibility of a U-235 bomb until more experiments had been performed. Chadwick's support for the technical content of the memorandum greatly enhanced its credibility. The committee now realised that an atomic bomb was feasible in principle. The physicists were electrified by the possibility.

Frisch and Peierls had understood the implications of what they had discovered. But there was never any question in their minds as to the morality of initiating a project that could produce a weapon of unimaginable destructive power, a weapon capable of 'killing large numbers of civilians'. Frisch later wrote:

Why start on a project which, if it was successful, would end with the production of a weapon of unparalleled violence, a weapon of mass destruction such as the world had never seen? The answer is very simple. We were at war, and the idea was reasonably obvious; very probably some German scientists had had the same idea and were working on it.

Fear drove them. Fear that such a weapon might be delivered into the hands of a regime capable of subjugating Europe, if not the world. A regime capable of perpetrating an evil beyond their darkest imaginings.

Spontaneous fission

Peierls and his Russian wife Genia moved into a more spacious house in Edgbaston in Birmingham and at their invitation Frisch moved from his cramped bedsit to lodge with them. Genia was a phenomenon. She ran the house 'with cheerful intelligence, a ringing Manchester voice and a Russian's sovereign disregard of the definite article'. She taught Frisch to shave every day and to dry dishes as fast as she could wash them up.

As they waited patiently for news of the reaction to their memorandum, Frisch became increasingly concerned about his own position. He was summoned by the police to answer a barrage of questions concerning his status – did he have any dependent relatives, was he studying for an examination, was he about to get a degree which would allow him to find work? Genia was convinced he was about to be interned, along with other enemy aliens, on the Isle of Man. She made him buy some cotton shirts that even a bachelor could launder. Through his scientific contacts he tried to get the message to the authorities that he was involved in important war work. This seemed to do the trick. He never heard from the police again and, fortunately, never had to launder his cotton shirts.

The small apparatus Frisch had set up to separate U-235 was going to require a considerable amount of time and patience to produce quantities of the isotope sufficient to make measurements. This was patience he did not possess. He therefore worked out an alternative 'quick-and-dirty' method of making the necessary measurements. This involved bombarding natural uranium with slow neutrons which, according to Bohr and Wheeler, would cause fission only in U-235. He decided to revert to a relatively old method of generating neutrons, using gamma radiation from radium to dislodge neutrons from a beryllium target. This was a method that had been overtaken by the development of more modern high-tension equipment, and Lawrence's cyclotrons.

Armed with a small amount of highly radioactive radon gas that he had extracted from a sample of radium stored deep underground in Derbyshire's Blue John Cavern, he produced neutrons from beryllium which he used to bombard about a gram of uranium enclosed in an ionisation chamber. He made a series of measurements over a 36-hour period, taking naps on a camp bed in the laboratory between measurements. He made two important discoveries.

What had first appeared a puzzling experimental artefact turned out to be spontaneous fission in natural uranium. The U-235 'liquid drops' are so unstable that every now and then one will fall apart all by itself, producing fission fragments and secondary neutrons. The second discovery was that Frisch had overestimated the rate at which U-235 nuclei are fissioned by slow neutrons. This meant that he had underestimated the critical mass of U-235 required to support a chain reaction. Fortunately, in the meantime Peierls had worked out that the critical mass could be reduced by surrounding it with a substance that would reflect any escaping neutrons back into the fissile material. So they were back pretty much where they started.

Peierls had also been thinking about isotope separation. He had consulted Franz Simon, a first-rate chemist who had been born to a Jewish family in Berlin and had won the Iron Cross, first class, during the First World War. He had been rescued from Germany and brought to Oxford by Lindemann in 1933. Frisch had opted for the Clusius–Dickel thermal diffusion method because it had seemed the simplest, but Simon and Peierls were not convinced this method would work. Simon thought that ordinary gaseous diffusion through a porous barrier might be more effective. Peierls wrote to Thomson, urging the committee to consult with Simon, and then wrote to Lindemann. Simon and Peierls approached Lindemann in person in June 1940. Peierls could not translate Lindemann's grunts, but thought he had been convinced by their arguments.

Maud Ray Kent

In Stockholm, Meitner had noted the paper by Segrè, published in *Physical Review*, in which he had reported his failed attempt to find rhenium-like chemical properties in the mysterious substance with the characteristic 2.3-day decay time. Segrè had speculated that the mysterious substance was nothing more than a fission fragment. Meitner was convinced it was element 93.

To prove this she needed access to a source of neutrons. Through late winter she waited patiently for Siegbahn's cyclotron to become operational, before giving up and going to Copenhagen to use the cyclotron in Bohr's institute. She arrived on the afternoon of 8 April 1940.

Bohr himself was in Norway on the final leg of a lecture tour. He dined with King Haakon VII that evening. It was a gloomy occasion. The King and his government officials were depressed at the prospect of an impending German invasion. Bohr travelled back to Copenhagen on the night train, and was woken by Danish police with the news that Denmark had been invaded, too. Meitner woke in Copenhagen to the roar of German aeroplanes overhead.

Denmark was granted a semblance of self-rule. In return for co-operation with the German occupying force, the Danish government negotiated protection for Denmark's 8,000 Jews, a condition that infuriated Hitler. Meitner was not threatened in occupied Denmark, and she stayed for three weeks before returning to Stockholm. Before she left, Bohr asked if she would send a telegram from Stockholm to British physicist Owen Richardson, a Bohr family friend, to advise him that he and his wife Margrethe were in good health, if not in good spirits.

The telegram read:

MET NIELS AND MARGRETHE RECENTLY BOTH WELL BUT UNHAPPY ABOUT EVENTS PLEASE INFORM COCKCROFT AND MAUD RAY KENT

Richardson passed the message to Cockcroft, who puzzled over its meaning. It was, perhaps, easy to be paranoid under the circumstances. German forces were tearing the heart out of continental Western Europe. German physicists were busy working on a secret project to build an atomic bomb. Germany had no cyclotron, but radium and beryllium were known to serve as a useful source of neutrons. Cockcroft became convinced that the last three words of Meitner's telegram were, in fact, a crude coded message. MAUD RAY KENT was surely an anagram – albeit imperfect – for 'radium taken'. It was a message consistent with the idea that the Germans were taking control of all the radium they could get their hands on. It all pointed to a concerted effort by the Germans to develop their nuclear technology.

Cockcroft voiced his concerns to Chadwick, and Thomson decided to use the message as the basis for a new name for the advisory committee that had been assembled in response to the Frisch–Peierls memorandum. He called it the M.A.U.D. Committee, a name deliberately obscure and meant to throw any German intelligence agent off the scent.[6] It was, perhaps, also meant to serve as a reminder for all those involved and aware of Meitner's telegram that they were now inexorably locked in a race with the Nazis to develop an atomic bomb.

Nobody thought to go back to Meitner to seek clarification.

Subsequent publication of the McMillan–Abelson paper confirmed what Meitner had suspected – the substance with the 2.3-day decay time was indeed element 93. Had not the war intervened, she would have been able to demonstrate this much herself, ending a search for transuranic elements that she had begun with Hahn in 1934. Among her many disappointments, this was most bitter.

'Mad Jack' Howard

Working at the Collège de France in Paris, Frédéric Joliot-Curie, Hans von Halban and Lew Kowarski had been one of the first research groups to demonstrate experimentally the possibility of a self-sustaining nuclear chain reaction in uranium. From that point they had moved quickly. By August 1939 they had recorded increased fission in blocks of uranium oxide immersed in ordinary water, although this activity was insufficient to support a chain reaction.

Halban, who had worked on neutron absorption in deuterium with Frisch at Bohr's institute in Copenhagen in 1937, suggested that heavy water would make a much better moderator. Of course, this was a

---

[6] This did not stop inventive onlookers from trying to decipher what the acronym stood for. A popular choice was: Military Applications of Uranium Disintegration.

suggestion that was confirmed by the German physicists' subsequent interest in stocks of heavy water from the Norsk Hydro plant at Vemork. He and Kowarski carried out calculations on all the possible candidates, including heavy water and ultra-pure graphite, and concluded that heavy water would work best.

Together they worked on the theory of chain reactions. By now wary of publishing their results in the open literature, they wrote a paper primarily to establish a claim to any original discovery. They lodged the paper, sealed in an envelope, with the French Academy of Science for safekeeping.

The 185 kilos of heavy water rescued by Allier from the Norsk Hydro plant were stored in an air-raid shelter at the Collège de France. However, the scientists had little opportunity to use these materials to put their theories to the test. As the Germans marched on Paris in early June 1940, Joliot-Curie received instructions from the French armaments minister not to allow the uranium and heavy water to fall into enemy hands.

Halban and Kowarski left Paris with their families and headed south. Joliot-Curie and his wife Irène, daughter of Marie Curie, followed. Halban loaded the 26 jerrycans of heavy water into a car with his wife Else and young daughter and drove to Le Mont-Dore, a spa town in central France. There they were joined by Allier. The heavy water was stored initially in the local women's prison, then in the death cell in neighbouring Riom's state prison. The condemned prisoners themselves carried the heavy water into the cell. Next morning the prison governor, perhaps nervous of the impending arrival of new masters, refused to release the heavy water until Allier threatened him with a loaded revolver.

Halban and Kowarski were busy setting up new laboratory facilities in the Villa Clair Logis at Clermont-Ferrand when, two days after the fall of Paris, the word came from Allier to evacuate France altogether. They headed for Bordeaux, where Charles Henry George Howard, twentieth Earl of Suffolk and thirteenth Earl of Berkshire, was waiting for them.

Howard was a character straight out of a Wodehouse novel. Having become the Earl of Suffolk and Berkshire at the age of eleven, 'Jack' Howard left Radley public school to work as a deckhand aboard the wool clipper *Mount Stewart*, and sailed for Australia. After a short spell in the

Scots Guards he returned to Australia and managed a sheep station in Queensland. Back in Britain, he studied for a degree in pharmacology at the University of Edinburgh and in 1937 was elected a Fellow of the Royal Society of Edinburgh. As the war began he joined the Ministry of Supply's Department of Scientific and Industrial Research (DSIR) and was dispatched to Paris to act as liaison with the French Ministry of Armaments.

Howard's mission was to 'rescue' valuable machine tools, millions of dollars' worth of industrial diamonds, the heavy water, and about 50 French scientists. When bankers were reluctant to release the diamonds from their vaults to a man possessing only a letter of introduction from the French Minister of Armaments, he would nonchalantly let his jacket fall open to reveal Oscar and Genevieve, a pair of .45 automatics nestling in shoulder holsters. That tended to overcome the bankers' inhibitions, and the diamonds were swiftly given up into his care.

His rather imaginative approach to his missions had earned him the nickname 'Mad Jack', and Halban and Kowarski quickly began to appreciate why. With the port under attack, and hundreds of thousands of refugees flocking to the docks seeking safe passage, chaos reigned. Howard, unshaven and covered in tattoos, simply got the crew of the British coal ship SS *Broompark* too drunk to set sail until he had completed his mission. With Halban, Kowarski, their families and the heavy water aboard, the ship sailed from the dock down the Gironde estuary on 19 June. A nearby ship hit a mine and sank. Joliot-Curie later claimed to the German invaders that it was on this ship that the heavy water had been stowed.

There were 25 women on board the *Broompark*, including Howard's private secretary, Eileen Marden. When some of the women began to complain of sea-sickness, Howard administered his remedy of choice – champagne. Kowarski quickly learned to have absolute confidence in Howard's abilities: 'His infectious good humour made the entire trip seem like a schoolboy adventure.'[7]

---

[7] Howard went on to specialise in defusing unexploded bombs. He worked closely with his cockney driver, Fred Hards, and his secretary Eileen Marden, who would take notes while Howard examined the devices. They became known as the 'Holy Trinity'. Tragically, all three were killed by a bomb on 12 May 1941. Howard was awarded a posthumous George

Joliot-Curie and Irène decided to return to Paris. Their reasons are obscure, but it may be that Irène refused to abandon French soil. Joliot-Curie's concern for his own academic status in England or concern for the continuity of French science may also have been factors.

The *Broompark* docked at Falmouth on 21 June 1940. Howard delivered the diamonds into the hands of Harold Macmillan, then an Undersecretary at the Ministry of Supply. The French physicists and their precious cargo reached London, where the heavy water was temporarily stored at Wormwood Scrubs prison, before being transferred into the care of the librarian at Windsor Castle. Halban and Kowarski joined the growing ranks of MAUD physicists and relocated to Cambridge's Cavendish Laboratory, where they formed a research group to develop a uranium–heavy water reactor.[8]

Frisch and Chips

Now that a decision had been taken to act on Frisch and Peierls' memorandum, the émigré physicists were finally allowed to contribute, not directly to the MAUD Committee but to a technical sub-committee. Frisch welcomed the opportunity: 'not only had our report started the whole thing, but we had also thought about many of the additional problems that would arise.'

Among the most pressing of problems was the separation of U-235, and it was by now clear that the facilities at Birmingham could not support work on both radar and the atomic bomb simultaneously. Frisch discussed the options with Chadwick in Liverpool, which at this time was not involved in war work. He returned to Oliphant with a proposal. Radar had to take priority in Birmingham; Liverpool offered greater access

---

Cross for 'conspicuous bravery'. Harold Macmillan later wrote: 'I have never known in a single man such a remarkable combination of courage, expert knowledge and indefinable charm.'

[8] The French physicists' adventure was turned into a feature film in 1949, titled *La Bataille de L'Eau Lourde* (The Battle for Heavy Water). It starred the physicists, playing themselves.

to laboratory facilities and a cyclotron, and although as an enemy alien Frisch was in principle prohibited from entering the port city of Liverpool, Chadwick had agreed to accommodate him there. Oliphant agreed, and Frisch moved to Liverpool in July 1940.

It was here that Frisch had his first experience of life under threat of German bombs. The Luftwaffe had failed to subdue Fighter Command in the Battle of Britain. Churchill, frequently given to colourful turns of phrase, was not exaggerating when he declared that never had so much been owed by so many to so few. Hitler suspended Operation Sealion indefinitely in mid-September 1940. Although this was Germany's first defeat of the war, it did not substantially alter the reality of German supremacy in Western Europe. Hitler switched his attention to the Atlantic – he intended to cut off vital supply routes and simply starve Britain into submission.

In November 1940 the Luftwaffe targeted Britain's major industrial cities and ports. Coventry was 'blitzed' on 14 November by more than 500 German bombers. Liverpool received some of the heaviest sustained bombing outside London, with over 300 air raids before the end of the year. Frisch was huddling under the staircase of his boarding-house during one particularly heavy raid, when a bomb blew in most of the windows. The landlady promptly quit and left without collecting the outstanding rents. Frisch also decided that the time had come to seek sanctuary in the suburbs.

In the laboratory Frisch worked with John Holt, a young student assigned to him by Chadwick. Frisch charged about energetically, with Holt trailing in his wake, and the pair earned the nickname 'Frisch and Chips'. They quickly discovered that uranium hexafluoride is one gas for which the Clusius–Dickel method produces no discernible separation of isotopes. Peierls and Simon had been right to be suspicious.

Simon had been co-opted onto the MAUD Committee in the middle of 1940 and worked full-time at Oxford on the problem of separating U-235 by an alternative gaseous diffusion technique. In this method, isotope separation is achieved by virtue of the fact that gases diffuse through a porous barrier at a rate that depends on their atomic or molecular weight. Lighter gases diffuse faster than heavier gases.

By December, Simon had worked out the details of a full-scale industrial plant that could separate as much as a kilo of U-235 per day. By his estimate, such a plant would cost about £5 million to construct, it would cover 40 acres and require some 60,000 kilowatts of electricity. He summarised his work in a detailed report and, not wishing to trust the report to the wartime postal service, he braved the bombing and journeyed to London shortly before Christmas 1940 to deliver it by hand to Thomson.

The comfort of restful sleep

Of all the MAUD Committee scientists, Chadwick probably had the widest perspective on the work that was now in hand. Although the British programme was committed to an approach based on U-235, the possibility of using other potentially fissionable materials such as element 94 had also been recognised by MAUD physicists. Production of element 94 in quantities sufficient for a bomb depended on the ability of the team at Cambridge to build a working nuclear reactor based on their uranium oxide–heavy water design. Chadwick had been so incensed by the publication of the McMillan–Abelson paper on element 93 that he had asked the British embassy to lodge an official protest. Lawrence was on the receiving end.

There were few in Liverpool to whom Chadwick could turn to discuss the visions that were now starting to haunt him. Frisch and Rotblat were getting along famously but they were not British citizens and Chadwick felt unable to confide in them. The other physicists working on the project were too young to be burdened with his dark thoughts. Weighed down by the inevitability of creating a weapon of inconceivable destructive power, the comfort of restful sleep began to elude him. He started to take sleeping pills.

He took sleeping pills for the rest of his life.

Chapter 4

A VISIT TO COPENHAGEN

October 1940–September 1941

The Virus House was ready in October 1940. In addition to a new laboratory, it also housed a special circular brick-lined pit about six feet deep into which a reactor vessel could be lowered. The pit could then be filled with water to act as both neutron shield and reflector, directing stray neutrons back into the reactor core, thereby prolonging any chain reaction that was initiated. It was in this pit that the world's first experimental nuclear reactor was to be assembled.

In December, Heisenberg, Weizsäcker, Wirtz and two other Uranverein physicists packed a domed aluminium cylinder with alternating layers of uranium oxide and paraffin wax, the latter to be tested for its efficacy as a moderator. The cylinder was a little under 150 centimetres in diameter. They lowered the cylinder into the pit and inserted small quantities of radium and beryllium into the centre of the reactor vessel to provide a source of neutrons to initiate a chain reaction. None of the Uranverein physicists was quite sure what to expect.

The experiment, designated B-I, failed. The physicists were looking for evidence of neutron 'multiplication', increased production of neutrons at different points in the reactor which would signal progress towards a self-sustaining chain reaction. But they found that the number of neutrons was actually diminished, rather than increased. The same arrangement was used in a repeat experiment (designated B-II) some weeks later. This time a little over six tons of uranium oxide was used, again with a paraffin wax moderator. The results were similar.

In Leipzig, Robert Döpel worked under Heisenberg's direction to assemble an experimental pile with a different, concentric, arrangement of uranium oxide and paraffin wax. This experiment, designated L-I, was also negative. This merely served to confirm that a reactor could not be constructed using carbon and hydrogen as a moderator, at least in the form of paraffin wax.

In the meantime, experiments conducted by Bothe's team in Heidelberg confirmed that heavy water would function quite effectively as a moderator. But what of graphite? Bothe had tended to dismiss his team's earlier results on neutron absorption by graphite as the result of impurities and, through German Army Ordnance, he sought and obtained a quantity of pure graphite from the Siemens company. Siemens supplied a 100-centimetre sphere of electro-graphite claimed to be of the highest purity. In January 1941 he and Peter Jensen now found, much to their surprise, that purification had worsened, not improved, its performance. Bothe very much doubted that impurities could be to blame, and the physicists concluded that unless the uranium could be greatly enriched in U-235, graphite simply could not be used as a moderator in a nuclear reactor.

These results were undoubtedly affected by impurities – most probably boron contaminants arising from the manufacturing process which Szilard had the previous year taken great pains to exclude from samples of graphite used to make similar measurements at Columbia University. Göttingen physicist Wilhelm Hanle was suspicious of Bothe and Jensen's results, however. Although he was not part of the Uranverein, Hanle carried out measurements of his own to show that impurities were indeed to blame and that graphite would function as an effective moderator. He reported his results to Army Ordnance.[1]

---

[1] See Mark Walker, *German National Socialism and the Quest for Nuclear Power, 1939–1949*, p. 26.

In the meantime, opinion had swung firmly in the direction of heavy water as the moderator of choice. The cost of producing graphite of sufficient purity was deemed prohibitive, compared to the cost of an assumed abundant supply of heavy water from the captured Vemork plant.

Yet by the end of 1940 the German physicists had obtained only eight litres from Vemork. Construction of a heavy water plant in Germany was again debated and deemed uneconomic. Wirtz was despatched to inspect the Vemork plant and discuss ways in which production could be increased.

Critical to future progress

There was more bad news to follow. Harteck and fellow Uranverein scientist Hans Jensen in Hamburg had finally come to accept that the Clusius–Dickel thermal diffusion technique did not work for uranium hexafluoride. They had constructed separation tubes on a larger scale than Frisch in Liverpool, including an apparatus eighteen feet tall installed at I.G. Farben's Leverkusen works, but the results were identical. It had taken seventeen days to produce a single gram of uranium hexafluoride which contained twice the normal amount of U-235, a separation of a miserly 1 per cent. At temperatures for which uranium hexafluoride remained stable, the separation factor was found to be essentially zero. Separation might have been improved by operating at higher temperatures, but the uranium hexafluoride tended to decompose at higher temperatures. This was clearly not a technique that could be used to separate U-235 or enrich uranium on the scale they were going to need for a reactor, or a bomb.

The meeting of the Uranverein held in March 1941 was a gloomy affair. Harteck subsequently reported to the German War Office that the scientists faced two urgent problems. They needed to produce heavy water for use as a moderator and they needed to come up with an alternative way of separating U-235. Of these two problems, the first was deemed to be the more tractable – a sufficient quantity of heavy water would allow a reactor to be constructed from natural uranium. If sufficient quantities of heavy water were not available, then uranium enrichment would be required to allow a reactor to be constructed using ordinary water as a moderator. Harteck judged enrichment to be relevant 'only for special applications in which cheapness is but a secondary consideration'.

In other words, separation of U-235 was viable only if a bomb was to be built. The problem of isotope separation was not abandoned, and a couple of radical alternatives were discussed. Bagge raised the possibility of an electromagnetic separation method, exploiting the subtle differences in the flight paths of the different uranium isotopes as they are propelled in a confined 'atomic beam' through an electromagnetic field. Directing the beam through two shutters rotating at different speeds would allow that part of the beam rich in U-235 to pass; that part rich in U-238 would be blocked.

A second suggestion was made by Harteck's Hamburg colleague Wilhelm Groth. This was based on the use of an ultracentrifuge. Wirtz and Uranverein physicist Horst Korsching had made some promising advances with the application of thermal diffusion methods to liquids. Nobody in the Uranverein thought to suggest gaseous diffusion through a porous membrane, as Simon and Peierls had done in Britain.

The Uranverein physicists had put all their faith in the Clusius–Dickel method, so had no choice but to start working from scratch on alternatives. Access to heavy water was now absolutely critical to any future progress.

They should accelerate the thing

Even in these extraordinary times, Fritz Houtermans was rightly regarded as an extraordinary character. Born in Danzig, he grew up with his part-Jewish mother in Vienna. He rejected his privileged background (his father was a wealthy Dutch banker) and became a political radical. His programme of psychoanalysis with Sigmund Freud was terminated when he admitted that he had been inventing his dreams. He was expelled from school for publicly reciting *The Communist Manifesto* to fellow pupils on May Day.

Drawn to physics, he studied in Göttingen in Germany with James Franck and met a succession of notables passing through the German university town during this period, including Heisenberg, Fermi and Oppenheimer. In the 1920s and early 1930s, Houtermans established his scientific reputation primarily through his work on the physics of energy production in stars. While at Göttingen he met and courted German physicist Charlotte Riefenstahl[2] (she was also courted by Oppenheimer). He married Charlotte while attending a physics conference in Odessa in August 1931. Rudolf Peierls was a witness.

When Hitler came to power Houtermans had already developed a passionate contempt for Nazism. A brush with the Gestapo and Charlotte's insistence prompted their departure for Britain, but Houtermans was restless. He busied himself helping Szilard find places for the stream of displaced physicists emerging from Germany. He also worked out how to reproduce pages of *The Times* small enough to be hidden beneath postage stamps and sent these to friends in Germany in an attempt to combat the disinformation which formed the staple diet of the German media.

His Communist sympathies led him to be persuaded to take a job at the Ukrainian Physics Institute in Kharkov, just as the Great Purge was about to begin. Within a few years he was living the Stalinist nightmare. The Kharkov Institute was accused of harbouring German spies and Houtermans was arrested on 1 December 1937. With the help of Soviet physicist Peter Kapitza, Charlotte escaped with their two children first to Copenhagen, then to America.

Houtermans was in prison for two and a half years. Sent first to the notorious Lubyanka prison at NKVD[3] headquarters in Moscow, he was transferred to Butyrka, then Kholodnaya Gora in Kharkov, then the central NKVD prison in Kharkov, where he was tortured. He later vividly

---

[2] Charlotte was no relation to Leni Riefenstahl, the German film-maker and propagandist.
[3] Narodnyi Komissariat Vnutrennikh Del (NKVD), or the People's Commissariat for Internal Affairs, was the main directorate for state security in the Soviet Union. In addition to managing all of the USSR's prisons (including the Gulag – the network of forced labour camps), the NKVD was also responsible for administering the Soviet Union's foreign intelligence service and covert operations overseas. It was renamed the Ministerstvo Vnutrennikh Del (MVD) in 1946 and, after the arrest and execution of Lavrenty Beria, became the Komitet Gosudarstvennoy Bezopasnosti (KGB) in 1954.

described the various methods that had been used by his NKVD interrogators. One form of torture had required him to place both feet on the floor of his cell and lean forward against a wall with his full weight pressing on his fingertips. The pain in his fingertips quickly grew intolerable. And yet, he refused to break.

But when his interrogators threatened to arrest his wife and children (Houtermans did not know that they were by this time safe in America) he agreed to sign a confession, implicating colleagues who he thought were already out of the country and safely out of reach. He was released into the hands of the Gestapo in April 1940, a beneficiary of the Nazi–Soviet pact. He was promptly rearrested as a suspected Soviet spy and imprisoned in Berlin.

His colleague and close friend Max von Laue helped free him in July. He learned about the Uranverein, and was shocked to discover the roles that Heisenberg and Weizsäcker were now playing in the German nuclear project. But Houtermans himself was about to be drawn into work on nuclear fission.

Despite his release from prison he was still under suspicion and under Gestapo surveillance. He was banned from taking an academic position or working on government research projects. Laue found him a position in the research team of Manfred von Ardenne, an independent scientist and entrepreneur who had used an inheritance to found his own laboratory in Lichterfeld, a suburb of Berlin. Ardenne had managed to secure funds for independent research on nuclear fission in uranium from the German Post Office. Both Ardenne and Wilhelm Ohnesorge, head of the Reich Postal Ministry who agreed the funding, were aware of the possibility of atomic bombs based on uranium fission. Ohnesorge had even indirectly advised Hitler.

Houtermans was assigned the task of working out the theory of nuclear chain reactions. By the end of 1940 he had independently come to the conclusion that Weizsäcker, McMillan and Turner had each arrived at over a year before. Resonance capture of a neutron by U-238 would eventually lead to the production of a new fissionable element with 94 protons. If a nuclear reactor could be constructed, it could be used to produce element

94 which could be separated relatively easily from the spent reactor materials and used to make a bomb. Houtermans was horrified.

Ardenne was not an academic scientist. He had studied physics, chemistry and mathematics for only four semesters before leaving university to educate himself. He had set up his laboratory to carry out research on radio and television technology and electron microscopy, but he operated largely outside academic circles. The Uranverein physicists had no choice but to tolerate his activities but tended to keep their distance. Houtermans was an altogether different prospect, however. Unlike Ardenne he understood the physics and its implications. He expressed his concerns about the possibility of building an atom bomb based on element 94 to both Heisenberg and Weizsäcker in early 1941.

The precise details of the exchange between these three physicists were never made clear. Houtermans was not working on the 'official' uranium project and the Uranverein physicists would have been well aware of the interest that he still attracted from the Gestapo. It seems that Houtermans came to understand that Heisenberg and Weizsäcker were seeking to 'make use of warfare for physics', but, of all those involved in nuclear research, Houtermans also understood only too well how quickly moral resolve could collapse under brutal tyranny.

However, Houtermans also picked up signals that suggested Heisenberg and Weizsäcker were actively trying to play down the significance of element 94. This was a dangerous impression to leave with someone under the Gestapo's watchful eye. It also contradicted the impression left by Weizsäcker's own eagerness to communicate the possibility of building a bomb using element 93, which he had reported to the Army Weapons Research Bureau in July 1940. Furthermore, if Weizsäcker was actively trying to downplay the significance of element 94, then a patent application which he drafted sometime during 1941 which spells out how element 94 can be produced in a reactor, separated and used to make a bomb 'about ten million times' more powerful than any existing explosive material is, perhaps, difficult to understand.

In any event, Houtermans' concern spilled over into action. He was alerted by Laue to an opportunity to send a message to America via Fritz

Reiche, a Jewish physicist who had managed to secure the necessary travel permits and visas and in mid-March 1941 was about to depart for New York.[4] Houtermans asked Reiche to commit a message to memory. As Reiche later recalled, Houtermans had asked that he:

> Please say all this: that Heisenberg will not be able to withstand any longer the pressure from the government to go very earnestly and seriously into the making of the bomb. And say to them, say they should accelerate, if they have already begun this thing ... they should accelerate the thing.

Irrespective of Heisenberg's true motivations, his very involvement in the Uranverein sent all kinds of signals to physicists working in Britain and America, particularly those who were also refugees from Nazi Germany. Now Houtermans was signalling that Nazi desire for a super-weapon could soon overwhelm any internal resistance – real or not – from the German physicists themselves.

But no amount of element 94 could be produced without first building a working nuclear reactor. And no reactor could be built in Germany without first obtaining a reliable supply of heavy water. There would be no progress of any kind until this problem was solved.

Blood is thicker than heavy water

Jomar Brun was head of hydrogen research at Norsk Hydro when in 1933 he realised the potential for large-scale production of heavy water at the Vemork plant, whose primary function was to produce ammonia for use in nitrogen fertiliser. He worked with inorganic chemist Leif Tronstad from the Norwegian Institute of Technology in Trondheim to draw up plans for a heavy water production facility involving hundreds of electrolysis, combustion and condensation cells. It was an incredibly speculative proposal,

---

[4] He was just in time. All Jews were banned from emigrating from Germany in May 1941.

but Norsk Hydro had given the go-ahead and the facility met its first order for heavy water – from Birkbeck College in London – in August 1934. Tronstad and Brun published important results on the physical properties of heavy water in the British journal *Nature* in 1935.

Uranverein physicist Karl Wirtz had corresponded with Brun before the war and, as a result of Wirtz's visits to the plant, they had become friends. Together with Harteck, Wirtz made a further visit to the Vemork plant in May 1941. The German scientists discussed their requirements and proposals to expand production capacity using a new catalytic process devised by Harteck and his colleagues in Hamburg, but they were evasive about what they wanted the heavy water for.

Brun and Tronstad's suspicions do not seem to have been overly raised by these discussions. However, when Tronstad joined Operation Skylark, based in Trondheim, his suspicions were raised considerably. Skylark was part of a wider network of Norwegian resistance groups that had been established by the British Secret Intelligence Service (SIS)[5] to gather intelligence on the movements of German battleships along the Norwegian coastline. Skylark had been in radio contact with the SIS since February 1941. In April, the Norwegians had received the message:

FOR HEAVENS SAKE KEEP THIS UNDER YOUR HAT STOP TRY TO FIND OUT WHAT THE GERMANS ARE DOING WITH THE HEAVY WATER THEY PRODUCE AT RJUKAN STOP PARTICULARLY FIND OUT WHAT ADDRESS THEY SENT IT TO IN GERMANY STOP

The message likely originated with Lieutenant-Commander Eric Welsh, a veteran SIS operative and accomplished dissembler. Welsh spoke fluent Norwegian and had worked for many years in Bergen, where his expertise in industrial paints had been put to good use in the design of special corrosion-resistant floor tiles used at the Vemork heavy water plant. Welsh knew Brun, and was broadly familiar with the layout of the plant.

---

[5] Also known as Military Intelligence – Section 6, or MI6.

The reason for Welsh's enquiry is somewhat less clear. The MAUD Committee physicists were obviously well aware of the German interest in heavy water as a result of the briefing that Thomson, Oliphant and Cockcroft had received from Allier in April 1940, and it is perhaps possible that a request for more intelligence had been passed through to the SIS. It is also quite possible that Welsh had been tipped off by Paul Rosbaud who, after ensuring that his Jewish wife and daughter were safe in Britain in 1938 had returned to Berlin to continue to meet his editorial responsibilities for *Die Naturwissenschaften* and to spy for Britain. Rosbaud had helped Lise Meitner to escape and in January 1939 had rushed into print Hahn and Strassman's paper describing their results on uranium. His continued friendly relations with the German physicists involved in the Uranverein meant that he was aware of the significance of heavy water. Welsh was Rosbaud's spymaster.

The Norwegians' response, when it came, was rather puzzling:

IF YOU CAN ASSURE US THAT IT IS OF IMMEDIATE IMPORTANCE TO THE PRESENT WAR WE WILL GET THE INFORMATION YOU REQUEST AT ONCE STOP IF IT IS ONLY FOR THE ICI THEN PLEASE REMEMBER THAT BLOOD IS THICKER THAN HEAVY WATER STOP

Norsk Hydro and Britain's Imperial Chemical Industries (ICI) were commercial rivals. The Norwegian resistance simply wanted to ensure that they weren't engaged in industrial spying on behalf of British commercial interests. Tronstad was subsequently identified as the author of this message, although it seems that he did not join Operation Skylark until after the response was sent.

Tronstad did not immediately see the significance of Welsh's request. He had trained in chemistry in Berlin, Stockholm and Cambridge and had been appointed professor of inorganic chemistry at the Norwegian Institute of Technology in May 1936. Although he had been intimately involved with the development and operation of the Vemork facility, and few scientists knew the physical properties of heavy water better than he, he was not a nuclear physicist and was probably unaware of the substance's potential as a moderator in a nuclear reactor.

In any case, he didn't have much time to ponder. The Gestapo uncovered and shut down Operation Skylark in September 1941, and Tronstad was obliged to escape with his family to Britain via neutral Sweden the following month.

In Britain, Tronstad met with Welsh, who briefed him about the significance of heavy water and reassured him that details of the design of the Vemork plant were required in order to ensure that production was stopped, not because ICI were interested in obtaining Norsk Hydro's industrial secrets. Tronstad was convinced, and told the SIS all he knew.

Barbarossa

Hitler's sustained aerial bombardment of Britain began to wind down in May 1941. At the end of this period over 40,000 civilians had been killed and one million homes damaged or destroyed. Britain was not beaten, but it was bowed. Some among Hitler's military advisers urged that Britain should be finished off completely, but the country was on its knees and hardly likely to pose a threat in the immediate future. Besides, Hitler himself had become impatient. He had written in *Mein Kampf* of his intention to invade Russia and he didn't want to put this off any longer.

He launched Operation Barbarossa on 22 June 1941, deploying about three million German soldiers against a similar number of Soviet troops. The Nazi–Soviet pact was shown to be what everyone knew it to be the day it had been signed by Molotov and Ribbentrop – a sham.

Stalin had anticipated a German invasion, and had used the Nazi–Soviet pact to buy time. The Red Army was still recovering from the consequences of the Great Purge, and had not fared particularly well in the invasion of Finland in the winter of 1939–40. The contrast with the ruthless efficiency of the German war machine in its conquest of the rest of mainland Europe was stark, and plain for all to see. Stalin needed time to restructure his armed forces and rearm them, but had been wary of committing to a full mobilisation for fear of precipitating the very attack he was hoping to postpone.

It may be that he had at least anticipated a warning, an ultimatum from Berlin that would have allowed time to mobilise his forces.

There was no warning. The Red Army fell back in confusion as German forces swept through the Baltic States and into Russia itself. The German invasion was organised around three army groups supported by the Luftwaffe. Army Group North forged north-eastwards from East Prussia towards Leningrad. Army Group Centre headed for Moscow. Army Group South headed south-eastwards from Poland towards the Crimean Peninsula and the oilfields of the Caucasus.

Stalin was utterly shocked by the speed and ferocity of the attack. He had justified his brutal purges as necessary preparation for the defence of the Soviet Union. Now that defence had been tested, and found wanting. He waited for twelve days before broadcasting an unprecedented appeal to the Soviet people, calling them brothers and sisters and urging them to rally to the call of the party of Lenin and Stalin.

The Soviets declared a Great Patriotic War. The invasion led to the opening of the Eastern Front, set to become the largest theatre of war in human history.

I'm afraid it went badly wrong

Progress had been slow for the Uranverein through the summer and autumn of 1941. The German War Office had placed an order for 1,500 kilos of heavy water but by the end of the year only 360 kilos had been delivered. The German physicists now had plenty of powdered uranium metal, obtained from the Degussa company, but Heisenberg elected to use the first of the heavy water that was now becoming available in conjunction with uranium oxide.

In the late summer of 1941 Heisenberg and Döpel set up experiment L-II using the same reactor configuration they had used almost a year before, this time in an aluminium sphere about 75 centimetres in diameter containing a little over 140 kilos of uranium oxide and about 160 kilos of heavy water. The results were again negative, but after correcting for the absorption of neutrons by the aluminium their calculations suggested the merest hint of neutron multiplication. The physicists sensed they were on the right track. This was little more than intuition, a 'gut feeling', but Heisenberg later stated that: 'It was from September 1941 that we saw an open road ahead of us, leading to the atomic bomb.'

Heisenberg probably did not doubt that German science was already far ahead of anything that physicists might have done in Britain or America. This left him in a unique and, possibly, quite uncomfortable position. It was Weizsäcker who urged him to consult his former mentor, Niels Bohr. But Heisenberg's reasons for seeking such a meeting may have been quite complex.

Bohr had preferred to remain at his Institute for Theoretical Physics in Copenhagen following the German occupation of Denmark in April 1940. Of Jewish descent, Bohr was considered a 'non-Aryan' by the occupation forces but, like all 8,000 Danish Jews, was protected – at least temporarily – by the agreement reached with the Danish government designed to preserve the fiction that the Nazis were there by invitation.

That Heisenberg and Weizsäcker were concerned for Bohr's welfare is not in doubt. Growing moral qualms about the work the Uranverein was doing may also have played a part in Heisenberg's decision to visit Bohr in Copenhagen. In his memoir, written almost 30 years after the event, Heisenberg recalled Weizsäcker's proposal. 'It might be a good thing,' Weizsäcker had said, 'if you could discuss the whole subject with Niels in Copenhagen. It would mean a great deal to me if Niels were, for instance, to express the view that we are wrong and that we ought to stop working with uranium.'

According to Heisenberg, his primary purpose was to seek guidance from Bohr on the morality of working on scientific problems that could have 'grave consequences in the technique of war'. As Heisenberg's Uranverein colleague Peter Jensen later put it: Heisenberg, the high priest of German theoretical physics, sought absolution from his Pope. Or, perhaps, as Peierls later suggested: '[Heisenberg] had agreed to sup with the devil, and perhaps he found that there was not a long enough spoon.'

Heisenberg looked forward with great eagerness to the prospect of talking to his former mentor. To him, and indeed to many other physicists of his generation, Bohr had long been something of a father-figure. As Heisenberg's wife Elisabeth later wrote:

In Tisvilde, the beautiful vacation home of the Bohrs, he had played with their children and taken them for rides on a pony wagon: he had gone for long sailing trips on the ocean with Bohr, and Niels had visited him in his ski cottage; together they had grappled with the problems of physics, and he thought he could talk about anything with Bohr.

But there may have been other reasons for seeking a meeting. It is possible that, alarmed by reports of American efforts to build an atomic bomb that had appeared in the Swedish press, Heisenberg and Weizsäcker may have also wanted to discover what Bohr knew.

Having decided to pay Bohr a visit, Heisenberg now faced a number of practical hurdles. Although Germany dominated much of continental Western Europe, travel was restricted and the authorities were initially reluctant to let Heisenberg make the trip. Weizsäcker proposed a potential solution. He had already given several lectures in occupied Copenhagen, the most recent being a lecture on the philosophical implications of quantum theory at Bohr's institute. At Weizsäcker's prompting, an invitation was issued to both Heisenberg and Weizsäcker to participate in a symposium on astronomy, mathematics and theoretical physics at the newly-formed German Cultural Institute in Copenhagen.

The proposal was initially rejected by the Reich Ministry of Education but, after some arm-twisting by the German Foreign Office (whose officials suggested that State Secretary Ernst von Weizsäcker – Carl Friedrich's father – might intervene), approval was granted, provided Heisenberg kept a low profile and stayed only a few days.

Despite this proviso, Heisenberg arrived in Copenhagen early on Monday, 15 September 1941, four days before the conference was due to start. As he described it in a letter to his wife that he composed at various times during his trip, it was a journey into the recent past:

Here I am once again in the city which is so familiar to me and where a part of my heart has stayed stuck ever since that time fifteen years ago. When I heard the bells from the tower of city hall for the first time again, close to the window of my hotel room, it gripped me tight inside, and everything has stayed so much the same as if nothing out there in the world had changed. It is so strange when suddenly you encounter a piece of your own youth, just as if you were meeting yourself.

Such was his eagerness to see Bohr that he made his way to Bohr's Carlsberg residence that first night, walking through the darkened city under a clear and starry sky. He was relieved to discover that Bohr and his family were doing fine. Their conversation quickly turned to the 'human concerns and unhappy events of these times'. In his letter to his wife he expressed some dismay that 'even a great man like Bohr can not separate out thinking, feeling, and hating entirely', but then added: 'But probably one ought not to separate these ever.'

But Heisenberg remained stubbornly insensitive to the perceptions and feelings of his former colleagues. In occupied Copenhagen in September 1941, with much of Europe conquered by Axis forces, with the German Army Group North just seven miles from Leningrad, with Kiev encircled and the assault on Moscow about to begin, it was not difficult to conceive a German victory. There was a certain startling inevitability about the impending Nazi domination of Europe, and everything that this implied.

Heisenberg, the pragmatist, had made his bargain long ago. And here he now was, a representative of German culture, in Copenhagen at the behest of the German Cultural Institute, promulgating what many Danes may have perceived to be thinly-veiled Nazi propaganda. Surely, Heisenberg may have reasoned, in the face of the inevitability of Nazi victory it was in the best interests of themselves and of physics for his former colleagues in occupied countries to make their bargain too?

Bohr and his colleagues boycotted the formal conference proceedings, but Heisenberg still sought them out. He visited Bohr's institute and joined some of the physicists for lunch on a couple of occasions. Among those present were Christian Møller and Stefan Rozental. They later had bitter recollections of the discussion: '[Heisenberg] stressed how important it was that Germany should win the war ... the occupation of Denmark, Norway, Belgium and Holland was a sad thing but as regards the countries in Eastern Europe it was a good development because these countries were not able to govern themselves.'

It was during Heisenberg's second meeting with Bohr on the Wednesday evening that he raised the issue of atomic weapons. Their subsequent recollections of this highly-charged meeting were vague and contradictory. Heisenberg remembers that they took an after-dinner stroll, principally to avoid the risk of Gestapo surveillance. Bohr believed the conversation had taken place in his study. It would have made sense that Heisenberg would want to talk somewhere more secure, away from prying ears or listening devices, because when he raised the question of the military application of atomic energy with Bohr he was in principle committing an act of treason.

The conversation got off to a poor start. And then it got rapidly worse. Bohr had heard about Heisenberg's insensitive remarks, and became angry when Heisenberg not only defended Germany's invasion of the Soviet Union but further argued that it would be a good thing if Germany were to win the war. When Heisenberg finally raised the question of working on an atomic bomb, Bohr was completely and utterly shocked.

As far as Bohr understood, he had already shown in 1939 that achieving an explosive nuclear chain reaction would 'take the entire efforts of a country'. Yet here was his friend and former colleague, with whom he had shared some of the most thrilling moments of scientific discovery in his life, explaining with some impatience that a bomb was possible and that he was working on it for the Nazis. In a letter that Bohr composed to Heisenberg long after the war, but which he never sent, he wrote:[6]

---

[6] This should be read in its proper context. In drafting this, and subsequent letters and notes to Heisenberg, he was reacting – many years after the event – to Heisenberg's version of events that had been incorporated in the 1957 Danish translation of Robert Jungk's popular book, *Brighter Than a Thousand Suns*.

... in vague terms you spoke in a manner that could only give me the firm impression that, under your leadership, everything was being done in Germany to develop atomic weapons and that you said that there was no need to talk about details since you were completely familiar with them and had spent the past two years working more or less exclusively on such preparations.

Heisenberg may have even drawn a sketch to explain his work, though this now seems quite doubtful. When Bohr produced this some years later it appeared to be a sketch of a reactor, but whether Heisenberg intended it or Bohr misunderstood, Bohr assumed it was a sketch of an atomic bomb. Even worse was to come. Heisenberg appeared to be probing Bohr for any information about an Allied bomb programme. Was this, after all, an intelligence mission? On whose authority was Heisenberg now acting?

After the war, Heisenberg claimed he was trying – through Bohr – to establish a commitment from nuclear scientists not to develop atomic weapons. Whether this was really his intention or not, Bohr interpreted his efforts as those of a brashly confident representative of an aggressive occupying power bent on delivering the ultimate weapon to his masters. In another letter to Heisenberg that was never sent, Bohr wrote:

It had to make a very strong impression on me that at the very outset you stated that you felt certain that the war, if it lasted sufficiently long, would be decided with atomic weapons ... You added, when I perhaps looked doubtful, that I had to understand that in recent years you had occupied yourself almost exclusively with this question and did not doubt that it could be done.

Heisenberg later recalled Bohr's observation that it would be hopeless to try to influence the activities of physicists now working in various countries, 'and that it was, so to speak, the natural course in this world that the physicists were working in their countries on the production of weapons'.

Despite this exchange, they appeared to have parted amicably. Heisenberg met with Weizsäcker on the Langelinie, the picturesque walk near Copenhagen harbour. Heisenberg admitted: 'You know, I'm afraid it went badly wrong.'

Heisenberg and Weizsäcker participated in the conference on 19 September. It was attended by just five local astronomers from the observatory in Copenhagen. In his obligatory report on the visit, Heisenberg commented that 'our relations with scientific circles in Scandinavia have become very difficult'.

After attending a lunchtime reception at the German embassy, Heisenberg joined Weizsäcker for a last visit to the Bohr household. This final visit was free of political or scientific discussions. Bohr read aloud and Heisenberg played a Mozart sonata.

Faustian bargain

If Heisenberg had really been trying to stop the development of an atomic super-weapon, he had failed. If he had been seeking absolution, he had failed. But the manner of his failure was to have profound implications. The wartime efforts of nuclear scientists in Britain and America were driven by a deep-rooted fear of what the Nazis might do with such a super-weapon. Against this background of fear, the interpretation of the *intent* of the German physicists, especially of Heisenberg, and the determination of the Nazi military authorities provided a critical moral justification for the work that Allied physicists were now doing.

In the end, it did not matter much precisely what Heisenberg had really intended to say to Bohr. As a result of the meeting, one of the most respected and revered of nuclear physicists, a Danish half-Jew living under Nazi occupation, had been left with the firm impression that Heisenberg was working earnestly to deliver an atomic bomb to Hitler's arsenal.

Heisenberg had been unable to see all possible ends. His dangerous Faustian bargain was beginning to have unexpected consequences.

Chapter 5

TUBE ALLOYS

March–December 1941

The work of the MAUD Committee had proceeded apace through the last few months of 1940. Much more was now understood about the critical mass of a U-235 bomb and the challenges of separating the less abundant isotope from natural uranium on an industrial scale. And yet, the entire project still rested on little more than an intelligent guess.

All the physicists' calculations depended on an assumed rate of fission of U-235 nuclei by fast neutrons. It had not proved possible to separate even a small amount of U-235 on which to make some direct measurements. Chadwick's team in Liverpool used the cyclotron to measure how the rate of fission of natural uranium varied with the energy – or speed – of the neutrons fired at it. These rates depend on the combined fission rates for both U-235 and U-238, and their variation with neutron energy gives clues to the underlying behaviour of the individual isotopes. The experimental results closely fit theoretical predictions. 'The first test of theory had given a completely positive answer and there is no doubt that the whole scheme is feasible', Peierls wrote in March 1941.

The results confirmed independent measurements available from physicists working at the Carnegie Institution in Washington. The possibility of spontaneous fission in U-235, which Frisch had discovered nearly a year earlier, was potentially worrying as it was thought that it might lead to premature release of neutrons and hence premature detonation of any putative bomb. However, the rate of spontaneous fission was found to be insufficient to threaten the bomb's practical feasibility.

By April 1941 Simon's team in Oxford had experimented with a half-scale model of a single stage of the proposed gaseous diffusion plant, and a full-scale model was under construction. The results were sufficiently encouraging for Simon to propose building a twenty-stage pilot plant. Towards the end of May, Metropolitan-Vickers was awarded a contract for the design of this plant, to be constructed by the end of the year at the Valley Works site in Rhydymwyn, near Mold in North Wales.[1] ICI was contracted to supply quantities of uranium hexafluoride and to provide chemical engineering support for the plant's construction and operation.

It seemed that a method was now available for the large-scale separation of U-235 and it seemed certain that a bomb could be fashioned from a relatively small, super-critical mass of fissionable material. Attention turned to further questions about how best to bring such a mass together and what kind of explosion would result.

The most obvious way of creating an explosive super-critical mass of U-235 is to bring two sub-critical pieces together. This, it was reasoned, would need to be done very quickly. Assembled too slowly, the mass would release neutrons and detonate prematurely – it would simply blow itself apart – yielding much less than the explosive force potentially available. The proposed solution was to shoot a small sub-critical mass of the active material into another sub-critical mass, in a process that was later to become known as the 'gun method'. Thomson was assured by British armaments experts that constructing such a gun would be entirely practicable.

---

[1] This site had been purchased by the Ministry of Supply in 1938 and was used for the manufacture and underground storage of chemical weapons. The Valley Works factory produced mustard gas shells until April 1945. The site was decommissioned in 1959 and the stockpile of chemical weapons was disposed of over the following ten years. The site is now a nature reserve.

What kind of damage would such a bomb do? The calculations began to crystallise. The physicists estimated that a U-235 bomb consisting of just 25 pounds of active material would explode with a force equivalent to 1,800 tons of TNT. There was just one precedent for such an explosion. During the First World War the French munitions ship SS *Mont Blanc* had entered the harbour of Halifax, Nova Scotia. On deck were about 2,300 tons of wet and dry picric acid, 200 tons of TNT, ten tons of gun cotton and many drums of high octane fuel. On 6 December 1917 the *Mont Blanc* collided with a Norwegian ship, the SS *IMO*. The fuel spilled on the deck of the *Mont Blanc* and was soon ignited.

The resulting blast completely destroyed the ship and its surroundings over an area of nearly a square mile. Structural damage extended out a further half a mile. A mushroom-shaped cloud rose several miles into the sky as debris was thrown a distance of up to four miles and windows were shattered up to ten miles from the blast. The ship's gun landed over a mile away, near Alboro Lake. About 1,600 people were killed immediately, the death toll rising to 2,000 from secondary effects of the blast.

The physicists had little doubt that the effort required to build bombs capable of delivering such devastating effects could be justified. No nation would want to be caught without such a decisive weapon in its arsenal.

In Cambridge, Halban and Kowarski continued their research on a uranium–heavy water reactor. They confirmed that a nuclear reactor was no longer just a possibility, it was a near certainty. Although this was work not directly related to the design or production of a bomb, the MAUD physicists were by now aware of the potential for production of element 94. Perhaps surprisingly, they tended to play down its significance. Some argued that element 94 would be unsuitable for a bomb. Besides, it was clear that this material could be produced only in a working nuclear reactor, and a working nuclear reactor would need large quantities of heavy water (the physicists believed they would need several tons of both uranium oxide and heavy water to make a working reactor). There was no heavy water plant in Britain, and sourcing this substance in such quantities from the Vemork plant in occupied Norway was clearly out of the question. Separation of the small amount of U-235 needed for a uranium bomb seemed a much more practical and immediate route to an atomic weapon.

Work on various reactor designs continued, but the MAUD Committee judged that this was work that promised peacetime, rather than wartime, dividends. Halban and Kowarski were studying different reactor configurations, possible cooling systems and control systems. This work was obviously too important for it to be shelved for the duration of the war. And yet, when set against the wartime imperatives that the MAUD Committee physicists now faced in Britain, it was equally clear that the materials and resources required to support it properly could not be afforded.

Discussions about the future of Halban and Kowarski's work grew into a wider debate about what the MAUD Committee could hope to achieve in a country at war and under direct attack. It became clear that the next step for the committee physicists was to draft a compelling report and seek support for its conclusions from the British government.

Penny-in-the-slot

Peierls had continued with his MAUD Committee work through the winter of 1940–41, but he had greatly missed his associate Frisch. With pressing problems to address concerning the physics of gaseous diffusion, he decided that he needed a new assistant. He recruited a quiet, somewhat reserved and unassuming émigré who had been working at Edinburgh University. This was a physicist with some considerable skill in mathematics, just the kind of talented theorist that Peierls needed. He joined the group in Birmingham in May 1941 and took the room in the Peierls' home in Edgbaston that Frisch had vacated nearly a year before.

His name was Klaus Fuchs.

Fuchs had arrived in Britain on 24 September 1933, part of the first wave of émigrés looking to escape Nazi Germany. However, in Fuchs' case, this was not emigration forced by anti-Semitism and new Nazi regulations. Fuchs was a Roman Catholic. He was also a Communist. As a young boy he had adopted the socialist leanings of his father Emil, a clergyman in the Lutheran Church, but in 1932 had come to reject the Socialist Democratic

Party in favour of the harder line sponsored by the German Communist Party. He had sensed that the disciplines imposed by party membership and party activism were the only meaningful response to the Nazi threat.

As Hitler began to use the emergency powers granted to him as Chancellor in 1933, Fuchs stepped up his activism. Slightly built and ascetic, he would have hardly posed much of a threat to the Nazi thugs now stalking the streets, yet he did not lack courage. Challenging the Brownshirts who were protesting against the rector of Kiel University, he was promptly beaten up and thrown in the river. When the Nazis began arresting Communists in significant numbers, Fuchs first went into hiding in Berlin before accepting the advice of his party comrades to leave the country. He was just 21 years old.

He found refuge in England in the home of Ronald and Jessie Gunn, a wealthy English couple who were also Communist sympathisers. The Gunns introduced Fuchs to Nevill Mott, professor of physics at Bristol University, who had just been appointed head of the physics department. Mott agreed to find a position for Fuchs as an assistant in his research team.

Where in Germany Fuchs had been an outspoken activist, characterised by an almost brash self-confidence and self-belief, in Britain he became shy and withdrawn, rarely speaking unless spoken to. He did not openly discuss politics. But the fact that Fuchs was no longer outspoken on political matters did not mean he had abandoned his political beliefs. After a time, he quietly registered his presence in Britain with Jurgen Kuczynski, a fellow member of the German Communist Party who had come to Britain in 1936. Unlike Kuczynski, Fuchs did not openly declare his affiliation to the party. Not that this mattered, perhaps, as the British police had already been informed that Fuchs was a Communist by the German Consulate in Bristol.

Mott soon found that Fuchs was a talented and persistent – if not dogged – theoretician. Fuchs worked diligently and secured his doctorate, on the application of quantum mechanics to metals, four years later. Realising that Fuchs' personality and style would make for a poor lecturer, Mott sought and secured a post-doctoral research position for Fuchs with Max Born, another German émigré physicist, at Edinburgh University.

Born had helped to found the new quantum physics at Göttingen in the 1920s and early 1930s. Together with his assistant Pascual Jordan, he had in 1925 published an elaboration of a theory developed by Heisenberg which was to become known as matrix mechanics (an early version of quantum mechanics). Oppenheimer had studied for his doctorate under Born, and Fermi, Teller and Wigner had worked at different times as Born's research assistants. Although a Lutheran, Born was classified by his ancestry as a Jew, and he lost his professorship in 1933. He had taken up the offer of a lectureship at Cambridge University before becoming a professor of natural philosophy at Edinburgh University in 1936.

Born warmed to Fuchs and, inasmuch as it was possible to be a friend with someone who never really opened up, Born and Fuchs became friends. As Fuchs' reputation grew and he began to establish the beginnings of an academic career, he became somewhat less reserved. But he still kept his distance.

The Nazi–Soviet pact and the Soviet Union's subsequent invasion of Finland were both shocks to Fuchs' system, but he quickly managed to rationalise Stalin's decisions as necessary preparation for the war between Germany and the Soviet Union which would surely follow. His application for British citizenship was overtaken by the outbreak of war in September 1939, and he became an enemy alien. Initially classified as a minimum risk to security because of his anti-Nazi stance, he was nevertheless interned on the Isle of Man in June 1940 as the Wehrmacht swept through Europe. From the Isle of Man he was moved swiftly to an internment camp at Sherbrooke, near Quebec City in Canada.

Most of the internees at Sherbrooke were Jewish, as they constituted the majority of the German émigré population in Britain at the time. Their Canadian guards were puzzled, as Jewish émigrés (including a few rabbis) hardly seemed to pose much of a threat to the security of a country at war with Nazi Germany. However, some among their number were not Jews, and among these were genuine Nazis. Fuchs resented being imprisoned alongside them.

The camp developed a thriving culture. Fuchs gave lectures on physics. Now back in a German community, he once again openly acknowledged his Communist beliefs and regularly attended discussion meetings organised by fellow Communist internees. He was in touch with his younger sister Kristel, who had emigrated to America in 1936 and was now married and living in Cambridge, Massachusetts. Through one of her contacts she was able to arrange for Fuchs to receive some magazines. These were sent to Fuchs at Camp Sherbrooke by Israel Halperin, a young professor of mathematics at Queen's University in Kingston, Ontario. Fuchs and Halperin had never met, but Halperin was a member of the Canadian Communist Party.

Born lobbied the British authorities and just six months after arriving at Sherbrooke, Fuchs was released and returned to Edinburgh. He left Sherbrooke on Christmas Day, 1940. Five months later he received a letter from Peierls inviting him to join a project that couldn't be specified in detail, but which was expected to be important to the war effort. Fuchs accepted without hesitation.

There remained the question of Fuchs' clearance to work on such a sensitive wartime project. MI5 had only two items of intelligence on him. There was the report from the German Consulate in Bristol, and a more recent report from one among the community of German refugees. Both reports declared that Fuchs was a Communist. Peierls was told that he could hire Fuchs provided he was told only what he needed to know in order to carry out his work, but Peierls claimed he could not work with Fuchs this way. The restriction was dropped and Fuchs was cleared to join the MAUD Committee team of physicists.

Fuchs quickly settled into his new surroundings in Birmingham. He set to work on two main tasks, the theory of nuclear chain reactions in U-235 and the theory of gaseous diffusion as a method for separating U-235 from U-238. Fuchs' quiet reserve led Genia Peierls to coin him a new nickname – 'penny-in-the-slot'. As she would explain to others: 'Put a question in and you get an answer out. But if you don't put anything in, you don't get anything out.'

The MAUD report

On 15 July 1941, the MAUD Committee submitted two reports, one on the use of uranium for a bomb, the second on the use of uranium as a source of power. The first of these reports was unequivocal:

We have now reached the conclusion that it will be possible to make an effective uranium bomb which, containing some 25 lb of active material, would be equivalent as regards destructive effect to 1,800 tons of T.N.T. and would also release large quantities of radioactive substance, which would make places near to where the bomb exploded dangerous to human life for a long period.

The report went on to make three recommendations:

(i) The committee considers that the scheme for a uranium bomb is practicable and likely to lead to decisive results in the war.
(ii) It recommends that this work be continued on the highest priority and on the increasing scale necessary to obtain the weapon in the shortest possible time.
(iii) That the present collaboration with America should be continued and extended especially in the region of experimental work.

The committee physicists declared that they had started work on the project 'with more scepticism than belief', but, perhaps somewhat modestly, stressed that 'the lines on which we are now working are such as would be likely to suggest themselves to any capable physicist'.

The report was optimistic about the likely time to production, suggesting that an atomic bomb could be available as soon as the end of 1943. Among the MAUD Committee physicists, only Blackett thought this overly optimistic. He very much doubted that so novel a project could be implemented without unforeseen difficulties likely to cause delays.

The MAUD Committee reports produced a flurry of activity in Whitehall. Lindemann had maintained a watching brief on the physicists' deliberations and had attended many of the meetings of the technical sub-committee. He set great store by the judgement of Thomson and Simon and had been impressed by Peierls. Lindemann was now the British government's leading scientific adviser. He had been ennobled as Lord Cherwell in June.

Cherwell had counselled Churchill in August 1939 that atomic weapons would not be available for 'several years'. Although he had not been proved wrong, his initial scepticism about the possibility of an atomic bomb had now given way to great concern. Knowing Churchill liked briefing documents to occupy no more than half a page, Cherwell thought this one so important that he let it run to two and a half. But he still hedged his bets: 'I would not bet more than two-to-one against, or even money,' he wrote, 'But I am quite clear we must go forward. It would be unforgivable if we let the Germans develop a process ahead of us by means of which they could defeat us or reserve the verdict after they had been defeated.'

Churchill sought the views of his Chiefs of Staff: 'Although personally I am quite content with the existing explosives,' he claimed, 'I feel we must not stand in the path of improvement, and I therefore think that action should be taken in the sense proposed by Lord Cherwell: and that the Cabinet Minister responsible should be Sir John Anderson. I shall be glad to know what the Chiefs of Staff think.' John Anderson, Lord President of the Council, was a former Home Secretary and physical chemist. He had completed his doctorate on the chemistry of uranium at Leipzig University.[2]

The MAUD Committee reports were formally reviewed by the Defence Services Panel of the Scientific Advisory Committee. This was chaired by Lord Hankey and included the physicist Sir Edward Appleton, the pharmacologist Sir Henry Dale, Nobel laureate and President of the Royal Society, and Sir Edward Mellanby, who had discovered vitamin D. The panel met with MAUD Committee physicists on 16 September. They discussed

---

[2] Anderson had been responsible for preparing defensive measures against air raids prior to the outbreak of war. He initiated the development of a prefabricated air-raid shelter that came to be called an 'Anderson shelter'.

fuse mechanisms, the twenty-stage gaseous diffusion pilot plant, ICI's commitment to source quantities of uranium hexafluoride and types of membranes that could be used in the diffusion plant that were available in America.

This was a meticulous review, which concluded:

> We have been impressed by the unanimity and weight of scientific opinion by which the proposals are supported. The destructive power of the weapon which would thus be created, and the ultimate importance of the issues at stake, need no emphasis. Moreover, we have to reckon with the possibility that the Germans are at work in this field and may at any time achieve important results ... For all these reasons we are strongly of the opinion that the development of the uranium bomb should be regarded as a project of first class importance and all possible steps should be taken to push on with the work.

On 20 September the Chiefs of Staff agreed, recommending that no time, labour, materials or money be spared. The panel completed its report, which also covered many points of policy and organisation, and submitted it to Anderson on 25 September.

Responsibility for the British project was passed to the Department of Scientific and Industrial Research. A senior ICI executive, Wallace Akers, was appointed to head it. Anderson and Akers gave the project a suitably misleading name – Tube Alloys – and the new organisation headed by Akers was called the Directorate of Tube Alloys. Akers was assisted by a deputy, Michael Perrin, also from ICI, and in October 1941 set up offices at 16 Old Queen Street in London, close to the headquarters of the SIS in Queen Anne's Gate.

Leif Tronstad was one of the first to visit.

Of great value to the Soviet Union

On precisely the same day of the panel's report to Anderson, NKVD spy Anatoly Gorsky (alias Soviet embassy attaché Anatoly Gromov) filed a report from London to Moscow Centre. In Moscow the report was circulated by Yelena Potapova, an NKVD officer with good English and a grounding in science. The report read:

VADIM has relayed a report from [LISZT] about a meeting of the Uranium Committee of September 16, 1941. The meeting was chaired by BOSS.

VADIM was Gorsky's codename. BOSS was a reference to Lord Hankey. Gorsky's source was John Cairncross, Lord Hankey's private secretary and a Soviet spy. Cairncross had been recruited to the Soviet cause in May 1937 by Anthony Blunt and Guy Burgess, as a replacement-in-waiting for another Soviet spy in the Foreign Office, Donald Maclean.[3] Cairncross's decision to betray atomic secrets to the Soviets was justified, in his own mind, as a way of providing practical support to an Allied power in the fight against the mutual threat of Nazism. Cairncross's favourite composer was Franz Liszt.

The report went on to describe everything that had been discussed at the meeting, and so summarised the present state of the British atomic bomb project. The report concluded:

The Chiefs of Staffs Committee at their meeting on September 20, 1941, made a decision to immediately launch construction in Britain of a plant to manufacture uranium bombs.

The Soviet Union thus knew of Britain's decision to build an atomic bomb literally within days of the decision being made.

Soon the Soviets would know much more. While on a visit to London in late 1941, Fuchs decided to visit his friend Kuczynski. Perhaps Fuchs had guessed that Kuczynski was also working for Soviet intelligence. In

---

[3] Cairncross is often referred to as the 'fifth man' in the notorious Cambridge spy ring. However, although Cairncross knew of Blunt, Burgess, Maclean and 'Kim' Philby, he later insisted that there was never a link between their activities as spies.

fact, Kuczynski was an agent of the GRU.[4] Fuchs had advised him that he had information about a secret project that could be of great value to the Soviet Union.

Fuchs became a Soviet spy.

Kuczynski put Fuchs in contact with Simon Kremer, secretary to the military attaché at the Soviet embassy in London. Kremer was also an agent of the GRU, known to Fuchs only by his cover name Alexander. Kremer gave Fuchs some rudimentary instruction in the tradecraft of a spy and they organised clandestine meetings at a house near Hyde Park. At these meetings Fuchs would hand over pages of carefully typed or handwritten script summarising the work in which he himself had been directly involved.

Despite his commitment to the Soviet cause and his decision to betray the secrets of the country for which he was now working, Fuchs adopted his own strict ethical code. He had relatively free access to the work of others, such as Peierls, and had sight of American research, but he refused to pass on documents relating to anything other than his own work. However, he did provide verbal summaries of what he knew.

After three meetings with Kremer over six months, Fuchs was passed to another agent, known to him only as Sonja. This was, in fact, Ruth Beurton, née Kuczynski, Jurgen's sister. Living in Kidlington, near Oxford, with a British husband, Ruth Kuczynski had no formal association with the Soviet embassy and was therefore less likely to attract unwanted attention from MI5. They arranged to meet in Banbury, a quiet market town between Birmingham and Oxford.

Meddling foreigners

'If Congress knew the true history of the atomic energy project,' said Leo Szilard after the war, 'I have no doubt that it would create a special

---

[4] Glavnoe Razvedyvatel'noe Upravlenie (Main Intelligence Directorate), created by Lenin in 1918. The GRU handles all military intelligence.

medal to be given to meddling foreigners for distinguished services, and Dr Oliphant would be the first to receive one.'

American work on nuclear fission had continued to stumble forward through the winter of 1940–41, but the programme was languishing. Research was under way at various institutions on the theory of fission reactions, isotope separation, the properties of element 94, nuclear reactors and the production of heavy water. But none of this work was directed towards any kind of war objective.

NDRC funds were being used to study separation of U-235 by gaseous diffusion at Columbia University, by high-speed centrifuge methods at the University of Virginia, and by electromagnetic methods of the kind that Nier had used on a small scale in Minnesota to obtain tiny quantities of the isotope. Nier concluded that the electromagnetic method would be unsuitable for separation of U-235 on a large scale but Lawrence saw an opportunity to put one of his redundant cyclotrons to good use. The electromagnetic separation method and the working principles of the cyclotron are very similar, and Lawrence used NDRC sponsorship to investigate the possibility of converting the 37-inch cyclotron at the Rad Lab into a large mass spectrometer for isotope separation.

Bush remained determinedly sceptical of the prospects for progress towards a bomb, and much of the high-level discussion centred around the use of nuclear fission as a means of generating power. On 17 March 1941 Lawrence decided to agitate. Conant had just returned from a visit to Britain, a visit which had afforded him an opportunity to discuss nuclear fission with a number of MAUD Committee physicists. These discussions had opened his mind to the possibility of a bomb, but he had simply assumed that when Bush was ready to hear about what the British had to say, he would go through the proper channels. Lawrence thought the time had come to 'light a fire under the Briggs Committee'. He asked Conant to pass on the message of frustration to Bush.

When Bush met with Lawrence two days later, Bush let him have both barrels. It was he who was running the show, he claimed, and he would back Briggs and the Uranium Committee unless there was a strong case for getting personally involved. In truth, Bush was still puzzled by the science, and he feared the possibility of spending great sums of money without a clear end result. He decided to turn to the National Academy of Sciences for help.

Bush asked the National Academy for an 'energetic but dispassionate review of the entire situation by a highly competent group of physicists'. In April 1941 the Academy in turn asked Nobel laureate Arthur Compton to head the review group. Compton was a respected physicist whose work on the absorption and scattering of X-rays and gamma rays had led him to discover the 'Compton effect',[5] but he was not a nuclear physicist. Having initially expressed doubts about his own fitness for the role, he quickly and eagerly accepted. Lawrence was also appointed to the group, along with physicists John C. Slater and John H. Van Vleck.

Should further impetus have been needed, it arrived in the form of Houtermans' warning. On his arrival in America, Reiche had passed the message to Rudolph Ladenburg at Princeton. Ladenburg invited a number of distinguished physicists to dinner so that Reiche could repeat the message. Within this group, only Wigner was involved with the Uranium Committee and he made no comment. When, a few days later, an opportunity arose to alert Briggs himself, Ladenburg did so. Briggs expressed deep concern, asked that he be given any further information and promptly buried the matter in his files.

In the end, Compton's review group needed three tries to get the story straight. The group submitted its first report on 17 May. It reflected Briggs' conservatism, focusing on the promise of the controlled release of nuclear energy which would require years of development. No firm recommendations were made about the possibility of a bomb, but the report did suggest that any such weapon could not be expected before 1945. There was nothing at all in the report about fission of U-235 by fast neutrons, critical mass or bomb mechanisms.

Meanwhile, the prospects for a bomb based on element 94 were improving considerably. Glenn Seaborg was the son of Swedish immigrants (his

---

[5] And incidentally proving that light behaves as a collection of particles (photons) as well as a wave.

father's surname, Sjöberg, had been anglicised on Ellis Island). Born in Michigan, Seaborg had studied chemistry at the University of California, Los Angeles (UCLA) and had completed his doctorate at Berkeley, where he had become interested in the chemistry of radioactive substances. After hearing of Hahn and Strassman's discovery, he was inevitably drawn to the study of uranium, and the new elements 93 and 94. Together with his graduate student Arthur Wahl, he had succeeded in separating a microscopic amount of element 94, a discovery he had wanted to shout from the rooftops, but which he had instead quietly filed with the Uranium Committee and the editor of *Physical Review* for publication after the war.

At first Seaborg did not give the new element a name, referring to it in conversation using the codename 'copper'. When subsequent experiments required the use of real copper, the physicists had to refer to it as 'honest-to-God copper', to distinguish it from element 94.

The Berkeley scientists immediately began work to investigate the fission properties of the new element. McMillan had left Berkeley in November 1940 to work on radar at MIT.[6] So Seaborg and Emilio Segrè continued to use the 60-inch cyclotron to produce quantities of element 94 so that its fission properties could be measured. On 18 May they recorded a slow-neutron fission rate for element 94 that was nearly twice the comparable rate for U-235. There could now be no doubt that element 94 would be a suitable active material for an atomic bomb.

Bush was busy reorganising the structure and management of government-funded science. The NDRC was appropriate for managing laboratory research efforts but had no authority over the kind of engineering effort that would be required to translate scientific research into armaments. He proposed to establish a new agency, the Office of Scientific Research and Development (OSRD), with executive authority over the NDRC and any engineering projects resulting from its work. Bush proposed himself as

---

[6] The MIT laboratory where wartime research on radar was carried out was named the Radiation Laboratory to disguise its purpose. To avoid confusion with the Berkeley Radiation Laboratory, in this book I will always refer to 'radar at MIT' and reserve the name Radiation Laboratory (or Rad Lab) for the Berkeley facility.

chairman of the OSRD, reporting directly to Roosevelt. Conant replaced him as chairman of the NDRC.

The Nazis launched their invasion of the Soviet Union on 22 June, forcing the pace still further and changing the status of a decision on the future of the American fission programme from 'urgent' to 'desperately urgent'. Conant decided that the review group needed an injection of engineering pragmatism, and engaged the services of engineers from General Electric, Bell Laboratories and Westinghouse. But their second report, delivered on 11 July, was another step backwards. The prospects for nuclear power were again positively evaluated, but there was hardly any mention of element 94, or a bomb.

Compton, who had been in South America at the critical time, feared that the government might abandon fission studies altogether. Lawrence had missed the meeting at which the report had been drafted because his daughter Margaret had been ill. He decided to draft a separate letter to the review group spelling out the importance of element 94. 'If large amounts of element 94 were available,' he wrote, 'it is likely that a chain reaction with fast neutrons would be produced. In such a reaction the energy would be released at an explosive rate which might be described as a "super bomb".'

Bush received an unofficial copy of Thomson's draft of the MAUD Committee report shortly before it had been approved in July. This led to some high-level discussions and a slightly greater sense of urgency, but it appears that Bush resolved to wait until he had received a copy of the report through official channels before taking any further action.

Enter Oliphant.

It had become transparently clear that Britain could not possibly contemplate building an atom bomb on its own. The country lacked the money and the resources for the job and despite the fact that Germany's attentions had turned eastward, Britain remained a nation under siege. Oliphant flew to America in late August 1941 to find out what was happening and, if necessary, stir things up.

He found that the MAUD Committee report had been passed to Briggs and that 'this inarticulate and unimpressive man had put the reports in his safe and had not shown them to members of his Committee'. Oliphant was greatly distressed. When he met with members of the Uranium Committee he talked openly and persuasively about the possibility of a bomb. At least one member of the Committee was shocked. Oliphant had come to the meeting and 'said "bomb" in no uncertain terms … I thought we were making a power source for submarines', he said.

Oliphant met Lawrence in Berkeley on 21 September. Lawrence drove them up Charter Hill, to the construction site where the 184-inch super-cyclotron was being built. Oliphant summarised the MAUD Committee report and Lawrence enthused about electromagnetic separation of U-235 and the fission properties of element 94. Back in Lawrence's office, they were joined by Oppenheimer, who heard for the first time about the tentative steps now being planned to build an atomic bomb.

Oliphant had further, unsatisfactory, meetings in New York with both Conant and Bush, and left to return to Britain wondering if his visit had had any real impact. He need not have worried. Lawrence was now fully committed. He contacted Compton, telling him that he now believed that an atomic bomb could be built that might determine the outcome of the war. Compton suggested they both meet with Conant a few days later at the University of Chicago's 50th birthday celebrations, at which both Lawrence and Conant were due to receive honorary degrees.

They met at Compton's house. Lawrence summarised the British work, the results on element 94 and the prospects for separating U-235. He expressed his dissatisfaction with Washington's complacency in the face of compelling evidence of German interest in atomic energy. Conant's initial reluctance gave way to conviction as Compton rallied to the cause. Then Conant turned to Lawrence: 'Ernest, you say you are convinced of the importance of these fission bombs', he said. 'Are you ready to devote the next several years of your life to getting them made?' Lawrence's jaw dropped, but he did not hesitate. If Conant told him this was his job, he would do it.

Bush received an official copy of the MAUD Committee report on 3 October 1941, two weeks after its content had been discussed in Moscow. Bush took the report to Roosevelt on 9 October. America, criticised for its policy of isolationism, was not yet at war. Yet when presented with the evidence that an atomic bomb was possible within the likely timeframe of the war in Europe, Roosevelt committed to action, without consulting Congress. He also reserved the matter of nuclear policy for himself and a small group of advisers, later to become known as the Top Policy Group, consisting of Bush, Conant, Vice President Henry Wallace, Secretary of War Henry L. Stimson and Army Chief of Staff George C. Marshall.

A third report was requested from the National Academy review group. Conant asked his Harvard colleague, chemist and explosives expert George Kistiakowsky, to participate. With Compton's blessing, Lawrence asked Oppenheimer to support the theoretical work. Compton had known 'Oppie' for fourteen years and was glad to receive his helpful suggestions.

Compton presented the third, and final, report personally to Bush on 6 November 1941. Like the MAUD Committee report, this was now unequivocal:

A fission bomb of superlatively destructive power will result from bringing quickly together a sufficient mass of element U-235. This seems to be as sure as any untried prediction based upon theory and experiment can be ... The mass of U-235 required to produce explosive fission under appropriate conditions can hardly be less than 2 kg nor greater than 100 kg ... The possibility must be seriously considered that within a few years the use of bombs such as described here, or something similar using uranium fission, may determine military superiority. Adequate care for our national defense seems to demand urgent development of this program.

The American physicists were much more conservative in their estimates than their counterparts in Britain, but their conclusions were broadly similar. In the rush to embrace U-235, element 94 had once again slipped off the radar. Bush delivered the report to Roosevelt on 27 November. Roosevelt in effect sanctioned a decision that had already been made.

A new committee, called the S-1 Committee, was convened, reporting to Bush's OSRD. Bush had thought to appoint Lawrence to chair this committee, but he had grown increasingly concerned about Lawrence's inability to work under strict secrecy – Conant had reprimanded him for telling Oppenheimer about the project without authorisation. In the event, Conant became chairman. Briggs was left in place as chairman of the S-1 Section, devoted to physical measurements, and also served as a member of the S-1 Committee.

There appears to be no formal document signalling the decision to begin the American atomic bomb project, merely a short note on White House paper accompanying the returned report of the National Academy review group. The note is dated 19 January 1942 and reads: 'V.B. OK – returned – I think you had best keep this in your own safe. FDR.'

Winter in Moscow

The German encirclement of Moscow stumbled to a halt on 15 November 1941, as a cold, hard frost bit deep into the mud. By the end of November it was snowing heavily and the temperature had dipped to minus 20° Celsius. Machinery froze. Without gloves, winter shoes and clothing thick enough to cope with a hard Russian winter, men froze too. Many froze to death. The Russians, however, did not.

On 6 December General Georgei Zhukov ordered his troops to launch a massive counter-attack on German positions across a 200-mile front. The Soviet Union had lost about four million men, two-thirds of its coal production capacity, and three-quarters of its iron production. The sheer scale of the Soviet counter-offensive in the depths of winter was startling. Hitler ordered his troops to maintain their positions, but they could not. They were overwhelmed.

Pearl Harbor

Japan was a small island nation with limited natural resources but with big ambitions. Steeped in a warrior culture, Japanese writers and poets of the late nineteenth century had elaborated the vision of a Japanese Asian empire, presided over by Japan's emperor, a direct descendant of the sun god Amaterasu. In the first decades of the twentieth century, Japan had begun to build the foundations for such an empire using the unique blend of manufacturing know-how, commercial capability and violence borrowed from the successful model of British imperialism.

In 1937 Japan had renewed hostilities with its cultural nemesis, China. In August 1940, it had announced the Greater East Asia Co-Prosperity Sphere, intended as an economic bloc designed to achieve self-sufficiency for Japan, China and the Far Eastern British, Dutch and French colonies. It was not a new concept, and was perceived largely as an attempt to free Asia from European imperialism, replacing the European invaders with a Japanese empire. On 27 September 1940 Japan signed the Tripartite Pact in Berlin, binding the country to Germany and Italy, and undertaking to 'assist one another with all political, economic and military means if one of the Contracting Powers is attacked by a Power at present not involved in the European War or in the Japanese–Chinese conflict'. It was a not-so-veiled threat to America.

America responded with economic sanctions and boycotts. As Japan increased pressure on Malaya, the Dutch East Indies and French Indo-China in the first few months of 1941, America tightened the screw. On 25 July Roosevelt imposed an oil embargo and froze Japanese assets. The embargo tipped Japan's delicately balanced economy into crisis.

Roosevelt moved the American Pacific Fleet to Hawaii and ordered a build-up of forces in the Philippines, hoping that this show of strength would deter any further Japanese aggression in the region. Japan's military leaders considered their options. Admiral Isoroku Yamamoto, commander of Japan's Combined Fleet, had won support for a plan to attack American forces in the spring of 1941. He had been building up *matériel* and training pilots through the summer. The attack was eventually authorised on 1 December.

At 07:58 on 7 December an urgent message was relayed to a startled world from the Ford Island command centre:

AIR RAID PEARL HARBOR. THIS IS NOT DRILL.

The Japanese had launched their attack. The Allies declared war on Japan the next day. Germany and Italy declared war on America on 11 December. American Admiral William Halsey declared: 'When this war is over, the Japanese language will be spoken only in Hell.'

The war in Europe had now become the Second World War.

PART II

WEAPON

Chapter 6

A MODEST REQUEST

March–November 1942

Einar Skinnarland was a ruddy, blond, jovial Norwegian in his early twenties. Born and raised with seven brothers and sisters in the small town of Rjukan, he worked as a construction superintendent at the Møsvatn (Mos Lake) dam which provided power for the region, including the Vemork plant. His brother Torstein also worked as an engineer at the dam. He was highly resourceful, fluent in English, a champion skier and intimately familiar with the local landscape and the community that lived within it. In March 1942 he advised his employers of his intention to take some leave.

But instead of taking a holiday, he joined a small group of resistance fighters formally known as Norwegian Independent Company, No. One.[1] The unit had been formed a year before by the British Special Operations Executive (SOE) to carry out commando raids in occupied Norway. Under the leadership of Odd Starheim,[2] on 15 March the group seized the

---

[1] After the war this group became known as *Kompani Linge*, named for its first leader, Captain Martin Linge, who was killed by a German sniper during a raid in December 1941.
[2] Starheim was one of the first Norwegian resistance fighters recruited by the SOE. He was parachuted back into Norway in January 1942, charged with organising resistance in

600-ton coastal steamer SS *Galtesund* and sailed it to Britain. After two days crossing the North Sea in ferocious weather, they arrived in Aberdeen, Scotland.

The SOE had been established by Churchill on 16 July 1940 to conduct war by other means – facilitating espionage and sabotage behind enemy lines. It was also intended to serve as the core of a resistance movement in the event that Britain itself was invaded. It was formed initially from three departments: Section D of the SIS, Military Intelligence Research (a War Office department), and a propaganda organisation known as Department EH. The SOE went by various alternative names, including the 'Baker Street Irregulars',[3] the 'Ministry of Ungentlemanly Warfare' and 'Churchill's Secret Army'. Its activities remained strictly secret throughout the war.

Inevitably, there was a certain amount of internal rivalry between the SIS and SOE. The SIS was charged with the task of gathering intelligence and exerting influence through its network of agents and therefore favoured a quiet, thoughtful and calm approach to its missions. In contrast, the SOE thrived on creating chaos, consistent with Churchill's charge that it should 'set Europe ablaze'. A successful SOE operation frequently resulted in a crackdown by the Gestapo, risking the exposure and the very lives of valuable SIS agents. As an organisation the SOE was not without restraint, however. It adhered to a rule that there should be no explosions without the prior approval of the Foreign Office.

Skinnarland was a gift to the SOE. He had extensive contacts at the Vemork plant and knew the area as well as anyone could possibly know it. On his arrival in Britain he was questioned by an expert on Norwegian affairs. It soon became apparent that if they could get Skinnarland back to Rjukan before he was missed, he would be able to pick up his life as though

---

southern Norway together with his colleague Andreas Fasting, known by their codenames 'Cheese' and 'Biscuit'.
[3] The SOE was headquartered at 64 Baker Street in October 1940. As the organisation expanded it occupied 82 Baker Street and the top floor of 83 Baker Street situated across the road. The Baker Street Irregulars was the name of Sherlock Holmes' fictional group of spies.

nothing had happened. Armed with some rudimentary skills, he could be a very useful British spy working close to a production plant critical to the German nuclear programme.

Skinnarland was briefed by Tronstad, now head of Section IV of the Norwegian High Command, responsible for intelligence-gathering, espionage and sabotage in collaboration with the SOE. He was also given a short but intensive training course in wireless/telegraph operation, explosives and intelligence-gathering at one of the SOE's Special Training Schools in the Scottish Highlands. There was time enough for only one practice parachute jump. On 28 March, only eleven days since arriving in Aberdeen, he took off from Kinloss bound for a dropping zone near Rjukan. Aside from a moment's panic as he attempted what was only his second parachute jump, he landed safely and the following morning reported for work.

He told his colleagues that he had enjoyed a relaxing break.

Nuclear physics as a weapon

Heisenberg had returned empty-handed from Copenhagen, but he had little time to ponder on what might have been or what this meant. In September 1941 the German armed forces had appeared near-invincible. They had annexed or conquered most of Western Europe and the Eastern European countries bordering the Soviet Union. But by December the tide of war was beginning to turn against Nazi Germany on its Eastern front.

Hitler put the German economy on a formal war footing. Difficult decisions had to be taken, with conflicting demands weighed in the balance of expediency. Inevitably, the Uranverein received notice that its work could continue 'only if there is a certainty of getting some benefit from it in the near future'.

Bothe, Hahn, Harteck and Heisenberg were called to report on the progress they had made at a conference held on 16 December 1941 at the headquarters of German Army Ordnance in Berlin. The conference concluded that nuclear fission was unlikely to offer any significant advantage to the German war effort, at least in the near future. Schumann went on to recommend that the army withdraw from nuclear research and from the Kaiser Wilhelm Institute for Physics, and pass responsibility for supervision of the programme to another organisation. The Reich Research Council, which had originally established the Uranverein in April 1939, was waiting eagerly in the wings.

If, as they later claimed, the German physicists were looking for ways to avoid work that might lead to delivery of a super-weapon for Hitler's arsenal, then this decision certainly worked in their favour. However, the Reich Research Council was a weak organisation and the army retained control over that part of the programme that it had itself initiated in 1939 under Kurt Diebner.

By this time Heisenberg and Weizsäcker's research priorities appeared to have undergone a subtle shift. On their return from Copenhagen, they had tended to focus their research efforts on a reactor and downplay the potential for a weapon. But Diebner did not share their pessimism. His group had continued to contribute to the nuclear research programme from a laboratory in the Gottow suburb of Berlin, and he had developed considerable enthusiasm for the prospects for a bomb.

The Uranverein drafted a report, delivered to German Army Ordnance in February 1942, which summarised the physicists' current thinking. It seems that optimism prevailed. The report clearly stated the potential for an explosive 'a million times greater than the same weight of dynamite' based on either U-235 or element 94, the latter produced in a nuclear reactor. It concluded that a bomb could be constructed using from '10 to 100 kilograms of fissionable material' and recommended that a major industrial effort be initiated. It is clear from this report that the Uranverein had arrived at broadly the same conclusions as the MAUD Committee in July 1941 and the National Academy review group four months later.[4] Within the space of just seven months, physicists in Britain, America and now Germany had all concluded that an atomic bomb was feasible in principle,

---

[4] The 25 pounds of active material mentioned in the MAUD Committee report corresponds to a little more than 11 kilograms. The third National Academy report suggested that an explosive mass of U-235 between 2 and 100 kilograms would be required. This suggests that the German physicists understood the principles of an atomic bomb based on the physics of fast-neutron fission reactions.

and they had all determined a similar range for the mass of active material that would be required.

But whereas the MAUD Committee and National Academy conclusions galvanised British and American efforts, the Uranverein conclusions appear to have become lost in the noise of economic re-prioritisation, as the German military made ready for a war of attrition on the Russian front. In January both Harteck and Weizsäcker were called up. Heisenberg had to pull out all the stops, through personal contacts in the military, to restore their status as 'indispensable' contributors to the nuclear research effort, and therefore exempt from direct military service.

Despite the optimistic assessment for the prospect of atomic weapons in the report to Army Ordnance, the decision to transfer responsibility for the Uranverein was not reversed. However, the opportunity was not yet lost to capture the attentions of senior figures in the German government and military.

The army's decision to relinquish control of the Kaiser Wilhelm Institute for Physics left the directorship of the Institute open. Weizsäcker and Wirtz garnered support for Heisenberg's appointment as director, but Schumann recommended Bothe. Heisenberg may have been wary of losing his position and his ability to influence events. When Schumann announced a second Army Ordnance conference to be held on 26–27 February 1942, it was clear that Heisenberg would need to be persuasive of the benefits of continuing the programme.

The February conference eventually became two conferences. The Reich Research Council decided to hold a special convention on nuclear research at its headquarters on 26 February. This was to be a series of popular lectures delivered by the physicists to an array of highly-placed and influential figures, such as Albert Speer, Heinrich Himmler, Hermann Göring, Wilhelm Keitel and Martin Bormann. The physicists would then proceed to Harnack House, the headquarters of the Kaiser Wilhelm Society, where a conference organised by German Army Ordnance would begin later that same day. This would be a much more in-depth meeting on the physical problems primarily for the benefit of the physicists themselves, featuring some 25 scientific papers.

However, when invitations to the two conferences were sent out on 21 February, a secretary at the Reich Research Council mixed up the agendas. The luminaries from the Nazi government were sent the agenda for the physics meeting. Instead of the titles of eight popular lectures beginning with Schumann's opening presentation on 'Nuclear physics as a weapon', they received a dense agenda consisting of 25 lectures with highly obscure titles. Himmler expressed his regrets: 'As I will not be in Berlin at the time in question, I regret I will not be able to attend the event.' Not surprisingly, they all declined.[5]

During his popular lecture, Heisenberg explained the essence of the problem they faced:

The behaviour of neutrons in uranium can be compared with the behaviour of a human population sample taking the fission process as analogous to 'marriage' and the capture process as analogous to 'death'. In natural uranium, the 'death rate' exceeds the 'birth rate', with the result that any given population is bound to die out after a short time.

To build a working nuclear reactor or explosive device, the physicists needed either to raise the number of offspring produced by each marriage – increase the rate of production of secondary neutrons – or reduce the death rate, meaning the capture of neutrons by U-238. The way to reduce the death rate was to enrich uranium in the rare isotope U-235. For pure U-235 the death rate would be greatly reduced and explosive population growth would result. Heisenberg emphasised that: 'Pure uranium-235 is thus seen to be an explosive of quite unimaginable force.'

A further way to reduce the death rate in natural uranium was to use a moderator. This would produce a reactor, not a bomb, but the reactor would have many military applications, for example in submarines. In addition, a reactor would over time produce a quantity of element 94,

---

[5] Even without the mix-up, it is doubtful that such senior figures would have accepted the invitation. However, members of their various staffs did attend.

a potentially more powerful explosive than U-235 and capable of being chemically separated from the spent reactor materials.

The physicists made a good impression. The new sponsors of the programme from the Reich Research Council were sufficiently encouraged to seek increased funding. Sights were set on building a working reactor. Although the possibility of making atomic weapons was not discounted, this was promoted as a longer-term possibility that would be investigated when a working reactor was available. Although no high-ranking government or military figures had been present to be similarly impressed, the word nevertheless spread quickly. A month later Josef Goebbels, Hitler's Reich Minister of Propaganda, learned of the latest research in the 'realm of atomic destruction'.

Heisenberg was formally appointed as director at the Kaiser Wilhelm Institute for Physics on 24 April 1942. This made him, in effect, the most senior physicist (if not actually the scientific head) of the Reich Research Council's nuclear programme. The personal animosity that had developed between Heisenberg (and his close associates Weizsäcker and Wirtz) and Diebner now spilled over into political confrontation. Diebner, dismissed as a mediocre physicist, was asked to leave the Institute. He and his team of young physicists retreated to the Army Ordnance laboratory in Gottow and continued to work on nuclear physics.

Plans were well in hand to construct a fourth test reactor in Leipzig, named L-IV. Attention focused yet again on the supply of heavy water from the Vemork plant. The increase in production to 140 kilos per month by the end of 1941 was still insufficient to meet the demands of the programme. A new contract for production and delivery of five tons of heavy water had been drawn up, but production in the first few months of 1942 had actually declined.

Passive resistance

Shortly after Skinnarland had returned to Norway in March 1942, Tronstad sent the first of several letters to his former colleague Jomar Brun. Referring to heavy water by the codename 'juice', Tronstad asked that Brun

keep him advised of heavy water production for the German nuclear programme. Brun himself had been summoned to Berlin earlier in January to
meet with the German physicists and discuss ways in which production of
heavy water could be increased. Although he was not told why, it was clear
that the heavy water was of significant importance.

In a further letter to Brun, probably inspired by Welsh, Tronstad asked
if it might be possible to conceive a plan to transport a considerable quantity of heavy water to Britain. Brun responded that this would be almost
impossible. Perhaps, Brun went on to suggest, if a plane could be landed
on one of the frozen lakes in the mountains near the plant it might be
possible 'with the aid of Norwegian compatriots, to transfer our juice
from the plant to the plane'. The plan was abandoned as impractical and,
as concern about German progress in nuclear technology grew in Britain,
Brun turned to sabotage.

Brun began adding castor oil to the electrolyte, causing heavy foaming
in the system which would halt production for several hours, sometimes
days. So severe was the foaming at times that Brun had to suppress its
effects to avoid suspicion. He did not realise it at the time, but he was not
the only one at the plant involved in this kind of sabotage. Others were
adding cod liver oil to the electrolyte.

In April 1942 the whole plant was shut down and no heavy water was
produced at all. In May the number of heavy water concentration cells
used at the plant was doubled, but the additional cells did not begin producing until mid-June. The lack of progress was put down to a 'certain
passive resistance' by the plant's Norwegian operators and engineers.

Preparations for L-IV were complete by the end of May. The experimental reactor consisted of powdered uranium metal, contained about 140
kilos of heavy water and weighed almost a ton. It was arranged in a spherical configuration, roughly 80 centimetres in diameter, with two concentric
layers of uranium separated by heavy water. The radium–beryllium initiator was inserted into the centre of this configuration via a sealed shaft.

This time there could be no mistaking the evidence for neutron multiplication, which Heisenberg and Döpel estimated at 13 per cent: 'So we
have at last succeeded in building a pile configuration that generates more

neutrons than it absorbs.' They estimated that a pile consisting of ten tons of uranium metal and five tons of heavy water would produce a self-sustaining chain reaction.

Meeting at Harnack House

Albert Speer had been Hitler's chief architect before becoming Reichsminister for Armaments and War Production following the death of Fritz Todt in a plane crash on 8 February 1942. Speer himself had been due to board the ill-fated flight but had preferred instead to get some sleep after an exhausting two-hour discussion with Hitler that had ended at three in the morning.

Of all Hitler's cabinet, Speer was probably personally closest to Hitler because of their shared architectural vision, and Hitler had always displayed a kind of fellowship towards him. But Speer's appointment to the second most important cabinet position after Hitler himself, made with haste to head off a bid for the position from Göring, carried their relationship to an entirely new level. Speer, an army, industry and Nazi Party outsider, had never been a soldier, had never fired a rifle and had never had anything to do with weapons of war. He protested that he was not up to the job, but Hitler insisted: 'I have confidence in you,' he had said, 'I know you will manage it. Besides, I have no one else.'

Towards the end of April he held one of his regular meetings with General Friedrich Fromm, chief of the reserve army and responsible for military training and personnel, over lunch in a private room at Horcher's Restaurant in Berlin.[6] Fromm remarked that the only way Germany could now win the war was with the aid of new weapons and that he was in touch with a group of scientists working on a weapon which could 'annihilate whole cities'.

Göring had recently issued a decree forbidding investment in research programmes that were likely to deliver results only after the war. Conscious

---

[6] This was one of Göring's favourite restaurants. He had exempted the restaurant staff from military service, receiving in return 70,000 bottles of port wine for the Luftwaffe.

of Fromm's argument, and of further complaints from other sources that nuclear research was being neglected, Speer decided to raise the matter directly with Hitler. At a meeting with Hitler on 6 May, he suggested that Göring be placed in charge of the Reich Research Council to emphasise its importance. Göring was subsequently appointed on 9 June.

Speer called for a meeting with the physicists to be held on 4 June at Harnack House. In addition to Speer, the meeting would be attended by Fromm, Fromm's subordinate General Emil Leeb, head of Army Ordnance, Admiral Karl Witzell, armaments chief of the navy, and Field Marshal Erhard Milch, State Secretary of the Air Ministry. It was the kind of illustrious military audience that the Reich Research Council had tried, but failed, to assemble in February. It was certainly the most senior military audience that the German nuclear physicists had ever been invited to address.

Heisenberg abandoned L-IV, still sitting in its water tank in Leipzig, and headed for Berlin. Among the physicists assembled for the meeting were Ardenne, Diebner, Hahn, Harteck, Hans Jensen, Fritz Strassman, Weizsäcker and Wirtz. In all, there were about 50 people crammed into the Helmholtz lecture room at Harnack House. Heisenberg delivered another popular lecture, but now with some notable departures from the version he had given in February. This was a military audience, and so it was inevitable that the question of a bomb would arise. Heisenberg launched into a discussion of the military applications of nuclear fission right at the start of his lecture. For some in the audience, this was the first they had heard of the possibility, their reactions mirroring those of the American Uranium Committee nine months previously, when Oliphant had said 'bomb' in no uncertain terms.

'Given the positive results achieved up until now,' Heisenberg said, 'it does not appear impossible that, once a uranium [reactor] has been constructed, we will one day be able to follow the path revealed by von Weizsäcker to explosives that are more than a million times more effective than those currently available.'

When asked how large an explosive charge would need to be to destroy a large city, Heisenberg replied that it would need to be 'about the size of a pineapple'. He stressed that theoretically there was nothing standing in the way of a German atomic bomb, but the technical prerequisites would take at least two years to develop. He estimated that even if the Americans forged ahead with a programme of their own, there would be no threat from an American atomic bomb before 1945 at the earliest.

When pressed by Speer for an estimate of the budget required to support further work, the German physicists, like their American counterparts more than two years before, were somewhat at a loss for an answer. Weizsäcker suggested 40,000 Reichsmarks, a significant sum by the standards of university research projects. But by the standard of wartime investments in armaments, it was extremely paltry. 'It was such a ridiculously low figure,' Milch subsequently commented, 'that Speer looked at me and we both shook our heads at the artlessness and naiveté of these people.' Milch was unimpressed. Just two weeks later he authorised the mass production of the first *Vergeltungswaffe*, or 'vengeance weapon', the V-1 flying bomb.

Later that evening, Heisenberg found himself sitting next to Milch at dinner, and asked him point blank how he thought the war would turn out. Milch's reflex answer was that if the war was lost they should all take strychnine, before recovering his poise and giving Heisenberg the party line about Hitler's well-laid plans. When Heisenberg asked the same question of Speer during a tour of the laboratory facilities after dinner, Speer did not respond. He simply stared at Heisenberg for several minutes before moving on, as though the question had not been asked. Heisenberg interpreted this as a tacit acknowledgement that the answer was well known, but could not be articulated.

Speer was seduced by the prospect of nuclear power and was not so easily put off by the physicists' 'artlessness'. After the meeting he urged them to inform him of the measures that needed to be taken, and of the sums of money and materials required to advance the nuclear programme. Fromm offered to release several hundred scientifically-trained personnel from the armed forces to serve as assistants. It was, in essence, the physicists' last chance. They had reached a turning point.

It is likely that a request for a substantial budget in terms of personnel, money and materials would have been favourably received. This need not have been directed purely at a bomb but could have been directed at a reactor that would produce fissionable material for a bomb. In the event, after careful consideration, the physicists submitted a request for just 350,000 marks (about $80,000), an increase of a mere 75,000 marks on the existing budget for the nuclear programme.

Speer got the message:

> Rather put out by these modest requests in a matter of such crucial importance, I suggested that they take one or two million marks and correspondingly larger quantities of materials. But apparently more could not be utilised for the present, and in any case I had been given the impression that the atom bomb could no longer have any bearing on the course of the war.

Convinced that a bomb was outside their reach within the likely timeframe of the war, the Uranverein physicists had settled for relatively modest funding to continue their work on a nuclear reactor.

The physicists' actions at this critical turning point in the German programme require careful interpretation. From the very beginning, some in the group – and Heisenberg in particular – had sought to 'make use of warfare for physics'. The Uranverein represented an opportunity to continue research on nuclear fission at the military's expense and with relatively little interference. Every attempt to build the project into a focused, structured programme directed towards military ends had been gently rebuffed. When pressed by Schumann, the physicists working outside Berlin had in September 1939 refused to relocate to the Kaiser Wilhelm Institute for Physics, preferring to continue their academic work at their various institutes around the country. Consequently, the programme had remained fragmented and relatively unfocused.

Like many in German industry and society, the physicists had embraced the Nazi war as a means to an end, the end in this case being the furtherance of their own academic careers and positions. Most in the Uranverein were not Nazis. They were ready to take whatever personal advantages could be gained from the war, but they had no real wish to be part of it. For these physicists, the Uranverein represented an opportunity to make what they saw to be a valid contribution to the German war effort, with the potential promise of a decisive weapon in some ill-defined future, and without the personal risks associated with direct military service.

A request to Speer for millions of marks would have meant a high-profile military project, and a commitment to deliver an explosive device that could have an influence on the course of the war to a regime that was not known to be forgiving in the event of failure. For Heisenberg this must have been quite a gamble. He had dangled the prospect of 'explosives more than a million times more effective that those currently available' in front of an audience of high-ranking military advisers and Hitler's Reichsminister for Armaments. He had then deftly downplayed the potential for a weapon, citing the huge technological challenges that stood in the way. Then, finally, he had bid for a level of funding that was perfectly reasonable for the next step in a long-term research project, but completely insufficient to fast-track a war project designed to overcome the technological challenges and build a new 'wonder-weapon'.

There were still risks, of course. Successful construction of a working nuclear reactor would mean that element 94 could be produced, though perhaps given the scale of the experiments envisaged, the material would not be produced in the kinds of quantities required to build a bomb. Heisenberg's gamble appeared to have paid off. The work would continue under the auspices of the Reich Research Council, a civilian organisation, and there would be little or no further reference to the possibility of a bomb, at least among the core group of physicists working under Heisenberg. The German atomic bomb project, if it had ever really existed as such, was now at an end.

Hitler had received several 'Sunday-supplement' accounts of the potential for an atomic bomb from various non-expert and often ill-informed advisers, and Speer was wary of having his master's imagination captured by an idea that was plainly far from becoming a reality any time soon. On 23 June Speer advised Hitler that the nuclear programme might produce some useful things in the long term, but there was no decisive super-weapon in the offing. The German military turned its attentions elsewhere.

Success with 'atomic fission'

On the day that Speer advised Hitler of the rather limited prospects for the German nuclear programme, the experimental pile L-IV exploded. Heisenberg and Döpel narrowly escaped serious injury, or death.

The pile had been sitting in water for more than twenty days, and had developed a leak. It began to emit a stream of bubbles which Döpel ascertained contained hydrogen gas, formed by a chemical reaction between water and the uranium metal inside the sphere. Döpel hauled the sphere out of the water tank. When a laboratory technician opened one of the sphere's inlet valves, air rushed in, igniting the uranium powder inside which then sprayed outwards.[7] The aluminium began to melt, igniting more uranium powder. Döpel and a couple of the technicians managed to put out the fire and gingerly lowered the sphere back into the water. Heisenberg was summoned to inspect the apparatus. Satisfied that everything was under control, he promptly left to deliver a seminar.

But the sphere was not under control. Some time later Heisenberg was called back to the laboratory and watched with Döpel as the sphere first shuddered, then visibly swelled. The physicists ran for the door, emerging from the laboratory seconds before an explosion tore it apart.

They had survived, but they had lost their laboratory, the powdered uranium metal and the heavy water. The chief of the local fire brigade remarked on Heisenberg's success with 'atomic fission'. Rumours spread, and evolved into reports that several German physicists had been killed in the accidental explosion of a uranium bomb.

Hurry up – we are on the track

Despite the loss of the Leipzig laboratory, Heisenberg had reasons to be satisfied. The reactor experiments were heading in the right direction. The meeting with Speer had concluded favourably in that the importance of

---

[7] The physicists had earlier discovered that powdered uranium metal is pyrophoric – it spontaneously catches fire on contact with air.

nuclear physics had been recognised and modest funding had been agreed. 'No orders were given to build atom bombs,' he later wrote, 'and none of us had cause to call for a different decision.' He had been appointed director at the prestigious Kaiser Wilhelm Institute for Physics, which he regarded as a personal victory over 'Aryan' physics. The continuation of the nuclear programme as a civilian research project left him free to continue his academic research and build his professional and social standing in Berlin.

Heisenberg's conscience may have been clear, but the bargain he had struck still exerted legacy effects. It is one of the great ironies of the first war of physics that, at precisely the moment the German atomic bomb project formally ceased to exist, the palpable fear of a German atomic weapon that had continued to build in Britain and the United States was about to be translated into action. The race to build the atomic bomb, such as it was, was soon to claim its first lives.

The most disturbing news came from Szilard in America. Through various contacts he had picked up the message that the German physicists had already succeeded in establishing a self-sustaining chain reaction which, by Szilard's own reckoning, put them a year ahead of the Allied programmes. Wigner later recalled receiving a telegram from Houtermans in Switzerland urging them to 'Hurry up. We are on the track', but the source of Szilard's message was not Houtermans, though it did come from Switzerland.[8] Szilard alerted Compton, and during July 1942 Compton wrote letters to Conant in Washington:

> We have become convinced that there is a real danger of bombardment by the Germans within the next few months using bombs designed to spread radio-active material in lethal quantities ... Apparently reliable information has reached us to the effect that the Germans have succeeded in making the chain reaction work. Our rough guess is that they may have had the reaction operating for several months.

---

[8] Wigner may have mis-remembered the warning that Houtermans had sent via Fritz Reiche the previous year.

The warning was passed to the Directorate of Tube Alloys via the American embassy in London. It contradicted the intelligence that had been gathered by the SIS, which indicated that German nuclear research was still very much 'in progress'. The British sources were varied, and included reports from Rosbaud in Berlin, conversations between Brun and Uranverein physicists Hans Suess and Karl Wirtz who had separately visited the Vemork plant in July 1942, and comments made by Hans Jensen to several physicists in Denmark and Norway, including Bohr. All the intelligence pointed to a continued effort to develop a reactor, and that the German physicists had yet to succeed in establishing a chain reaction.

Nevertheless, the possibility of warfare using weapons designed not necessarily to cause destruction but to render an area radioactive and therefore inhospitable had to be taken seriously. Alan Nunn May, a Cambridge-educated physicist now working with Chadwick, was asked to assess the feasibility of such a weapon. He concluded that the prospects for such radiation weapons were very limited.

Operation Freshman

The concerns that had been raised about German progress nevertheless brought into sharp relief discussions that had been going on within the Directorate of Tube Alloys and the British military authorities since April. The dependence on heavy water from the Vemork plant represented a significant vulnerability for the German nuclear programme. Sabotage by Brun and other engineers at the plant had certainly limited production (of the five tons that Heisenberg had estimated to be required, by June 1942 less than a ton of heavy water had actually been delivered). However, it was obvious that this kind of sabotage could not be sustained indefinitely. Better to restrict German access to heavy water altogether by taking the plant out of commission.

Discussions about plans to attack the Vemork plant had begun in the spring of 1942 and had thus far involved the War Office, Chiefs of Staff, Combined Operations, the SIS, the SOE, the Foreign Office and the Norwegian government-in-exile. The discussions had produced hundreds of memos and telegrams and little else. The idea of an early strike against Vemork was shelved in May when it was realised that, in the land of the midnight sun, there were at this time of year only a few hours of darkness available in which to carry out a sabotage operation.

Any attack would require a detailed understanding of the layout of the plant. At Tronstad's request, Brun obtained plans and photographs of the plant which were reproduced in microphotographs with the aid of a friend in Rjukan. The microphotographs were then concealed in tubes of tooth-paste and couriered to Tronstad via Sweden.

Churchill had been briefed on the discussions and, shortly after he returned from a strategy meeting with Roosevelt in Washington in June 1942, the Vemork plant was identified as a top priority target. In July the war cabinet issued a memo to Combined Operations, under the leadership of Lord Louis Mountbatten, requesting the examination of options for an attack on the plant sufficient to destroy all existing stocks of heavy water, the plant's electrolysis cells and the adjacent power plant.

Combined Operations turned to the SOE for help. In truth, there were few options. These included sabotage using Norwegian patriots already working at the plant, agents that could be infiltrated into the plant or a party of SOE saboteurs, a Combined Operations attack using British commandos delivered using gliders or hydroplanes, or a bombing raid mounted by the RAF. All of these options were highly precarious and fraught with considerable risks.

Tronstad argued strongly against a bombing raid. He feared that bomb-ing would be too indiscriminate. 'The valley is so deep,' he said, 'that throughout the winter the sun's rays never reach Rjukan's streets. If stray bombs were to hit the liquid-ammonia storage tanks at the bottom of that valley, the whole Rjukan population would be in the gravest danger.' Given the remoteness of the plant, any British commandos landed in the region would have great difficulty getting out again, turning a sabotage operation into a potential suicide mission. Any actions taken by locals risked German reprisals against the local population. Tronstad thought that Brun might be willing to scale up his sabotage activities, but Brun was known to be away from the plant.

Discussions dragged on through August and September. In the meantime, the ranks of Norwegian patriots recruited to the SOE's Norwegian section, under the command of Colonel John 'Jack' Wilson, were scoured for potential candidates for the raid. Ten men were identified, and the SOE began putting together plans for landing a small advance party onto Norway's notoriously inhospitable Hardanger Plateau – 3,500 square miles of virtually uninhabited, frozen wilderness about 3,000 feet above sea level, on the edge of which the Vemork plant was perched.

One of those selected was Knut Haukelid, who subsequently provided a detailed account of the training that he and his comrades received in various SOE Special Training Schools – referred to by the Germans as the 'International Gangster School'. 'We not only learned to force locks and break open safes,' he wrote, 'but we were taught the use of explosives in all circumstances ... We learned to use pistols, knives and poisons, together with the weapons nature had given us – our fists and feet.'

An accident with a loaded pistol during a field exercise ruled Haukelid out of the advance party, which was given the codename Grouse, led by Second Lieutenant Jens Anton Poulsson. The Grouse party included Claus Helberg, Knut Haugland and Arne Kjelstrup. Poulsson, Helberg and Kjelstrup were all natives of Rjukan (Poulsson and Helberg had been school classmates, Kjelstrup had been born in Rjukan but had lived most of his life in Oslo). All four were hardened 'hillmen', intimately familiar with the survival challenges of the Hardanger wilderness.

After much vacillation, a decision on the plan of attack – codenamed Operation Freshman – was finally made by Combined Operations on 13 October 1942. The Grouse party was to reconnoitre the area and identify a suitable landing site for a further party of glider-borne commandos, comprising Royal Engineers of the First Airborne Division, who would carry out the raid. After destroying the heavy water plant, the commandos were expected to make their way on foot to the Swedish border, about 250 miles away.

Both Wilson and Tronstad had argued strenuously that the plan was ill-conceived and susceptible to failure. Norway was unsuitable for a glider operation, over a towing distance longer than had ever been attempted, even in daylight. Success would demand very favourable weather conditions, something that could not be guaranteed. They were overruled. Tronstad sent word to Brun that he should leave for Britain without delay.

After a couple of aborted attempts during September, the Grouse party parachuted onto the Hardanger Plateau on 18 October 1942. They landed 30 miles from the designated drop zone. As Haukelid later remarked: 'It was always curious to the Norwegians how incapable the British and American pilots were at navigating over mountains and forests. They were looking for towns, navigable big rivers, railroad lines, big electrical conduit lines, etc. In the Norwegian mountains there is only wilderness.'

Although the weather had been fine at the time of the drop, the Grouse party was subsequently hit by several storms and took fifteen days to trek to their base of operations close to the Møsvatn dam. 'In good weather, it would have taken us a couple of days,' Poulsson later said, 'but because the snow was wet, the ground wasn't frozen, the streams and lakes were open [ice-free], it took us one hell of a long time with all that equipment.'[9]

The group made contact with Einar and Torstein Skinnarland and radioed their arrival back to the SOE on 9 November. On the same day Jomar Brun and his wife, who had fled to Sweden with the aid of Norwegian intelligence, boarded a plane bound for Britain. Three days later the advance party signalled that they had found a suitable landing site, three miles to the south-west of the Møsvatn dam.

The planning for Operation Freshman was finalised at SOE headquarters in Baker Street a few days later. Tronstad recommended that the attack be limited to the heavy water plant itself, which he believed could be put out of commission for up to two years. On 17 November, Grouse sent the following invitation:

Lake covered with ice and partly covered with snow. Larger lakes are ice free. Last three nights sky absolutely clear with moonlight. Temperature

---

[9] Television survivalist Ray Mears recreated this trek for a three-part documentary series first broadcast by the BBC in 2003.

about 23° Fahrenheit. Strong wind from the north has died down tonight. Beautiful weather.

Mountbatten advised Churchill on 18 November in a memo delivered via Cherwell. Churchill gave Operation Freshman a green light.

Chapter 7

THE ITALIAN NAVIGATOR

January 1942–January 1943

'Well have the chain reaction going here [in Chicago] by the end of the year', Arthur Compton declared from his sickbed on 24 January 1942. Ernest Lawrence bet him a thousand dollars that they wouldn't.

'I'll take you on that', Compton answered.

Lawrence backed off. 'I'll cut the stakes to a five-cent cigar', he countered.

Compton, who had never smoked a cigar in his life, agreed.

The United States had been at war for just seven weeks. Compton had hastily pulled together a plan for the S-1 programme, and had reached a critical decision point just as he had gone down with influenza. The programme would follow the conclusions of the third and final National Academy report and focus on the problem of uranium isotope separation and the physics of a U-235 bomb. However, although the possibility of building an atom bomb using element 94 may have all but disappeared from the final Academy report, Compton had not forgotten it.

Compton's principal task in the S-1 programme was to work on the physical principles of bomb design but, largely as an afterthought, he was also assigned the task of examining the potential for element 94. 'Except for this afterthought,' he later wrote, 'there might well have been no development of the nuclear reactor as a wartime project.'

From the beginning, the goal of the reactor project was understood to be the production of element 94. This was work that had to be consolidated in one location, and this was the decision that now confronted Compton. Szilard argued for Columbia University. Lawrence argued for Berkeley. Princeton and industrial laboratories in Pittsburgh and Cleveland were also considered. Compton advanced the case for Chicago. Like the Uranverein physicists in September 1939, nobody involved in the American programme wanted to relocate. Compton, who had already advised Conant that the work would be done in Chicago, exercised his executive authority. His bet with Lawrence was on.

Concern for secrecy led the laboratory at Chicago which was to house the project to be called the 'Metallurgical Laboratory', or just Met Lab, as deliberately obscure a name as MAUD and Tube Alloys. The only secret that Enrico Fermi's wife Laura discovered about Met Lab during the war was that there were no metallurgists employed there. 'Even this piece of information was not to be divulged', she wrote. 'As a matter of fact, the less I talked, the better; the fewer people I saw outside the group working at the Met Lab, the wiser I would be.'

According to Compton, Fermi agreed to the move from Columbia to Chicago at once. In truth, Fermi was reluctant. He and his small research team at Columbia had made significant progress with reactor configurations based on cubes of uranium oxide stacked in a lattice of graphite bricks. That they hadn't succeeded in creating the environment for a self-sustaining chain reaction was put down largely to problems of impurities in their materials. Now the team was to be dispersed as the physicists joined different parts of the S-1 programme.

Fermi travelled back and forth to Met Lab before settling in Chicago at the end of April 1942. His wife Laura joined him at the end of June, having first retrieved the cache of Nobel prize money that Fermi had won in 1938 and hidden in a lead pipe under the concrete floor in the basement of their New York home. This had been insurance against the possibility that – as enemy aliens – their assets might be seized.

By the time Fermi was settled in Chicago, element 94 had finally acquired a name. In a report on the chemical properties of elements 93 and 94 dated 21 March 1942, Seaborg and Wahl had decided to name element 94 *plutonium*. '[W]e considered names like extremium and ultimium', Seaborg wrote. 'Fortunately we were spared the inevitable embarrassment that one courts when proclaiming a discovery to be the ultimate in any field by deciding to follow the nomenclatural precedents of the two prior elements ... We briefly considered the form plutium, but plutonium seemed more euphonious.'

The report was typed by Lawrence's secretary at the Rad Lab, Helen Griggs. 'I like to say that she was so efficient as a secretary that I began to date her', said Seaborg. 'She doesn't like that characterization, and I have to admit immediately that she had other qualities.' Seaborg married Griggs in Nevada, on their way to Chicago.

Seaborg arrived in Chicago on his 30th birthday, 19 April 1942. If Fermi could get a reactor working before the end of the year, as Compton had wagered, then Seaborg's task was to work out how plutonium could be separated from the spent reactor materials. His biggest problem was that he needed to understand the chemistry of this new element before a working reactor was built. That meant finding another way to make the new substance in sufficient quantities for chemical analysis.

The best he could do was arrange for batches of uranium nitrate to be bombarded with neutrons in a cyclotron for weeks and months on end. This gave him quantities of plutonium measuring no more than millionths of a gram to work with, so he gathered around him a small team of experts in ultramicrochemistry – the science of chemical investigation using minute quantities of substances.

By 14 August, Seaborg's microchemists had succeeded in isolating a tiny quantity of plutonium.

This Napoleonic approach

The S-1 programme was investigating several different routes to an atomic bomb. In addition to the Met Lab project to construct a reactor and produce plutonium, projects were also in hand to separate quantities of U-235 using gaseous diffusion techniques, electromagnetic methods based on an adaptation of Lawrence's 37-inch cyclotron and centrifugal separation methods. The construction of a working reactor based on heavy water as a moderator was also being pursued in case the uranium–graphite pile failed for some reason. Construction of a heavy water plant was under way in Canada.

The MAUD Committee had strongly advocated collaboration with America in its July 1941 report, and yet the British were initially rather cagey in their approach to the Americans and somewhat preciously guarded what they believed was a lead in nuclear research. Cherwell had favoured continuing the project in Britain, or 'at worst in Canada'. Chadwick was hesitant.

Several Tube Alloys physicists travelled to America during the first few months of 1942, including Peierls, Simon and Halban. They visited all the major centres of S-1 activity, attended meetings of the S-1 Committee and enjoyed a completely free exchange of information.

While the British team could consider itself ahead in terms of research on the theoretical principles of bomb design, it was clear that the Americans were moving ahead rapidly on all the experimental fronts. Work on the pilot gaseous diffusion plant at Rhydymwyn in Wales was continuing, but the British team could make no contribution to the experimental work on electromagnetic or centrifugal separation of U-235. These separation methods had not been fully explored in Britain and Tube Alloys did not have the capacity to undertake such work. And the MAUD Committee, having backed U-235, had not sponsored much work on plutonium.

The American effort was now very impressive. 'One thing is clear,' remarked Akers shortly after arriving in the US, 'and that is that an enormous number of people are now on this work so that their resources for working out schemes quickly are vastly greater than ours.' The American project had languished until the MAUD Committee report and Oliphant's meddling had provided the impetus to establish the S-1 programme. However, by the end of spring 1942 it was clear that the American project was overtaking Tube Alloys.

The letters exchanged at this time between Vannevar Bush and John Anderson were cordial but deliberately vague, and spoke of more meaningful collaboration when the projects were ready to progress from the pilot plant stage to full production. But the visits by Akers and the Tube Alloys physicists had convinced them of the importance of pushing for a full Anglo-American programme, managed by a joint council and supported by joint technical committees. Akers approached Chadwick, and Chadwick set aside his reservations. A proposal for full collaboration was put forward to Anderson's policy council in June 1942.

The American entry into the war within a few days of the establishment of the S-1 programme meant that the constraints of the past few years had quickly vanished. Roosevelt stressed the importance of time at the expense of money – the Americans believed they were competing in a race in which the Germans already had a substantial lead.[1] Given the uncertainties they faced, it was impossible for the S-1 Committee to determine which of the various routes to an atomic bomb was most favourable. The committee determined to pursue them all. 'To embark on this Napoleonic approach to the problem would require the commitment of perhaps $500,000,000 and quite a mess of machinery', Conant concluded.

On 17 June Bush proposed to Roosevelt that the US Army Corps of Engineers get involved in the S-1 programme alongside the OSRD, taking responsibility for process development, large-scale engineering design, selection of suitable locations and procurement of materials. Roosevelt approved the proposal.

Anderson vacillated at precisely the wrong time. Bush advised him of the proposed changes to the organisation of the S-1 programme in June, but Anderson was not convinced of the value of full Anglo-American collaboration until the end of July, when he drafted a memo to Churchill:

---

[1] Compton had estimated that, with access to a uranium–heavy water reactor operating at 100,000 kilowatts for two months, the Germans could have enough plutonium for six atomic bombs by the end of the year – a frightening prospect.

We must, however, face the fact that the pioneer work done in this country is a dwindling asset and that, unless we capitalise it quickly, we shall be rapidly outstripped. We now have a real contribution to make to a 'merger'. Soon we shall have little or none.

American policy-makers had already concluded that they could manage quite well without British help.

Negotiations to move Halban's Cambridge team to America had foundered on the issue of security. This team had continued to work on a uranium–heavy water reactor, a project that was now seen as secondary to the main thrust of Tube Alloys. Although it now made sense to everyone involved to consolidate the S-1 and Tube Alloys reactor work at the Met Lab, foreign nationals were prevented from participating in secret American war projects. Bush thought that it might be possible to bend the rules for a team of British nationals, but Halban's team actually had few British nationals in it. This proved to be an insurmountable problem. In truth, personality clashes, differences of opinion about priorities and the perception that the American work had already moved far ahead were also factors.

It was decided instead to transfer Halban's team to Canada. The Canadian government eagerly accepted the suggestion, and terms were agreed in late autumn 1942. The work was to be administered by Canada's National Research Council. Laboratory premises were found in Montreal.

Leftwandering activities

With the experimental projects to develop the material for a bomb under way, Compton had turned his attention to the physics of fast neutron reactions and the implications for bomb design. He had appointed Russian-born physicist Gregory Breit to lead this effort, but Breit quickly became frustrated by what he perceived to be a lack of progress and an intolerable lack of security. He resigned on 18 May 1942 to rejoin a navy project that he had worked on before becoming involved in the S-1 programme.

Compton, who had invited Oppenheimer to work on the project under Breit's overall supervision, now asked him to lead the work.

Oppenheimer was a remarkable physicist, but a somewhat flawed human being. The son of Jewish immigrants grown wealthy in America, he had a tremendous capacity for learning. At the age of nine, he would challenge a cousin to ask him a question in Latin which he would then answer in Greek. This capacity was not, however, matched by a corresponding level of human empathy. He did not wear his learning lightly. As a young boy he would compensate his physical awkwardness and shyness by showing off. He could be boastful and patronising, and he developed an acid tongue. The impression left with school friends and, subsequently, with scientific colleagues and collaborators ranged from compassion to exasperation.

Oppenheimer was a polymath, with interests not only in science but also in psychotherapy and the arts. At Harvard University he had majored in chemistry, but had also studied Greek, architecture, classical literature and art. On graduating from Harvard he studied under J.J. Thomson at the Cavendish Laboratory in Cambridge, before moving on to Göttingen in Germany. In Göttingen he worked with James Franck and Max Born and met Heisenberg and English physicist Paul Dirac, among many other eminent theoretical physicists passing through during this period.

Compton had first met Oppenheimer in Göttingen in 1927. 'A specialist in the problems of nuclear physics,' Compton later wrote, 'he was one of the very best interpreters of the mathematical theories to those of us who were working more directly with the experiments.'

After gaining his doctorate, Oppenheimer returned to Harvard before moving on to the California Institute of Technology in Pasadena. He put several offers of academic positions on hold for a year while he returned to Europe to continue his postdoctoral studies. He went first to Leiden in Holland to work with Paul Ehrenfest, and then to Zurich in Switzerland to work with Wolfgang Pauli, who had just completed the first part of a joint work with Heisenberg on quantum electrodynamics. He returned to America, and a position at the University of California at Berkeley, in July 1929. Lawrence had been appointed associate professor at Berkeley just a year before.

Oppenheimer was certainly talented, but he was more technician than innovator. He could refine and extend the original ideas of others, but was less able to produce original ideas of his own. On 22 April 1942, Oppenheimer had celebrated his 38th birthday. It was a widely acknowledged fact of scientific life that a physicist's best work is done when young. Heisenberg had won the Nobel prize for work he had done in his late twenties. Einstein had won the prize for work he had published at the age of 26. Oppenheimer may have been acutely aware that his best work was now behind him. And this had been work that had not been judged worthy of the coveted prize.

Oppenheimer was also politically active. His privileged upbringing may have fostered in him a sense of guilt which was manifested as a heightened social conscience. This does not seem to have been directed towards particular individuals, at least outside his own family, but rather to political ideals and social causes. As he later explained:

I had had a continuing, smouldering fury about the treatment of Jews in Germany. I had relatives there, and was later to help in extricating them and bringing them to [America]. I saw what the Depression was doing to my students. Often they could get no jobs, or jobs which were wholly inadequate. And through them, I began to understand how deeply political and economic events could affect men's lives.

From the very beginning, what Lawrence referred to as Oppenheimer's 'leftwandering activities' generated suspicion and caused concern over his involvement in the American bomb programme. By his own subsequent admission, in the mid-1930s Oppenheimer had been involved with just about every Communist front organisation in California.

Motivated by his infatuation with, and on-off engagement to, Jean Tatlock, daughter of a professor of literature at Berkeley and a prominent Communist Party activist, he had become involved with fund-raising to support the fight against the growing threat of European fascism. When his relationship with Tatlock ended, he married Katherine (Kitty) Peuning, a descendant of European royalty. Although Oppenheimer himself never became a card-carrying member of the American Communist Party, his wife Kitty, brother Frank, his close friend Haakon Chevalier and a number of his research team at Berkeley were at one time or another party members.

The FBI had opened a file on Oppenheimer in March 1941, after he was observed the previous December arriving by car for a discussion group meeting at Chevalier's home. The targets for FBI surveillance on this occasion had been two other activists, but further surveillance connected Oppenheimer with Steve Nelson, also known by the name Steve Mesarosh, a key figure in the Communist Party apparatus in San Francisco's Bay Area.

Croatian-born Nelson had spent two years at the International Lenin School in Moscow, where he had been taught working-class history, Marxism, and the practicalities of proletarian dictatorship. During this time, he had been sent on clandestine missions to Europe, India and China. Those looking for links between the legitimate (or, at least, tolerated) activities of the American Communist Party and the illegal activities of Soviet intelligence were looking closely at Nelson.

Nelson also knew Kitty Oppenheimer through her second husband, also a Communist, who had died fighting in Spain in 1937. The Oppenheimer and Nelson families met socially on several occasions. Robert was Kitty's fourth husband.

Oppenheimer was a security nightmare, yet his contributions to the S-1 programme had thus far been extremely valuable. Compton was now asking that he take over responsibility for the work on fast-neutron reactions and the principles of bomb design. Lawrence had insisted that Oppenheimer cease his flirtation with radical left-wing politics, and Oppenheimer had acquiesced (though he was still making financial contributions to left-wing causes as late as April 1942). He had been granted temporary security clearance so that he could work in support of Lawrence and he had completed a security questionnaire in April which he had answered mostly honestly.[2] However, full security clearance was not immediately forthcoming.

Oppenheimer could not afford to wait. As he began to get to grips with the nature and scale of the task that Compton had now assigned to him, he realised that he needed the best brains in the country to work on the problem. In early July 1942 he gathered together a study group at Berkeley made up of the brightest theoretical physicists he could find. He dubbed them the 'luminaries'.

Any lead that the Tube Alloys physicists believed they possessed in the areas of fast-neutron reactions and bomb design was about to evaporate.

The luminaries

As the S-1 programme moved forward through the spring of 1942, Teller had little choice but to cool his heels. Szilard, extremely annoyed at Compton's rather undemocratic decision to consolidate reactor research at Chicago, had nevertheless packed his bags and moved there at the end of January 1942. Wigner completed a study on chain reactions with his group at Princeton before joining the Met Lab in April, where he would work on reactor design.

Of the original Hungarian conspirators, Teller had so far been excluded from the most recent developments. He suspected issues with security clearance: 'Although Mici and I were both [American] citizens, our families were behind enemy lines.' But Oppenheimer wanted him on the programme, and overruled whatever security issues were causing problems with Teller's clearance. Teller joined the Met Lab in early June.

It seemed that nobody had given any thought to what Teller should be doing, so he joined forces with a young physicist from Indiana, Emil Konopinski, who had arrived in Chicago at about the same time. Fermi had earlier suggested that the temperatures generated in an atomic explosion

---

[2] Oppenheimer omitted to mention his activities in support of the radical Federation of Architects, Engineers, Chemists and Technicians (FAECT) union, and the discussion groups he had organised with Chevalier. See Herken, p. 58.

might be sufficient to trigger a thermal reaction between atoms of deuterium, fusing the atoms together and releasing an even more extraordinary quantity of energy in the process. The sun is powered by fusion reactions of this kind.

In this case the trigger was simply the high temperatures – about 400 million degrees Celsius – generated by an atomic explosion. A fusion reaction would not depend on acquiring a critical mass of a rare isotope or a new element unknown in nature, or self-sustaining chain reactions. Such a 'thermonuclear' reaction would therefore proceed as long as the temperature was maintained and there was enough deuterium fuel available to burn.

Teller and Konopinski figured they could usefully spend their time proving that this was impossible. They set to work, and discovered that for every objection they could identify, they could also find a potential workaround. They realised that a thermonuclear bomb was, in fact, possible. By the time Oppenheimer called the study group together at Berkeley, they believed they knew how such a bomb could be built.

In addition to Teller and Konopinski, Oppenheimer had also invited German émigré Hans Bethe from Cornell, John H. Van Vleck, Swiss-born Felix Bloch and Oppenheimer's former student Robert Serber, who had returned to Berkeley from Illinois at Oppenheimer's request. They met in a couple of attic rooms in Le Conte Hall, an administrative building on the Berkeley campus in which Oppenheimer had an office.

Until this point Bethe had refused to get involved in the programme, for the simple reason that he did not believe a bomb would work. Bethe had obtained his doctorate with Arnold Sommerfeld in Munich and had worked at Cambridge and with Fermi in Rome before taking up a position at the University of Tübingen. Although he had been raised as a Christian, his mother was Jewish and he lost his academic position in 1933. He made his way first to England, where he worked for a time with Peierls, before moving to a professorship at Cornell University in 1935.

On their way from New York to Berkeley, Bethe and his wife stopped off in Chicago to pick up the Tellers. Teller took the opportunity to show Bethe around the Met Lab, and particularly the latest experimental nuclear reactor that Fermi and his team were assembling in a doubles squash court in the west stands of Stagg Field, part of the University of Chicago. Bethe was amazed, and realised that his reservations might be poorly founded.

At Berkeley, the study group began work on the theory of a fission bomb, starting with the MAUD Committee report and the results of the various groups that had studied the problems under both Breit and Oppenheimer. It quickly became apparent that the fission bomb was a 'sure thing', and Serber was left to work out the details with Eldred Nelson and Stan Frankel, postdoctoral research assistants in Oppenheimer's group. Teller and Konopinski persuaded the rest of the group to work with them on the possibility of a thermonuclear bomb, which during the summer became known as the 'Super'.

Serber recalls their reaction:

At this point something remarkable happened. Teller brought up the idea of the Super, a fusion weapon, not a fission weapon, which was to be a detonation wave in liquid deuterium set off by being heated by the explosion of an atomic bomb. Well, everybody forgot about the A-bomb, as if it were old hat, something settled, no problem, and turned with enthusiasm to something new.

The figures were astonishing. If a thermonuclear reaction could be triggered, just twelve kilos of liquid deuterium could be expected to explode with a force equivalent to one *million* tons of TNT. Then Teller realised that a fission bomb might trigger other kinds of fusion reactions as well. Specifically, he figured that an atomic bomb would heat the atmosphere so intensely that fusion reactions involving nitrogen, constituting 80 per cent of the earth's atmosphere, would be triggered. Simply put, exploding an atomic bomb would set fire to the air.[3]

---

[3] Speer noted that Hitler would occasionally remark that 'the scientists in their unworldly urge to lay bare all the secrets under heaven might someday set the globe on fire'. See Speer, p. 317.

Oppenheimer was sufficiently perturbed by Teller's conclusions to seek an emergency meeting with Compton, who was holidaying in Michigan. Bethe, however, was immediately suspicious. He quickly spotted the unjustified assumptions that had propelled Teller's calculations to these stark conclusions. Teller was mollified and the possibility of catastrophe receded.

The group then realised that deuterium fusion reactions would proceed too slowly to support an explosive release of energy and, as the summer wore on, alternatives were discussed. These included reactions of deuterium with tritium, the heaviest isotope of hydrogen which contains one proton and two neutrons, and of deuterium and an isotope of lithium, Li-6, which produces tritium on bombardment with neutrons. At the end of their deliberations, Teller was left with the firm impression that the Super represented the ultimate prize and that the fission bomb had been reduced to a mere 'engineering problem'. However, this was not an impression that was shared by Serber, Bethe and Oppenheimer. For them the Super was, perhaps, an interesting possibility, but as it required a fission bomb to initiate it, this was a possibility that could be explored once a fission bomb had been built.

By the end of the summer the study group had done much to sharpen the thinking about both the fission and fusion bombs. In August 1942 Oppenheimer reported that a U-235 bomb would require 30 kilos of the isotope but 'should have a destructive effect equivalent to the explosion of over 100,000 tons of TNT'.[4] This was much, much more than the 1,800 tons claimed by the MAUD Committee physicists the previous year. The study group further claimed that by surrounding a fission bomb with 400 kilos of liquid deuterium, the destructive force would be greatly enhanced, to ten million tons of TNT equivalent, which would devastate an area of more than 100 square miles.

---

[4] There was no formal written report from the summer study group, and the figures were scaled down somewhat between the reports from Oppenheimer to Conant, and from Conant to Bush.

On hearing of these results, the S-1 Committee was stunned. Of course, the discovery of the possibility of a thermonuclear fusion weapon did not change the immediate priorities of the S-1 programme, which was to build a fission bomb. But it did change dramatically the scale of the enterprise. The Committee alerted Bush, and Bush alerted Secretary of War Henry Stimson.

Whether it was just an interesting possibility or not, the Super was now firmly on the radar.

The biggest sonovabitch

Bush's decision to involve the Army Corps of Engineers in the S-1 programme led to an inevitable culture clash just a few months later. This clash was, in part, precipitated by a notable shift in the balance of power within the programme. Despite the exceptional circumstances, and the ever-present threat that German physicists might get there first, the work on the American bomb project had to this point proceeded more or less democratically. The physicists had agreed to work together, to pool their knowledge and resources, and to each take tasks befitting their areas of expertise.

Now the physicists were losing control. A more authoritarian organisation was in place, issuing orders from Washington, and they were no longer involved in making key decisions. The involvement of the army and the introduction of formal management structures had increased bureaucracy. Communication between the physicists was being restricted in the interests of preserving secrecy through 'compartmentalisation', ensuring that very few individuals involved in the S-1 programme had the full perspective of it. This was not the way the physicists were used to working, and Szilard was convinced that this was not the best way to conduct the project. He proceeded to make a nuisance of himself.

The clash spilled over into confrontation over the design of the reactor cooling system. The engineering contractors brought in by the Army Corps of Engineers were more used to designing roads and bridges. The physicists were alarmed by their level of ignorance and incompetence.

Compton's efforts to impose his authority by reading a parable from the Old Testament to an assembly of Met Lab scientists did not help.

On 21 September, Szilard summarised his discontent, and his concerns for peace after the war, in an eleven-page memorandum. After noting how easy life could be if they all followed orders and just carried out the tasks assigned to them, he wrote:

> Alternatively, we may take the stand that those who have originated the work on this terrible weapon and those who have materially contributed to its development, have, before God and the World, the duty to see to it that it should be ready to be used at the proper time and in the proper way.

It had become clear to Bush in August that the division of the S-1 programme between the civilian OSRD and the army was not working. He raised the issue with General Brehon Somervell, head of the Army Services of Supply. Bush had been trying to work out a solution that would allow him to retain some civilian oversight, but Somervell thought to put the Army Corps of Engineers in charge of the whole programme.

Things were about to change and, from the scientists' perspective, not for the better.

Somervell needed a dependable individual to lead what was about to become a military programme, and he thought he knew just the man for the job. At the time, Colonel Leslie Groves was 'probably the angriest officer in the United States Army'. A West Point graduate, he had just agreed to accept an assignment overseas, having grown weary of the bureaucratic headaches associated with directing military construction projects with budgets of tens of millions of dollars (he had recently overseen construction of the Pentagon). His superiors leaned on him heavily. 'If you do the job right,' Somervell advised him, 'it will win the war.'

Groves' spirits fell. He could only answer: 'Oh, that thing.'

The entire budget for the S-1 programme was less than Groves would typically spend in a week. He set about taking control of the programme in a forthright manner. One of his subordinates, Lieutenant Colonel

Kenneth Nichols, another West Point graduate with an engineering Ph.D., remembered Groves as 'the biggest sonovabitch I've ever met in my life, but also one of the most capable individuals … I hated his guts and so did everybody else but we had our form of understanding.' Groves' first meeting with Bush was not an auspicious one. 'I fear we are in the soup', Bush wrote.

Groves may not have been noted for his tact and diplomacy, but he did move quickly. At the behest of Joliot-Curie in Paris and Henry Tizard in Britain, in 1940 Union Minière had shipped to the US over a thousand tons of pitchblende ore rich in uranium oxide from its mines in the Belgian Congo, to keep them out of German hands. The ore had been sitting at Port Richmond on Staten Island for six months. Groves had heard about the intention to appoint him to head the S-1 programme on 17 September. The very next day he despatched Nichols to New York to buy the pitchblende ore. That same day he approved the requisition of what was to become known as Site X, 56,000 acres of land near Oak Ridge in eastern Tennessee, where large-scale facilities for the separation of U-235 and the production of plutonium were to be built. At this point, a self-sustaining nuclear chain reaction had to be demonstrated.

Newly-promoted to Brigadier General, Groves took up his post formally on 23 September, and the programme finally kicked into high gear. The Army Corps of Engineers had referred to its contribution to the S-1 programme as Manhattan Engineer District, after its North Atlantic Division headquarters on Broadway, near New York City Hall. With the Army Corps of Engineers now in charge, the name was adopted to describe the whole programme. What was eventually to become known as the Manhattan Project was born.

But the compulsion for the American project remained the threat of a German atomic bomb. If Compton's estimates were correct, then Szilard's concern that they might not be ready 'before German bombs wipe out American cities' was an alarming prospect. German progress depended critically on access to heavy water. Groves added his voice to the growing demand for Allied action against the Vemork plant.

The Grouse party had parachuted onto Norway's Hardanger Plateau on 18 October. Operation Freshman was launched a month later.

Gliding to disaster

Operation Freshman went wrong virtually as soon as it began. Combined Operations decided that the target was sufficiently important to warrant a doubling of the personnel for the mission. Two Halifax bombers left Skitten airfield in Caithness, Scotland, on the night of 19 November. Each bomber towed a Horsa Mk. 1 glider flown by two pilots in which an officer, a sergeant and thirteen other ranks drawn from the Royal Engineers of the First Airborne Division sat huddled. This made a total of 34 men, all volunteers. Three days later, the British newspapers reported the following:

On the night of November 19–20th, two British bombers, each towing one glider, flew into Southern Norway. One bomber and both gliders were forced to land. The sabotage troops they were carrying were put to battle and wiped out to the last man.

One of the Halifaxes had got into difficulties as it approached Egersund, about 125 miles from the landing zone, and the glider was unexpectedly released. The glider crashed into a mountain near Helleland, killing two on board and severely injuring a third. The Halifax managed to climb above this mountain range, but then crashed into the next, killing the crew of seven.[5] Two of the survivors of the glider crash managed to get to a nearby farmhouse and local civilians hurried to warn them of the approach of German troops. The survivors must have decided that there was little choice but to surrender and sit out the war in a German prisoner-of-war camp.

---

[5] They were Flight Lieutenant A.R. Parkinson of the Royal Canadian Air Force, Flight Lieutenant A.E. Thomas, Pilot Officer G.W.S. De Gency, Flying Officer A.T.H. Haward, Flight Sergeants A. Buckton and G.M. Edwards, and Sergeant J. Falconer. They were buried in Helleland churchyard.

But Hitler, infuriated by the success of British sabotage operations, had just a few weeks before issued a new order. No quarter was to be granted to saboteurs, on principle, even if they were in uniform at the time of their capture. They were instead to be interrogated and then shot.

The survivors were rounded up by German troops and taken to a camp near Egersund. They were first interrogated and then taken into nearby woods where they were executed by firing squad, one after another. The bodies of all seventeen were buried in a trench in the sand dunes of Brüsand.[6]

At the landing zone, Haugland heard the drone of at least one bomber overhead. The night sky was cloudy but the moon was out and it was not particularly dark. But the noise of the plane died away. Running low on fuel and unable to identify the landing zone, the pilot of the second Halifax had decided to return to base. The plane and its glider were approaching the Norwegian coastline when the tow line froze solid and snapped, sending the glider hurtling to the ground. It crashed into mountains near Fyljesdal, north-west of Stavanger. Seven on board were killed instantly. One survivor managed to crawl away from the crash site but died of exposure and loss of blood. When German troops arrived on the scene, the bodies were buried in shallow graves.[7]

Of the nine remaining survivors, four were badly injured. All nine were taken to Stavanger county jail, where a dispute erupted between the Wehrmacht and Gestapo as to which of them had jurisdiction over the prisoners. By this time the hasty execution of the commandos from the other glider had come to the attention of General Wilhelm Rediess, head

---

[6] They were Pilot Officers N.A. Davies and H.J. Fraser of the Royal Australian Air Force, Lieutenant A.C. Allen, Lance Sergeant G. Knowles, Corporal J.G. Thomas, Lance Corporals F.W. Bray and A. Campbell, Sappers G.S. Williams, E.W. Bailey, C.H. Grundy, H.J. Legate, T.W. Faulkner, H. Bevan, L. Smallman and J.M. Stephen, and Drivers J.T.V. Belfield and E. Pendlebury. Their bodies were exhumed and reburied at Eiganes near Stavanger, with full military honours, in July 1945.

[7] They were Sergeants M.F. Strathdee and P. Doig of the Army Air Corps, Lieutenant D.A. Methven, Lance Sergeant F. Healey, Sappers J.G.V. Hunter, W. Jacques and R. Norman, and Driver G. Simkins. Their bodies were reburied at Eiganes in August 1945.

of the Gestapo in Norway, who angrily noted that Hitler's order did not preclude full interrogation of the prisoners first.

The four injured were taken for questioning by the Gestapo. Three of them were beaten and strangled with leather straps, and their chests and throats were crushed. They were then killed by injecting air into their bloodstreams. The fourth was executed by a single shot to the back of the head. The bodies were taken out to sea, weighted with stones and thrown overboard. They were never recovered.[8]

The five uninjured were imprisoned at Grini concentration camp, north of Oslo. They were advised that their rights as British soldiers under the Geneva Convention would be respected. However, on 18 January 1943, after much questioning, the five were taken from their cells by a 'Special German Delegation', blindfolded and shot.[9]

From maps and other documentation salvaged from the crashed gliders and, no doubt, from the Gestapo's brutal interrogation of the crash survivors, the Germans learned all they needed to know about the target of Operation Freshman. They fortified their positions in Rjukan and laid a minefield around the Vemork plant.

Now in great jeopardy, the four-man Grouse party vanished into the depths of the Hardanger wilderness.

The new world

Laura Fermi organised a party for the 'metallurgists' at the Met Lab early in December 1942. As the guests began to arrive, they were effusive in offering

---

[8] They were Corporal J.D. Cairncross, Lance Corporal T.L. Masters, Sapper E.J. Smith and Driver P.P. Farrell. They are commemorated on the Brookwood Memorial in Surrey. Stabsarzt W.F. Seeling, Hauptscharfuehrer E. Hoffman and Unterscharfuehrer F. Feuerlein were found guilty of their murder at a war crimes tribunal in Oslo in December 1945. Seeling and Hoffman were executed. Feuerlein was given life imprisonment and handed to the Russians to answer charges relating to atrocities committed against Russian prisoners-of-war.

[9] They were Lance Corporal W.M. Jackson, Sappers F. Bonner, J.F. Blackburn, J.W. Walsh and T.W. White. Their bodies were recovered and reburied at Vestre Gravlund, near Oslo.

their congratulations to Enrico. 'Congratulations?' Laura asked, puzzled. 'What for?' Nobody took any notice of her.

Her persistent questions received evasive answers or no answer at all. 'Nothing special,' said one, 'he is a smart guy. That's all.' 'Don't get excited,' said another, 'you'll find out sometime.'

Just a month before, Fermi had come to a difficult decision. Workers employed by the army contractors responsible for the construction of a new reactor building had gone on strike, delaying construction indefinitely. Fermi proposed to Compton that, rather than wait any longer, they instead make use of the squash court which had thus far been used to house the experimental piles.

Everyone involved would feel the pressure created by another delay, yet to trial a completely untested and potentially dangerous technology in the midst of a bustling city was to take an enormous risk. The term 'meltdown' had yet to be applied to a reactor that runs out of control, but the physicists involved had no difficulty imagining the consequences.

Fermi was able to reassure Compton that he could keep the chain reaction under control, relying on the fact that a small proportion of the secondary neutrons were emitted with some delay following fission. By operating the reactor so that its rate of production of neutrons fractionally exceeded the rate of neutron capture, the delayed neutrons would give the physicists enough time to respond should the chain reaction start to run away. Compton agreed, but decided not to inform the president of the University of Chicago just yet. When Compton advised Conant at a committee meeting on 14 November, Conant went white. Groves immediately started searching for an alternative site. But nobody told Compton to stop.

The morning of 2 December 1942 dawned cold, with temperatures falling below freezing, and a chill wind blew. Around mid-morning Fermi ordered that all but one of the cadmium control rods be removed from the reactor. The last control rod was then drawn halfway out, as the physicists carefully monitored the neutron intensity and compared the results with their estimates. Between 25 and 30 people watched from the balcony, including Szilard and Wigner.

By 2:00pm Compton had arrived and the group of observers had swelled to 42. Fermi ordered a repeat of their earlier experiment, and all but one of the control rods was withdrawn once more. With the last control rod about seven feet out of the pile, the chain reaction was almost self-sustaining, the pile almost critical. Fermi ordered that the rod be withdrawn another foot. As the rate of neutron production climbed inexorably, the steady click-click-click of the neutron counters became faster and faster until the clicks merged into a roar.

Physicist Herb Anderson described what happened next:

[W]e were in the high intensity regime and the counters were unable to cope with the situation anymore. Again and again, the scale of the recorder had to be changed to accommodate the neutron intensity which was increasing more and more rapidly. Suddenly Fermi raised his hand. 'The pile has gone critical,' he announced. No one present had any doubt about it.

The intensity of neutrons was now doubling every two minutes. If Fermi had let it run uncontrolled for another hour and a half, the reactor would have pushed on towards a million kilowatts, killing everyone in the room before melting down. Fermi shut the reactor down after just four and a half minutes. There had been nothing to see, and aside from the clicking of the neutron counters there had been nothing to hear. The reactor had produced a mere half-watt, but its significance that day was much, much greater than its output. The physicists had shown that it was possible to engineer the controlled release of the enormous and inexhaustible supply of energy bound in atomic nuclei.

Compton called Conant to tell him the news: 'Jim,' he said, 'you'll be interested to know that the Italian navigator has just landed in the new world.'

He had won his bet.[10]

---

[10] Although it seems that he never did get the cigar from Lawrence.

As the physicists celebrated, Szilard found himself standing alone with Fermi. 'I shook hands with Fermi,' he recalled, 'and I said I thought this day would go down as a black day in the history of mankind.'

Chapter 8

LOS ALAMOS RANCH SCHOOL

March 1942–March 1943

G eorgei Flerov had become involved in research on nuclear fission not long after the publication of the Frisch–Meitner paper in early 1939. He had studied at the Leningrad Polytechnic Institute and had worked in Igor Kurchatov's laboratory at the Fiztekh. Together with Lev Rusinov, Flerov had independently verified the production of secondary neutrons and indirectly confirmed that Bohr and Wheeler were right in their assertion that the rare isotope U-235 is responsible for fission in uranium. With Konstantin Petrzhak, he subsequently discovered that U-235 undergoes spontaneous fission.[1]

The outbreak of war in 1939 and the subsequent German invasion of the Soviet Union in 1941 caused the energies of Soviet physicists to be diverted away from problems of nuclear physics and towards essential war work. Flerov joined the Leningrad Air Force Academy to train as an engineer. However, he had glimpsed the possibility of an atomic bomb based on a fast neutron chain reaction and wasn't altogether ready to abandon

---

[1] Otto Frisch made the same discovery (see Chapter 3), but did not publish his results until after the war.

159

it. He wrote to several of his colleagues suggesting that research on nuclear fission be continued, to no avail.

When in late 1941 Flerov was stationed with his unit at Voronezh, not far from the front line, he decided to take advantage of its university library to catch up on the latest publications by Western nuclear scientists. He was particularly interested to find out how his work on spontaneous fission had been received in the scientific press.

What he found surprised and alarmed him. As he flicked through the pages of Western physics journals, he saw that there were simply *no* publications relating to nuclear fission research. He couldn't believe that such an interesting and important topic had been abandoned. It certainly wasn't the case that the leading figures in nuclear physics research had dropped the subject in favour of something else. Their names were similarly conspicuous by their absence.

The dogs were no longer barking. Flerov deduced that research on nuclear fission had become classified, a sign that American, British and – much more worryingly – German physicists were working on atomic bombs.

He decided to raise the alarm. He wrote a letter to Kurchatov in February 1942, recommending that research on fission in uranium be restarted in the Soviet Union. A further letter and a series of telegrams to Sergei Kaftanov, the recently-appointed State Defence Committee's plenipotentiary for science, met only with stubborn silence.

In frustration, in April 1942 Flerov wrote a letter directly to Stalin.

Stalingrad

The MAUD Committee report on the feasibility of building an atomic bomb had arrived in Moscow in late September 1941, just as German forces were advancing on the city. The Soviet government evacuated to Kuibyshev in October and by December the Wehrmacht was dug in only 30 miles outside Moscow.

Consequently, nearly six months had passed when, in March 1942, Lavrenty Beria began properly to consider the material that had been provided by Cairncross. Beria was head of the NKVD and a prominent member of the State Defence Committee. In the late 1930s he had presided over the closing stages of Stalin's Great Purge of perceived enemies of the Soviet state, through show trials, executions and imprisonment of political dissidents in a network of brutal labour camps which Alexander Solzhenitsyn later called the 'Gulag Archipelago'.

Beria was extremely suspicious. His first assumption was that this was disinformation planted by British or German agents designed to influence Soviet thinking and encourage wasteful expenditure on an ultimately futile project. However, he changed his mind after consulting with a trusted physicist who evaluated the report.

In March, Beria drafted a detailed memorandum to Stalin on the subject. He summarised the MAUD Committee conclusions and the British war cabinet's decision. He emphasised the 'importance and urgency of the practical utilization of the nuclear energy of uranium 235 for the Soviet Union's military purposes', before recommending the establishment of a consultative body of experts and the sharing of the espionage materials with a few 'prominent specialists'.

In April the NKVD consulted with Soviet nuclear physicists about the prospects for a bomb, but did not share the MAUD Committee conclusions or the reports that were now arriving from Fuchs via the GRU. The physicists were inevitably cautious, but by the time Flerov's letter to Stalin had arrived on his desk, Beria had already taken the decision to restart Soviet research on nuclear fission.

This was not to be an all-out effort to build an atomic bomb, since the Soviet Union was still fighting for its survival. Besides, so far as the physicists themselves understood, there were no large natural uranium deposits lying beneath Soviet soil. Instead, work was to begin on the feasibility of a Soviet bomb and an assessment of the potential threat from a German weapon. Planning proceeded slowly through the summer of 1942.

Although Hitler had failed in his mission to capture Moscow, by the late spring of 1942 the Eastern front had stabilised. Army Group South pressed on towards the Caucasus and the strategically important Soviet oil fields. They were making good progress, but Hitler, in frustration at his army's failures, had decided to take more direct control of his forces. At this point he decided to split his forces into two. One group would continue on to the Caucasus, while the German Sixth Army and, eventually, the Fourth Panzer Army were diverted 300 miles away towards the River Volga and the city of Stalingrad.

It was a puzzling decision. Stalingrad was a critical industrial city and the Volga an important means of communication. The city represented a gateway to the Urals and the north. Yet it was strategically less important than the oil fields to the south-east. It seems that Hitler feared that the Soviets would be able to launch attacks on his flanks from Stalingrad. It is also possible that Hitler was by now obsessed with the idea of destroying the 'city of Stalin'.

By September, Stalingrad was virtually overrun. The city had been fire-bombed and turned to blackened rubble by the Luftwaffe. Stalin insisted there would be no step backwards, and prevented civilians from leaving the city in order to encourage greater resistance by Soviet forces. The average life expectancy of a Soviet private newly arrived in the city fell to under 24 hours. But there was no surrender. Under the inspired leadership of General Vasily Chuikov, the Soviet 62nd Army defended the ruins of every building and every factory. Every territorial gain the Germans made during the day would be taken back by the Red Army at night. The battle for Stalingrad was fast becoming the bloodiest battle in human history.

Site Y

Within a few weeks of taking up his new assignment in September 1942, Groves embarked on an inspection tour of the facilities engaged in the American atomic bomb programme. What he saw was quite disheartening.

His first stop was Pittsburgh, and the research laboratories of Westinghouse Electric and Manufacturing Company. Work on centrifugal isotope separation had continued at the University of Virginia and at the Standard Oil Development Company's laboratory at Bayway, New Jersey, where a pilot plant was eventually constructed. Westinghouse was charged with the task of making the large, high-speed centrifuges needed for full-scale production of U-235. It was not an auspicious place to start the tour. The researchers had encountered major engineering problems and the work did not appear to have high priority. On Groves' advice, the project was closed down shortly afterwards.

From Pittsburgh, Groves travelled to Columbia University in New York, where work on gaseous diffusion was being conducted under the overall supervision of chemist Harold Urey. While the scientists he met there were more optimistic about this separation method, the corrosive tendencies of uranium hexafluoride were still causing major headaches. A gaseous diffusion plant would require innumerable porous barriers. These barriers had to be made from a material capable of resisting corrosion. No such material had yet been found. Groves thought the work should continue, but was doubtful that anything would come of it.

From Columbia, Groves headed westwards to the Met Lab in Chicago, arriving on 5 October. The work on the experimental reactor supervised by Fermi appeared to be progressing well. However, Groves was struck by the scientists' lack of clarity on what he understood from an engineering perspective to be quite fundamental parameters. If a bomb was to be built in time, then the programme needed to have answers to certain key questions: How much? How big? How long? The physicists appeared still to be content dealing with orders of magnitude estimates. Groves reminded them that if they were charged with the task of catering for a wedding reception, then being advised to expect anywhere between ten and a thousand guests was not a basis on which any kind of proper planning could be done.

Surrounded as he was by 'eggheads', Groves felt it necessary to point out to his audience, which included several Nobel laureates, that he was not overawed by their intellect. His ten years of formal education, he contended, must surely be equivalent to about two Ph.Ds. He left them to ponder the significance of this. Szilard, who had been in the audience, did not need long to ponder. 'How can you work with people like that?' he exclaimed later to his colleagues. In Szilard's case, the feelings were mutual.

Groves identified Szilard as a troublemaker almost immediately, and tried hard to have him locked up for the duration of the war.[2]

From Chicago, Groves travelled further westwards to the Rad Lab at Berkeley, arriving on 8 October. Lawrence, the accomplished showman, made a very favourable impression. Here in California, Groves thought, he was at last going to get some good news. Lawrence promised him a demonstration of his latest machine. By now Lawrence had graduated from the 37-inch cyclotron to the new 184-inch super-cyclotron which had been completed and was ready for operation in June 1942. The cyclotrons adapted to separate U-235 had been renamed 'calutrons' in honour of the University of California.

The 184-inch calutron was housed in a large circular building on Charter Hill, behind the Berkeley campus. Lawrence sat at the controls of this massive machine and explained how it worked. Suitably impressed, Groves asked how long it would take to get a real separation. Lawrence admitted that they hadn't so far achieved any sizeable separation. The machine had not yet been run for more than ten or fifteen minutes at a time. It needed between fourteen and 24 hours' operation just to achieve a proper vacuum.

Now seriously depressed, Groves made his way to Oppenheimer's office in Le Conte Hall. This meeting did not go the way that might have been predicted. Oppenheimer, the underweight, ascetic, radical intellectual capable of brilliance and arrogance in near equal measure, contrasted starkly with the military careerist sitting opposite. Groves was a sweet-toothed, overweight, conservative son of a Presbyterian minister with an engineer's pragmatism and distaste for intellectualism. Hardly a match made in heaven. Yet the two hit it off almost immediately, principally because Oppenheimer was keen to impress and win Groves over. His work on the atomic bomb project had given him a new sense of direction, and

---

[2] On 28 October Groves drafted a letter in Secretary of War Henry Stimson's name requesting that Szilard be interned as an enemy alien. Stimson refused to sign it. See Lanouette, pp. 238–41.

possibly a new lease of scientific life, and he wanted to ensure he at least retained his position.

Groves was struck by Oppenheimer's obvious capabilities as a physicist, his grasp of the situation and his ability to explain the science intelligibly. Most importantly of all, Groves found Oppenheimer curiously reassuring. 'There are no experts,' Oppenheimer claimed, 'The field is too new.' Oppenheimer argued that all the scientists working on the principles of bomb physics and design be brought together at a single, dedicated laboratory where they could work to solve the many problems they faced.

Groves had been thinking along the same lines, and had already conceived of the dedicated laboratory as 'Site Y'. He had gone to Berkeley with the intention of asking Lawrence to head the new laboratory but had concluded that Lawrence was critical to the success of the electromagnetic separation project. Groves swiftly reached a conclusion. He agreed with Oppenheimer that the programme needed a central laboratory, to be run as a military establishment. He also judged that Oppenheimer was best placed to serve as its scientific director, which may have been precisely the conclusion that Oppenheimer was seeking. Groves offered him the position the following week, on 15 October.

To many involved in the project, Oppenheimer's appointment was 'improbable'. There were many objections. Oppenheimer was a noted theoretician with a typical theoretician's clumsiness near experimental apparatus. Although this was a project that demanded theoretical input, it was principally going to be an experimental and engineering project, managed and run on a scale the like of which few physicists had experience of, and of which Oppenheimer had none. 'He couldn't run a hamburger stand', was one fairly typical observation. The project team would no doubt number a good many Nobel laureates, yet Oppenheimer himself had not won a Nobel prize.

And, of course, there were the security issues raised by his former Communist associations. On 10 October an FBI bug in Steve Nelson's Oakland office had picked up a conversation referring to 'an important weapon that was being developed' and an important contact on the project whom the FBI surveillance team believed was Oppenheimer himself.

The objections fell on deaf ears. Groves had found his man and he railroaded the decision through various committees. Oppenheimer was appointed on 19 October 1942.

Attention turned to the search for Site Y, the location of the new central laboratory. A remote wooded canyon in New Mexico, Jemez Springs, was rejected by Oppenheimer as too gloomy and depressing and by Groves because the site lacked existing buildings. Oppenheimer knew this area quite well. It was here that he had recovered from tuberculosis in the summer of 1928, staying at a log cabin on wooded slopes opposite the Sangre de Cristo mountain range. The cabin was nicknamed Perro Caliente (Spanish for 'hot dog') and always referred to by Oppenheimer as a 'ranch'.[3] That summer, he and his brother Frank had explored this whole area on horseback.

At Oppenheimer's suggestion, the search party drove on from Jemez Springs to a mesa on the other side of the Jemez mountains, and a private boy's school that he remembered from his tours of the area. This was the Los Alamos Ranch School, which had been established in 1917 by a Detroit businessman to provide a healthy, outdoors environment and a classical education for the sickly, coddled children of the well-to-do. It was affiliated to the Boy Scouts (the students at the school belonged to the mounted Los Alamos Troop 22). Alumni of the school included William Burroughs and Gore Vidal. The school was well known to Conant – he had once considered sending his youngest son there.

The site had buildings and supplies of water and electricity but the road up to the mesa from Santa Fe, 30 miles south-east, was little more than a dirt track. Groves nevertheless liked the site's isolation. At this stage Oppenheimer had estimated that the laboratory would need to house no more than 30 or so scientists, plus supporting personnel. Groves agreed to the location on the spot, and negotiations for the purchase of the site began within a week. The purchase was completed quickly, as the school had never really recovered from the Depression. Its last graduates,

---

[3] Oppenheimer later purchased the freehold.

which included the grandson of the founder, received their diplomas on 21 January 1943.

Oppenheimer had started unofficially recruiting scientists for the laboratory within days of his appointment. Now that Site Y had been identified, he and Lawrence began recruiting in earnest. Many that they approached were reluctant, however, some citing the remoteness of the location as a reason. Szilard, for one, declared: 'Nobody could think straight in a place like that. Everybody who goes there will go crazy.'

However, most of the scientists they approached were more concerned about the implications of working at a military laboratory and baulked at the idea of joining the army. Oppenheimer had not given this a second thought – he had happily reported for an army physical examination and had even been measured for a uniform. But he was persuaded by physicists Isidor Rabi and Robert Bacher, who were both busy working on radar at MIT, that the laboratory should retain 'scientific autonomy' and that joining the army should not be a prerequisite. Groves reluctantly agreed, provided the military retained authority and responsibility for security at the site.

The scientists working at Los Alamos could retain autonomy and continue as civilians, but at the cost of security measures that would give the laboratory the appearance of a concentration camp.

A game of murders

Despite the Canadian government's eagerness to embrace the proposal to relocate the Tube Alloys reactor team to Montreal, there remained concern about the number of non-British nationals who were to participate. As the scope of the project expanded, so too did the number of émigré scientists. Among these was George Placzek, already in America, who had agreed to become head of a new theoretical physics division at the Montreal laboratory. When it was suggested that the Italian physicist Bruno Pontecorvo join the project, the Canadians protested. But physicists of Pontecorvo's calibre (he had been part of Fermi's nuclear physics group in Rome) were in short supply, and many of those of British nationality were already accounted for by other projects. It was eventually agreed that Pontecorvo would join the team.

When in January 1943 the small party of scientists left Britain for Canada on a banana boat, Alan Nunn May was the only British-born scientist in the group. May was, like Fuchs, another quiet, withdrawn figure who rarely spoke unless spoken to and seemed to have few friends. He had been educated at Cambridge and, although not overtly politically active, served for a time on the editorial board of the *Scientific Worker*, the journal of the National Association of Scientific Workers, an organisation that had been infiltrated by Communists. He had joined Tube Alloys in April 1942, returning to the Cavendish Laboratory from Bristol, where he had been evacuated in the first months of the war.

Although there is no direct evidence that he was in contact with Soviet intelligence during this time, subsequent events would reveal that he was, in fact, a GRU spy, most probably recruited by one among the Cambridge spy ring.

To pass the time as they crossed the Atlantic, the party organised a few social events and games. One evening they played the game of murders, in which participants follow a series of clues to uncover the murderer among them. When one of the party took their turn to play the detective, she noticed that May stood rather aloof and detached from the proceedings. This was not untypical behaviour, and she quickly put him out of her mind as she turned her attention to the mystery at hand.

But it was May who was the murderer.

Raid on Vemork

Torstein Skinnarland was arrested in one of many German sweeps of the local population following the failed commando raid on Vemork. He was sent with his brother Olav to Grini concentration camp. Einar was forewarned of the raid and managed to escape onto the Hardanger Plateau, where he joined Poulsson and the other members of the advance party.

Einar Skinnarland was a fortunate addition to the team. He was able to procure supplies and his good-natured approach to their hardships helped keep their spirits up. With his help, the advance party – now renamed Swallow – managed to maintain wireless contact with SOE headquarters in Britain.

Back in London, the SOE was faced with a very difficult decision. Forty-one personnel had perished in the failed Freshman operation. Yet the high concentration cells at Vemork remained intact and continued to supply the heavy water required by the German atomic project. Nothing had changed: the destruction of the heavy water plant remained a top priority, although this was a task now made doubly difficult.

It was obvious that an operation like Freshman could not be repeated. Combined Operations passed the challenge back to the SOE, and an alternative plan was devised. This time, the SOE had the advantage of Brun's intimate – and recent – knowledge of the plant. Both Tronstad and Brun felt strongly that a small sabotage party could succeed where the larger-scale commando raid had failed.

A team of six Norwegians was drawn from the ranks of the Norwegian Independent Company. The team was led by 22-year-old Joachim Rønneberg, regarded as one of the best of the commandos from among those who had graduated from the SOE's Special Training Schools. Rønneberg selected Haukelid, who had by now recovered from the wound he had picked up in training, Kasper Idland, Fredrik Kayser, Birger Strømsheim and Hans Storhaug. All were accomplished skiers and out-doorsmen.

Unusually, all six were fully briefed by Tronstad about the fate of Operation Freshman, and Hitler's new commando directive. Tronstad and Wilson 'had thought it best to explain the whole situation to us', Haukelid later wrote. 'We must be prepared to receive no better treatment than the British soldiers if we were taken prisoner.'

The new raid was codenamed Operation Gunnerside. This time the planning was detailed and meticulous. Under Tronstad and Brun's direction, a replica of the heavy water plant was constructed at the SOE's Special Training School in Hatfield, Hertfordshire (STS-17). The sabotage team practised laying charges in precisely the right places on each high concentration cell to cause maximum damage. 'None of us had been to the plant in our lives but by the time we left Britain we knew the layout of it as well as anyone', said Rønneberg.

Each member of the team was issued with suicide pills, small quantities of cyanide encased in rubber that, if bitten through, would ensure death in three seconds. On their last day at STS-17, Tronstad explained how their mission would live on in Norway's history in a hundred years' time.

From Hatfield they headed north to STS-61, a large eighteenth-century country house near Godmanchester in Cambridgeshire. Here the team could enjoy some rest and relaxation prior to the mission, entertained by women from the First Aid Nursing Yeomanry, who kept house, cooked and organised their social lives. The house was also known as Farm Hall. It belonged to the SOE but was used as a staging-post for SIS agents about to depart for occupied territories, and as a debriefing or interrogation centre for agents or captives coming into Britain. Eric Welsh had had Farm Hall wired throughout, with concealed listening devices in all the bedrooms and reception rooms. A listening post was installed in a service wing, behind doors secured by special locks.

While it was pleasant, the team's extended stay at Farm Hall frayed their nerves. In December 1942 Operation Gunnerside was delayed by bad weather. Rønneberg insisted on a return to more arduous training in Scotland. The mission was delayed again on 23 January 1943, when the RAF pilot and navigator failed to find the designated drop zone and, running low on fuel, turned around and headed back to Scotland.

On the Hardanger Plateau, the Swallow team was experiencing the worst weather in living memory, the temperature barely rising above minus 30° Celsius. Though they were still in good spirits, their food rations were now very meagre and their health was deteriorating fast. When the Gunnerside team got airborne again on 16 February, the advance party had been holding out in Europe's most inhospitable wilderness for nearly four freezing months. They were in bad shape.

The advance party had been advised by wireless that the Gunnerside party had landed, but a severe storm had since descended and after several days without contact they began to fear the worst. A week later, the Gunnerside team finally made contact with two of the advance party,

Helberg and Kjelstrup, who had been sent by Poulsson to search for them. Four months surviving on the Hardanger Plateau had taken its toll. Helberg and Kjelstrup looked like tramps, their clothes filthy and covered in reindeer blood, bearded, malnourished, their emaciated faces a sickly yellow.

Back-slapping and hearty congratulations were followed by a veritable feast of reindeer and fresh rations. After a couple of days' recuperation, the Norwegians were ready to mount their attack. Helberg was dispatched to Rjukan to source information on the Vemork defences from a contact in the town, an engineer at the plant called Rolf Sørlie. The team then set to work to figure out how they were going to carry out their task.

There were about 30 German troops based at the plant itself, with many more garrisoned in Rjukan. The plant could be reached from the road only by a narrow suspension bridge, about 75 feet long. The bridge spanned the deep ravine which now separated the saboteurs from their target, and it was closely guarded. Gaining access to the plant without being detected and without an exchange of fire appeared impossible.

But what had been considered impossible by the German defenders was considered quite feasible by the Norwegian attackers. Helberg discovered that it was possible to descend into the ravine, cross the frozen river Måna at the bottom and ascend the other side, where the saboteurs could access a railway line cut into the mountainside. This railway line, which ran from Vemork to Rjukan, was used only occasionally to transport heavy machinery to the plant. It was not guarded. They had found a way in.

Finding a way out was more problematic. The explosion would undoubtedly alert the German troops, and if they chose to retreat via the ravine they risked becoming trapped. Rønneberg and Poulsson favoured fighting their way out across the bridge, but the others were not convinced. Democracy prevailed, and a retreat via the ravine was agreed.

The party split into two teams. Rønneberg led the sabotage team which included Idland, Kayser and Strømsheim. Haukelid led the covering team, comprising Poulsson, Helberg, Kjelstrup and Storhaug. Haugland and Skinnarland were to remain in wireless contact with the SOE and report the results of the operation.

They set out at 8:00pm on Sunday, 28 February. They all wore British uniforms and carried British papers so that their action would be seen as a military operation, hopefully reducing the risk of reprisals against the local population. Although it was a steep climb, they crossed the ravine without incident and managed to get to the railway line. They walked along it, a strong south-westerly wind covering any noise they made. They reached a small building about 500 yards from the plant at about 11:30, and waited for the change of sentry on the suspension bridge which was due at midnight.

The group separated at 00:30. The sabotage team cut through the flimsy chain on the fence around the plant and headed for the heavy water concentration cells in the basement. They split into two pairs as they tried to find a way in. Rønneberg and Kayser eventually gained access via a narrow cable shaft, surprising the Norwegian nightwatchman inside. Kayser covered the nightwatchman with his gun, as Rønneberg started to place charges.

He was about halfway through when Strømsheim crashed in through a window. He and Idland had tried to get in through the door on the ground floor, but had found it locked. Unable to find any other way in, they had decided to risk a noisy break-in. Kayser instinctively swung his gun from the nightwatchman to the window. 'I almost killed him', Kayser said later. 'If there had been a bullet in the chamber of my gun, I probably would have. I recognised him just in time.'

Idland kept watch outside the broken window as Rønneberg and Strømsheim placed the last of the charges. They had set the fuses when they were interrupted by a Norwegian foreman. Rønneberg lit the fuses and Kayser suggested to their two captives that they head upstairs as quickly as possible. By his reckoning, they should be able to get to the second floor before the explosion. The saboteurs left by a cellar door, and were no more than twenty yards from the building when they heard a muffled explosion.

'The explosion itself was not very loud', Poulsson later recalled. 'It sounded like two or three cars crashing in Piccadilly Circus.' However, inside the building it was a different matter, with one plant engineer who had been up on the third floor remarking: 'The explosion was tremendous, the power of it reverberated throughout the entire building.'

The sabotage party took cover as Haukelid and the covering team prepared for the appearance of German troops from the nearby barracks. But outside, the explosion had not been loud, and small explosions from the plant's combustion equipment (called 'cannons' because of their shape) were not unusual. A single guard appeared, flashed a torchlight inches above Haukelid's head, and went back inside.

The covering and sabotage parties reunited. They began their retreat back along the railway line, and then back down the ravine. They were crossing the now rapidly thawing river at the bottom when they heard the first sounds of sirens. Rønneberg had feared they would be trapped in the ravine, picked out by searchlights with no means of escape. But the Germans were busy searching the plant itself, convinced the saboteurs were still somewhere on the premises. They knew that nobody had passed the sentries on the bridge, and as far as they were concerned that was the only way out.

The raid had been a success, and now the party's main concern was for their own safety. They scrambled up the other side of the ravine. The road from Rjukan was now busy with traffic, including trucks carrying more German troops. Across the ravine they could see flashlights darting through the night as Germans traced their retreat along the railway line. They didn't have much time.

They followed the power line towards Rjukan, and then climbed up along Ryes Road as it zig-zagged beneath a cableway. The cableway had been built before the war to allow the citizens of Rjukan an opportunity to escape the gloom of winter during its four long months of darkness. It was now discontinued for public use. The road led up to the top of the cableway at Gvepseborg and the edge of the Hardanger Plateau.

There had been no exchange of fire. Aside from a couple of Norwegian workers at the plant, nobody had seen the raiders enter or leave. Between four and five months' production of heavy water washed uselessly over the basement floor.

General Rediess decided that this was the action of British intelligence and the Norwegian resistance, and threatened to execute ten of Rjukan's leading citizens in reprisal. Arriving on the scene shortly afterwards, Generaloberst Nikolaus von Falkenhorst, commander-in-chief of the German forces in Norway, decided that this had been a military operation, carried out by uniformed British soldiers. He called it 'the finest coup I have seen in this war', and ordered that the Rjukan citizens be released.[4]

This admiration spilled over into the media. A report on Swedish radio on 1 March claimed that the sabotage of the heavy water facility at Rjukan was intended to disrupt production of high-quality explosives. A garbled account appeared in the *Daily Mail* on 2 March 1943, submitted by the *Mail*'s correspondent in Stockholm. This account contained no reference to heavy water. A further report in the *Svenska Dagbladet* linked the raid with a 'secret weapon' based on heavy water. *The Times* reported on the raid on 4 April. On the same day the story made its way into the *New York Times*, which made a muddled connection between heavy water and atomic energy. To Groves, who was not well-disposed towards the British, these press reports were further examples of lax security. Bush wrote an angry note attached to the clipping from the *New York Times*, arguing that it 'gives sufficient basis in itself for insistence that knowledge be passed only to those who really need to know it'.

Special resolution

Against the odds, the Soviet forces held on in Stalingrad. Stalin sent the saviour of Moscow, Marshal Zhukov, to organise a counter-attack. Zhukov assembled massive forces on the steppes to the north and south of the city. On 19 November 1942 the Red Army launched its counter-offensive, codenamed Uran. Uran is usually translated into English as 'Uranus', but

---

[4] Falkenhorst was tried for war crimes in July–August 1945. He was found guilty of seven (of nine) charges, including a charge relating to the murder of nine commandos of Operation Freshman. His initial death sentence was commuted to twenty years' imprisonment. He was released on the grounds of ill-health in 1953, and died in 1968. Rediess committed suicide in May 1945.

can also be translated as 'uranium'. The choice of codename may have been quite coincidental, but the counter-offensive was a turning point in the battle (and arguably the entire war). Within a day the Romanian Third Army, which protected the German Sixth Army's northern flank, was crushed. The next day the Romanians guarding the Germans' southern flank were swept aside. Soviet forces completed their encirclement two days later.

Hitler insisted there would be no surrender, but attempts to relieve the trapped German forces by air failed. With ammunition and rations fast running low, the Germans surrendered on 2 February 1943. In all, the battle of Stalingrad had lasted nearly 200 days and had claimed between 1.7 and 2 million lives.

Nine days later, on 11 February, the Soviet State Defence Committee passed a special resolution on research into atomic energy.

Igor Kurchatov had studied physics at the Crimea State University and ship-building at the Polytechnic Institute in Petrograd before joining Abram Ioffe at the Fiztekh in Leningrad. With Ioffe he had worked on radioactivity before receiving funding to set up his own nuclear research programme in 1932. During the first years of the war he had worked on ways to demagnetise ships to protect them from magnetic mines. After the Nazi invasion, he declared his intention not to shave his beard until the enemy was defeated. Consequently, he grew a flourishing beard that gave him the appearance of an Orthodox priest. Inevitably, he gained the nickname 'the Beard'.

As the State Defence Committee searched for a scientific director for its atomic bomb programme, it was Kurchatov who impressed the most. His appointment was officially announced by Viacheslav Molotov, deputy chairman to the committee, later in February 1943.

But although Kurchatov understood the physical basis for an atomic weapon, he was also very well aware of the many problems standing in the way. 'Then I decided to give him our intelligence materials', Molotov later recalled. At the beginning of March, Kurchatov sat for several days in Molotov's office in the Kremlin studying the conclusions of the MAUD Committee and various papers on the Tube Alloys project on gaseous diffusion that Fuchs had provided.

This material showed that the British and the Americans were taking the possibility of a bomb very seriously, and confirmed some of the thinking about detonation of a U-235 bomb that had been developed largely by Flerov. But the material also provided new information that would help save time and effort in the Soviet programme. The British had abandoned thermal diffusion in favour of gaseous diffusion. A uranium reactor could be used to produce element 94, potentially more powerful than U-235 and without the problems associated with physical separation of a rare isotope.

Kurchatov wrote two memoranda for the State Defence Committee summarising his interpretation of the espionage materials. In the second, dated 22 March 1943, he emphasised the importance of element 94: 'The prospects of this direction are unusually captivating', he wrote. When asked for his opinion on the materials by Molotov, he declared that they were wonderful: '[T]hey fill in just what we are lacking.' Molotov introduced him to Stalin, who promised every kind of support.

It seemed reasonable to suppose that further work on uranium isotope separation and element 94 had been done in America, and Kurchatov closed his second memorandum with an appeal: 'In this connection I am asking you to instruct Intelligence Bodies to find out about what has been done in America ...' He listed four questions concerning element 94 that he wanted answers to, and requested an update on the work being carried out with cyclotrons. He also provided a list of American laboratories that should be targeted.

Berkeley's Rad Lab headed the list.

The Russian diplomatic problem

As the battle of Stalingrad wound to its bloody conclusion, Colonel Carter Clarke, chief of the Special Branch of the US Army's Military Intelligence Division, grew concerned that Moscow and Berlin might seek to negotiate a separate peace. On 1 February 1943 he ordered the army's Signals Intelligence Service to begin a small programme to study the encrypted Soviet message traffic passing between the US and Moscow. He had hoped to find clues to any such negotiation. The effort was referred to simply as the 'Russian diplomatic problem'.

The Soviets operated a 'belt and braces' approach to coding their messages. A message from a Soviet diplomat – or a spy – resident in America would first be coded into sequences of four-digit groups, using a code book. These were then regrouped into five-digit groups. The five-digit groups were then enciphered using a so-called one-time pad. Each page of the pad consisted of 60 five-digit random number groups which constituted the encryption key. By adding the original five-digit groups to the appropriate sequence of groups on the key page, a cipher clerk would produce a new series of five-digit groups. These were then translated into five-letter groups. The end result was a message consisting of seemingly random five-letter groups, which was then cabled to Moscow.

The cipher clerk at the receiving station would identify the specific page of the one-time pad that had been used, and decipher the message using the key page and the code book. It was a virtually unbreakable cipher system.

However, the whole point of a key page in a one-time pad is that it should be used only once. But the Soviet message traffic had grown so vast during the second half of 1941 that the cryptographers were unable to produce new key pages quickly enough to meet demand. In early 1942 the cryptographers started to produce duplicate key pages, doubling their output for no extra effort. About 35,000 duplicate key pages were shuffled and bound into different one-time pads during 1942, effectively making them two-time pads.

It also made the messages vulnerable to cryptanalysis.

Chapter 9

ЭНОРМОЗ

January–August 1943

By January 1943, the collaboration between British and American nuclear scientists had ground virtually to a halt. The Americans were stalling, and the British were deeply suspicious of the reasons.

Akers had met with Groves in Washington in early November 1942 and had been advised that concern for security was behind the Americans' growing tardiness. Groves had been alarmed by the extent to which British physicists had been in contact with their American counterparts across very different parts of the programme. Why, Groves enquired, had Peierls, supposedly busy with the physics of gaseous diffusion as part of Tube Alloys, been in contact with American physicists involved with the study of fast-neutron chain reactions?

Groves favoured the preservation of security through compartmentalisation, breaking the various parts of the programme down into self-contained projects in which the scientists and engineers involved knew absolutely nothing of the rest. The behaviour of British physicists, who were in any case too few in number to be compartmentalised this way, was therefore at odds with Groves' requirements.

Akers feared that there was more behind the Americans' unwillingness to co-operate than concerns for security. His fears were justified. Groves – a committed Anglophobe – Conant and Bush were all firmly agreed that full exchange of information with the British meant providing Tube Alloys with details of processes and plants to which the British had made no contribution. These were processes and plants being developed and built by American science funded by the American taxpayer. To gift information to the British, they concluded, would do little to aid the American (or, indeed, the Allied) war effort and could only be of value in supporting Britain's post-war ambitions.

Matters finally came to head in January 1943 when Conant shared a draft memorandum listing proposed general rules and regulations governing Anglo-American collaboration on the atomic bomb. The British were horrified.

The Conant memorandum 'derives from the basic principle that interchange on design and construction of new weapons and equipment is to be carried out only to the extent that the recipient of the information is in a position to take advantage of this information *in this war*'. This meant no further information was to be supplied by the Americans on electromagnetic separation, the production of heavy water, fast-neutron chain reactions and the manufacture of uranium metals and compounds such as uranium hexafluoride. Information transfer on the subject of gaseous diffusion was to be controlled by Groves.

The memorandum extended the proposed rules to information exchange with the group that had in the meantime been established in Montreal. It stipulated that while information on chain reactions could be exchanged with Met Lab physicists in Chicago, there was to be no further information transmitted on the properties or production of plutonium.

When Anderson saw the memorandum he was furious. He urged Churchill to discuss it with Roosevelt. But, unknown to the British, Roosevelt had already endorsed the stance the Americans were now taking. Churchill's telegrams to Roosevelt's aide, Harry Hopkins, went unanswered. A review of the cost to Britain of going it alone on the atomic bomb only served to confirm earlier conclusions. Without the Americans the bomb was out of Britain's reach, certainly within the likely timeframe of the war, and probably for many years thereafter.

In May 1943 Anderson discovered that the Americans had been quietly buying up refined uranium oxide from Ontario's Eldorado mine, and had secured rights to production for years to come. This had happened despite commitments from the Canadian government that the British would have joint control over the supply of uranium. The Americans had also secured a supply commitment from the heavy water plant which was to be built by the Consolidated Mining and Smelting Company in Trail, British Columbia.[1] Without access to uranium or heavy water, the Montreal project could not continue, and work ground to a standstill in June.

Churchill raised the matter with Roosevelt directly, following the conclusion of the two-week Trident conference in Washington in May. Roosevelt gave his personal assurance that the exchange of information with Tube Alloys would resume. But any hopes that this would happen were scuppered by Cherwell who, at a meeting held simultaneously in Washington on 25 May with Bush and Stimson, explained the real reason for Britain's desire to resume the exchange of information. This was necessary, he argued, so that Britain could pursue an independent atomic bomb programme after the war. The post-war vision for Britain was one of empire and atomic power. Cherwell's gaffe merely confirmed what Bush had always suspected. Bush informed Roosevelt of the discussion.

Having given his personal assurances to Churchill, Roosevelt was now perturbed by this evidence of British intent to hold on to its position on the world stage. Roosevelt was secretly committed to dismantling the British empire, turning former British colonies into sovereign nation-states with an American-style constitution. He made no final decision and the impasse dragged on.

---

[1] Consolidated Mining and Smelting signed an agreement with the OSRD on 1 August 1942, accepting millions of US dollars to build and operate a heavy water plant at Warfield, in a construction project named 'Project 9'.

Espionage assault

Stalin had realised that the defeat of Germany and Japan would leave three major world powers – the Soviet Union, the United States and Great Britain. Of these, Britain was the weakest, the sun surely setting on Churchill's dreams of empire. The Soviet Union already had a strong intelligence network operating in Britain, reliant on prominently placed, ideologically motivated individuals across all walks of political, military and scientific life. However, until the German invasion of the Soviet Union in 1941, America had not been regarded as an important target for Soviet espionage.

The German invasion forced the Soviet Union into an uneasy alliance with Britain and America, and the Soviets began to benefit from American 'Lend-Lease' military aid.[2] To manage the procurement and transport of Lend-Lease *matériel* – weapons, aircraft, vehicles and machinery – the Soviet Union greatly expanded its diplomatic presence in America. Among the diplomats arriving on American soil were many NKVD and GRU spies, as the Soviet Union launched an unprecedented espionage assault on its ally.

Pavel Fitin, head of the NKVD's First Directorate, established dedicated *rezidenturas* – bases for espionage operations – at the Soviet embassy in Washington, the Soviet consulate in New York, and the consulate in San Francisco. The Washington *resident*, or station chief, Vasily Zarubin (cover name Vasily Zubilin) had arrived in America in October 1941. Supported by his wife Elizabeth, who was also an NKVD agent, he was charged personally by Stalin to gather intelligence on American intentions towards Germany.

In late 1941/early 1942, Soviet vice consul and NKVD agent Anatoly Yatskov (cover name Anatoly Yakovlev) and trainee Alexander Feklisov (cover name Alexander Fomin) had been installed at the Soviet consulate

---

[2] 'Lend-Lease' aid to Britain, China, France, the Soviet Union and other Allied nations began in March/April 1941. The Allies were supplied with over $50 billion worth of *matériel* (equivalent to almost $700 billion in 2007) in exchange for (in Britain's case) the establishment of American military bases.

at 7 East 61st Street in New York, between Madison and Fifth.[3] The New York NKVD *rezident*, Gaik Ovakimyan, had been exposed as a spy by the FBI in April 1941 and had been obliged to leave the country. His role was filled temporarily by Pavel Pastelnyak, before Zarubin was assigned responsibility for the New York station.

Feklisov later explained why he had not been troubled at the thought of spying on his American ally. He had noted the political double-standards of the Roosevelt administration. The Soviet Union was an ally only because Soviet Communism was temporarily a lesser evil than German National Socialism. With his privileged access to detailed information on 'Lend-Lease' aid, he was able to surmise that the Americans were deliberately providing only defensive, not offensive, weapons to the Soviet Union. It seemed to him that America and Britain preferred to see the Soviet Union weakened as much as possible by its war with Nazi Germany, making it all the easier for them to control post-war events. 'When you know you are being taken advantage of,' Feklisov wrote years later, 'you have every right to be clever.'

The San Franciso *rezidentura* had been established in November 1941 and was staffed by vice consul Gregori Kheifets and third secretary Pyotr Ivanov. They were charged with development of an intelligence network targeted against Germany and, subsequently, Japan. Kheifets and Ivanov immersed themselves in the activities of the local American Communist Party and progressive unions such as the Federation of Architects, Engineers, Chemists and Technicians (FAECT).

Although the necessary intelligence infrastructure was in place in America as Kurchatov issued his appeal for more information about the American atomic programme, the Soviet spies did not have direct access to America's nuclear scientists. Fitin therefore decided to establish a *rezidentura* in New York, designated 'XY', dedicated to the task of gathering scientific and technical intelligence. Leonid Kvasnikov, an NKVD agent

---

[3] Although Feklisov's passport bore Fitin's signature, he was not protected by diplomatic immunity.

with some knowledge of engineering, was assigned to head it. He arrived in New York in January 1943 and took up the position of deputy *rezident*.

Fitin gave the operation the codename ЭНОРМОЗ (ENORMOZ), meaning, literally, 'enormous'.

The Chevalier incident

Kurchatov had put the Rad Lab at the top of his list of targets for Soviet espionage, and Kheifets and Ivanov had already begun recruiting intelligence contacts from among the group of radical physicists at the Berkeley laboratory. Given Oppenheimer's 'leftwandering', it was inevitable that they would try to approach him too.

Kheifets possessed an outgoing, friendly personality and spoke good English. He was familiar with the milieu of physicists, having once targeted Fermi and Pontecorvo as potential recruits to the anti-fascist cause while deputy *rezident* in Rome in the 1930s. He met the Oppenheimers at a couple of fund-raising events in December 1941 and 1942 (Oppenheimer knew Kheifets only as 'Mr Brown'). Shortly after his first meeting with Oppenheimer, Kheifets reported to Moscow of a private lunch conversation in which Oppenheimer had talked of Einstein's letter to Roosevelt and his frustration at the lack of progress on what Kheifets understood to be a secret atomic weapons project.

Late in 1942, Ivanov had approached George Eltenton, a British chemical engineer who had worked for a time at the Leningrad Institute of Chemical Physics with the noted Soviet chemists Yuli Khariton and Nicolai Semenov. During this period both Eltenton and his wife Dolly had become dedicated Communists. Eltenton was now working at the Shell Development Laboratory in Emeryville, about eight miles from Berkeley, and organised the local FAECT chapter. Ivanov explained that he knew that the work at the Berkeley Rad Lab was connected with atomic energy. 'Do you know any of the guys or any others connected with it?' he asked Eltenton. Eltenton volunteered to make an approach to Oppenheimer through their mutual friend, Haakon Chevalier.

Chevalier was a native of New Jersey, born of French and Norwegian parents. He was professor of French literature at Berkeley, an accomplished translator, and author of a biography of Anatole France. A visit to France in 1933 and the outbreak of the Spanish Civil War in 1936 had encouraged his left-wing sensibilities in the direction of Communism. He was introduced to Oppenheimer at a meeting of the Teachers' Union. They became good friends.

'Eltenton's manner was somewhat embarrassed', Chevalier later wrote. 'He seemed not too sure of himself. Through his roundabout phrases it gradually became clear to me that what the people behind him were really interested in was the secret project Oppenheimer was working on.' According to Chevalier's later account, he rejected the idea of spying for the Soviet Union but decided to mention Eltenton's approach to Oppenheimer.

The opportunity to do this arose during a dinner hosted by Oppenheimer and Kitty at their home on Eagle Hill.[4] The Oppenheimers were preparing to leave Berkeley for Los Alamos, and wanted to share a farewell dinner with their closest friends. With Kitty and Chevalier's wife Barbara playing a duet at the piano, Oppenheimer headed for the kitchen to fetch mixers and ice to make his legendary martinis. Chevalier followed.

Interpretations of what happened next vary. According to Chevalier, he simply described the approach that had been made to him by Eltenton. Oppenheimer agreed that Chevalier had been right to tell him about it and looked visibly shaken. Nothing more was said.

In later versions of the incident, Oppenheimer claimed that Chevalier had said that, through Eltenton, he had 'means of getting technical information to Soviet scientists', an overt suggestion that Oppenheimer pass atomic secrets to the Soviet Union.

Chevalier's marriage was already in trouble, and some 40 years later his embittered ex-wife Barbara made plain Chevalier's motives that night: 'I was not, of course, in the kitchen when Haakon spoke to Oppie, but I knew

---

[4] The timing of this meeting is not clear, but it occurred sometime in the winter of 1942–43.

what he was going to tell him. I also know that Haakon was one hundred per cent in favour of finding out what Oppie was doing and reporting it back to Eltenton. I believe Haakon also believed that Oppie would be in favour of co-operating with the Russians. I know that because we had a big fight about it beforehand.'

According to the version told many years after the event to Oppenheimer's secretary Verna Hobson, Kitty, not wanting Oppenheimer and Chevalier to be alone, had also entered the kitchen and had heard Chevalier's proposal. It was she who pointed out that passing information to Soviet scientists would be treason.

Irrespective of who said what, it is clear that Oppenheimer rejected this attempt to recruit him as a Soviet spy, if that was indeed what it was. However, it was not Oppenheimer's response that was to cause him problems. The 'Chevalier incident' was soon to return to haunt him in other ways.

Although the FBI had no knowledge of the burgeoning atomic programme, they uncovered the first hard evidence for Soviet espionage against it on 29 March 1943. The FBI had planted bugs in Steve Nelson's home, as well as in his Oakland office. Nelson returned home late from a union meeting to find someone, identified only as 'Joe', waiting patiently to talk to him. Joe explained that the Rad Lab physicists working on the bomb programme were soon to be relocated, and that he would be joining them.

They discussed 'the professor', who appeared to have forsaken his commitments to the Communist cause. 'To my sorrow, his wife is influencing him in the wrong direction', Nelson said. Nelson then pumped Joe for information, and though initially reluctant, Joe eventually described aspects of the Rad Lab work on electromagnetic separation and the site in Tennessee where a large-scale separation plant was already under construction.

The direct link with Soviet intelligence was subsequently confirmed by FBI surveillance. Nelson was observed meeting with Ivanov at San Francisco's St Joseph Hospital on 6 April. In a further bugged conversation between Nelson and a man later identified as Zarubin, recorded on

10 April, Zarubin was heard counting out wads of money, 'like a banker'. The two then discussed the Soviet intelligence network in America.

Whoever 'Joe' was, he was a Rad Lab physicist passing atomic secrets to the Soviets. And he was about to move to Los Alamos.

Los Alamos primer

Oppenheimer's tenure as head of the Los Alamos laboratory got off to a shaky start. It seemed that those who had declared Oppenheimer unfit to run a hamburger stand were going to be proved right.

The first few months of construction of the new facility were chaotic, the result of a general lack of leadership and direction. Despite cajoling, Oppenheimer had failed to produce an organisation structure for the laboratory. Even the very idea of an organisation structure was new to him. As far as he was concerned, about 30 physicists were heading for New Mexico where they would build an atomic bomb. What could be simpler? When confronted about the chaos at a dinner party at his Eagle Hill home, Oppenheimer exploded with rage.

But rage was followed by a calm appraisal of the situation, and Oppenheimer learned quickly. By March 1943 he had produced an organisation chart and revised his estimate of the size of the Los Alamos population from 100 to 1,500. He began to take managerial control. Oppenheimer, Kitty and their young son Peter (nicknamed 'Pronto' because of his very prompt appearance, less than seven months after the Oppenheimers were married) arrived in Santa Fe on 16 March 1943. A few weeks later they moved up to 'the Hill', as the fledgling laboratory had become known.

They moved into a modest cabin, one of six original Ranch School buildings which were fitted with bathtubs, in contrast to the accommodation now being hastily erected by the Army Corps of Engineers, which were fitted only with showers. There were trucks and bulldozers everywhere, as 3,000 construction workers erected the main buildings, including five laboratories, a machine shop, a warehouse and barracks. The spring thaw had turned the unpaved roads to thick mud. It was a vision straight out of the wild west. On his arrival, Bethe was shocked by the isolation and the shoddy buildings.

By early April, about 30 physicists had gathered on the Hill. Oppenheimer had moved quickly to recruit the 'luminaries' of the previous year's summer school – Bethe, Bloch, Teller and Serber among them. He had also recruited a young Princeton physicist called Richard Feynman, whose bongo-playing left an indelible impression on Teller, sitting in the next room.

Feynman was young (he would celebrate his 25th birthday on 11 May), precocious and passionate about physics. He was now meeting for the first time colleagues familiar to him only by their names, which he had seen in the pages of *Physical Review*. But his was not the kind of personality that was easily over-awed. He quickly gained a reputation for his voluble debating with Bethe, who needed and welcomed the stimulation, and was made a group leader under Bethe. His wife Arline was suffering from tuberculosis and had at Oppenheimer's direction been installed in a clinic in Albuquerque so that Feynman could make frequent visits.

Fermi's work on uranium–graphite reactors in Chicago was too important for him to abandon in favour of joining the team of physicists assembling at Los Alamos. And yet Fermi's grasp of the experimental problems and his insight were too important to forgo. Oppenheimer settled for a compromise. Fermi would serve as a visiting consultant to the Los Alamos laboratory.

Oppenheimer had wanted to recruit Rabi as associate director of the laboratory, but Rabi was working on radar at MIT and argued that this was more important than working instead to turn three centuries of gloriously successful physics into weapons of mass destruction. Although Rabi's reasons were different, Oppenheimer was able to persuade him to accept the same compromise solution as Fermi. With some reluctance, Rabi too became a visiting consultant to Los Alamos.

On 15 April, the Los Alamos physicists assembled for their first meeting, in an empty library, and the first of a series of inaugural lectures to be delivered by Serber. Groves' opening remarks were downbeat. It seemed as though he was already planning for failure and anticipating what he would say to the congressional committee that would undoubtedly be convened to find out how the money had been squandered.

Serber then delivered his first lecture, summarising the output of the summer study group and the work on fast-neutron fission that had been done in the last year. He was not a great speaker, but this was an occasion on which the substance was much more important than the style of delivery. 'The object of the project,' Serber said, 'is to produce a *practical military weapon* in the form of a bomb in which the energy is released by a fast-neutron chain reaction in one or more materials known to show nuclear fission.'[5] For many in the audience, compartmentalisation had so far prevented them from understanding the full implications of the work they had been doing. Some had guessed at the details. Others had heard rumours. Now they started to become absorbed by the larger problem.

On the surface, making an atomic bomb seems relatively straightforward. Take two pieces of U-235 or plutonium whose masses are subcritical when kept apart, but which when brought together form a mass in excess of the critical mass, which then explodes. But there are some not-so-straightforward hurdles to be overcome.

The first is concerned with efficiency. The critical mass of U-235 had by this stage been fixed at around 200 kilos – a little impractical for a weapon to be dropped from an aeroplane. The study group had proposed to raise the efficiency of the device and so reduce the amount of active, fissionable material required by surrounding it with a 'tamper' – made of U-238 or gold – to reflect neutrons back into the active material. For U-235 this brought the critical mass down to fifteen kilos. For plutonium the critical mass was estimated to be just five kilos with a uranium tamper.

But the critical mass is the minimum mass required to support a chain reaction, not the mass required to make an effective bomb. It was already clear that a bomb would need an amount of active material greater than

---

[5] Serber's lectures were faithfully noted down by Edward Condon, serving as one of Oppenheimer's associate directors. The lecture notes were then bound together to make an introductory manual known as the 'Los Alamos Primer'. This document was declassified and published in 1992.

the critical mass, called a 'super-critical' mass.[6] The first step was therefore to assemble a super-critical mass from several sub-critical components. The components would come together, creating a 'divergent' chain reaction, meaning that more neutrons are produced during the reaction than are consumed.

Timing is critical. It was estimated that a kilo of U-235 would fission in about a millionth of a second, producing an explosive force equivalent to 20,000 tons of TNT and generating initial temperatures that can be thought to be equivalent to about a thousand suns. At these temperatures, the uranium is rapidly vaporised and the vapour expands, making it more and more difficult to keep the chain reaction going. At some point, the vapour reaches 'second criticality', where neutrons produced by fission balance those escaping to the surrounding environment. Beyond this point, there is no further explosive release of energy.

Assemble the components too slowly and the mass would pre-detonate – blow itself apart prematurely – yielding much less than the potential explosive force. One proposed solution was to shoot a cylindrical plug of active material, called a 'shy', into a corresponding hole in a sub-critical sphere, thereby producing a combined mass in excess of the critical mass. Following the work of the Tube Alloys physicists, this method of assembling a super-critical mass became known as the gun method.

This was also the one aspect of early bomb design that was least understood. The pieces of active material had to be brought together fast enough to prevent pre-detonation, which meant a relative velocity of about 100,000 centimetres per second or more. The highest muzzle velocity available from conventional armaments was 3,150 feet per second (about 96,000 centimetres per second) with a projectile weighing 50 pounds. This was a US Army gun with a bore of 4.7 inches and a barrel 21 feet in length, weighing five tons. If, as expected, the shy in a U-235 bomb needed to be twice as heavy,

---

[6] In fact, the efficiency of a bomb – a measure of the amount of fissile material actually fissioned in an explosion compared to the total amount of fissile material – is proportional to the cube of the difference between the bomb mass and the critical mass.

then a correspondingly heavier gun would be needed. A U-235 bomb would therefore need to incorporate a gun weighing ten tons.

Then there was the question of initiating or triggering the bomb. Despite all the concern about pre-detonation, or 'fizzles', caused by stray neutrons coming from spontaneous fission in uranium or dislodged by cosmic rays (charged particles that shower the earth's atmosphere from space), it could not be assumed that simply assembling a super-critical mass would in itself be sufficient to initiate a chain reaction. The right kinds of neutrons had to be available at just the right time to get the initial chain reaction going. The core of the bomb could be shielded from cosmic rays and the problem of pre-detonation caused by neutrons from spontaneous fission could be resolved by using high muzzle velocities.

Serber suggested using small amounts of polonium and beryllium to trigger the chain reaction. Like radium, polonium is radioactive, producing alpha particles which can then go on to liberate neutrons from beryllium, the method that had been used to produce neutrons long before the first cyclotron.[7] The idea was to shield the polonium and beryllium from one another until the shy was fired. The blast from the gun would expose the two components of the initiator, producing a burst of neutrons just as the super-critical mass was assembled.

Although the gun method was undoubtedly the simplest, Serber also included alternative, rather more esoteric arrangements for assembling a super-critical mass. These had yet to be properly analysed. 'For example,' Serber said, 'it has been suggested that the pieces might be mounted on a ring ... If explosive material were distributed around the ring and fired the pieces would be blown inward to form a sphere.' Serber's sketch showed wedges of active material and tamper mounted on a ring. Forcing the ring in on itself would bring the wedges together to form a super-critical mass.

It was during a subsequent lecture by an expert on ordnance that Seth Neddermeyer, a young physicist from the US National Bureau of

---

[7] Frisch had used a mixture of radium and beryllium in the experiments in which he discovered spontaneous fission in uranium.

Standards, made the connection. He raised his hand. He was somewhat vague on the details because this was an area outside his expertise, but he thought that one way of assembling an explosive mass of active material without the need for a gun would be to make use of an *implosion*. His idea was basically to construct a hollow sphere consisting of separate plugs of active material which is then forced together at its centre by conventional explosives packed around the outside. Driving the sphere in on itself in this way would assemble an explosive mass of active material very quickly indeed.

His proposal met with objections from all sides. The most significant of these concerned the need for a near-perfect spherical shockwave generated by the conventional explosives if the explosive mass of active material was to be assembled properly. Oppenheimer himself was quite critical of the idea. However, he had also been wrong before, about several things. In conversation with Neddermeyer after the lecture, he agreed that this was something that should at least be investigated further. He promptly set up an implosion experimentation group in the Ordnance Division, and appointed Neddermeyer as group leader.

Serber concluded his lectures by outlining the challenge ahead:

From the preceding outline we see that the immediate program is largely concerned with measuring the neutron properties of various materials, and with the ordnance problem. It is also necessary to start now on techniques for direct experimental determination of critical size and time scale, working with large but sub-critical amounts of active material.

Any thoughts about what might happen if this weapon was actually used were pushed to the backs of the physicists' minds. They focused instead on the immediate challenges, which concerned neutron backgrounds, pre-detonation, critical masses, gun designs and implosive shockwaves. Many embraced these challenges with great eagerness.

Fermi, for one, was puzzled. The Italian Navigator saw his work on the bomb as a duty born of the necessities of wartime. He told Oppenheimer, with a note of some surprise in his voice: 'I believe your people actually *want* to make a bomb.'

Threats to security

The bugged conversation that had taken place in Steve Nelson's office on 10 October 1942 had stirred the FBI's interest in Oppenheimer. There followed an inevitable tussle between the FBI and the army's military intelligence organisation, G-2, over which of them had authority over counter-intelligence related to Oppenheimer and the Rad Lab physicists. The further bugged conversation between Nelson and 'Joe' on 29 March 1943, subsequent surveillance of Nelson and the confirmed link to Zarubin promptly ended the tussle. This was clear evidence of Soviet espionage against a secret military weapons programme.

The FBI was advised in outline about the military project now under way. It was agreed that G-2 would focus its attention on Rad Lab staff under contract to the bomb programme, while the FBI would focus on suspected Communists with links to the laboratory. J. Edgar Hoover, director of the FBI, sent a memo on 7 May to Roosevelt aide Harry Hopkins concerning the taped conversation between Nelson and Zarubin.

John Lansdale, who had been appointed head of security for the atomic programme, had by now prepared a counter-intelligence strategy. A graduate of the Virginia Military Institute and Harvard Law School, Lansdale had joined a law firm first in Cleveland, then in Washington, and was a lieutenant colonel in G-2 based in Washington. He had been called in by Conant in February 1942 to investigate potential security lapses at the Rad Lab and was very familiar with the territory.

Lansdale had installed Lieutenant Lyall Johnson in an office on the Berkeley campus. Johnson, a former FBI agent now working in G-2, had proceeded to recruit informants and place agents on the research staff. Lansdale also set up an office across the bay in San Francisco, supervised by Colonel Boris Pash. Pash was a native of San Francisco and a Russian expert. He was the son of Theodore Pashkovsky, a Russian Orthodox priest sent to California by his Church in 1894. The family had been recalled to

Russia in 1912, and Boris had fought in the White Russian navy during the October Revolution. He had returned to America in 1921.

Pash had already seen enough to be convinced that Oppenheimer was a real security threat. On his own initiative, he asked Peer de Silva, a young West Point graduate and Lieutenant in G-2, to begin his own investigation.

Despite the growth in surveillance by both G-2 and the FBI, the identity of 'Joe' was uncovered by accident. 'Joe' had been glimpsed leaving Nelson's home by an FBI agent, but this glimpse was insufficient to provide an identification. When in June a commercial photographer by chance took a photograph of four Rad Lab physicists walking arm-in-arm at Sather Gate, one of the entrances to the Berkeley campus, an undercover agent observing them purchased the negative. The 'Joe' glimpsed leaving Nelson's home was recognised in this photograph. He was Joseph Weinberg, hired by Oppenheimer to work on theoretical aspects of calutron design and operation.

The four in the photograph were Weinberg, Giovanni Rossi Lomanitz, David Bohm and Max Friedman. All were students of Oppenheimer. They had different backgrounds but were united by their firm friendship, their work on aspects of the electromagnetic separation method, which had begun the previous summer, and their political activism through organisations such as the Young Communist League. Weinberg had been a member of the American Communist Party since 1938. Bohm had joined the party in November 1942. Lomanitz had established a local chapter of the radical FAECT union at the Rad Lab, and Friedman was its organiser. All four were immediately put under surveillance.

To Pash, Oppenheimer's links with this group of young radical physicists and his recruitment of Weinberg to the project was yet more evidence of Oppenheimer's guilt. Pash was preparing to confront him with his allegations when he received news that Oppenheimer was planning an unexpected visit to Berkeley.

As he had been preparing to leave Berkeley for Los Alamos, Oppenheimer had said his farewells to Nelson over a quiet lunch. He had also received an urgent request for a farewell visit from Jean Tatlock, to which he had chosen not to respond.

Oppenheimer had first met Tatlock at a fund-raiser for Spanish Loyalists in the spring of 1936. She was young, just 22 years old, with an attractive figure, long dark hair, red lips, and hazel-blue eyes beneath heavy lashes. She had studied English literature at Vassar, before turning to psychology at Stanford University in California. Theirs had been an intense, and tempestuous, relationship. They shared interests in literature and psychology, and Tatlock's heightened social conscience, which had led her to join the American Communist Party in 1933–34, had encouraged similar leanings in Oppenheimer. It was Tatlock who had introduced Oppenheimer to Haakon Chevalier.

Their relationship had collapsed towards the end of 1939. They had come close to marriage several times, but Tatlock could be moody and introspective, and suffered bouts of severe manic depression. She would disappear for weeks and months at a time, only to return with taunts about who she had been with, and what they had done. It was Tatlock who ended the relationship. Oppenheimer met 29-year-old Katherine (Kitty) Puening Harrison just a few months later.

Tatlock went on to secure her medical degree from Stanford in June 1940. She had then worked as an intern in a psychiatric hospital before becoming a resident physician at San Francisco's Mount Zion Hospital. She had continued to see Oppenheimer after he had married Kitty, and it is clear that their feelings for each other had remained strong. By the time Oppenheimer left for Los Alamos, Tatlock was a qualified doctor, working as a paediatric psychiatrist at Mount Zion.

Oppenheimer may have felt guilty about not saying goodbye. In June 1943, on the pretext of recruiting a personal assistant, he arranged instead to meet Tatlock in Berkeley. Pash, his suspicions already firmly raised, arranged for military intelligence agents to follow Oppenheimer's every move. What they observed would only deepen suspicion.

After dinner at the Xochimilco Café on Broadway, Tatlock drove them to her apartment in San Francisco. The surveillance team noted the intimacy between them. Sitting in a parked car on the street below, they observed

the lights in Tatlock's apartment go out at 11:30pm. They saw nothing further until both Tatlock and Oppenheimer emerged from the apartment at
8:30 the next morning. They met for dinner again the following evening.
After dinner, Tatlock drove them to the airport so that Oppenheimer could
catch a flight back to New Mexico. Groves had insisted that all laboratory
directors travel by means other than air, for fear that an accident could set
the programme back many months. Oppenheimer's dalliance meant that
he would now have to break Groves' rule.

As far as Pash could tell, Tatlock was undoubtedly suspect, a prime candidate for a Soviet spy with access to the head of the Los Alamos laboratory, whose very existence was a state secret. Two weeks later, Pash drafted
a memo to the Pentagon recommending that Oppenheimer be denied
security clearance and be removed from the programme. He also wrote to
Lansdale, suggesting that if Oppenheimer could not be removed, he should
be threatened with the legal consequences of his actions.

But Lansdale's own appraisal of Oppenheimer was rather less hys-
terical. He judged that Oppenheimer's own ambition, driven behind the
scenes by Kitty, would guarantee his loyalty to the programme. Lansdale
believed that Oppenheimer was sincere. He suggested that Oppenheimer
be apprised of the evidence of Soviet espionage against the programme
and that he be encouraged to supply names. Groves, who already considered Oppenheimer to be irreplaceable, agreed. He pushed through
Oppenheimer's security clearance on 20 July.

Lomanitz was promoted to the position of group leader in Lawrence's
team on 27 July, with the intention to relocate him to Oak Ridge to supervise the work on electromagnetic separation. Three days later he was told
that he was to be drafted instead. He had been fired from the programme.
Oppenheimer sprang to his defence, but was advised by Lansdale that
Lomanitz was a lost cause. In discussion, Oppenheimer mentioned his
anger at Lomanitz's activism at the Rad Lab. He insisted that there was
a direct conflict between loyalty to the Communist cause and loyalty
to the atomic bomb programme, and America. He therefore wanted to
ensure that there were no Communist Party members working on the

programme. Lansdale judged that whatever connections Oppenheimer had had with the party were now firmly in the past.

Oppenheimer may have been somewhat shaken by the action taken against Lomanitz. He decided to come clean, and mention the approach by Eltenton. He mentioned the incident to Groves in August, but he withheld Chevalier's name.

On 25 August 1943 Oppenheimer discussed the Lomanitz situation with Lyall Johnson in the latter's office in Berkeley. He went on to suggest that Eltenton may have tried to acquire information about the work at the Rad Lab and should therefore be watched. Johnson called Pash, and Pash asked that Oppenheimer return for further discussions the next day.

When Oppenheimer returned to Johnson's office, he was surprised to discover that Pash was present. A small microphone hidden in the base of Johnson's telephone secretly recorded the ensuing conversation. Oppenheimer thought he had been asked to return for further discussions about Lomanitz, but Pash interrupted. He wanted to know about other groups interested in the work at the Rad Lab.

Oppenheimer was unprepared for this discussion. In his own mind he had already identified the guilty and the innocent and, in his arrogance, now sought to protect the innocent from the people whose job it was to judge these things for themselves.

Chapter 10

ESCAPE FROM COPENHAGEN

January–November 1943

Despite receiving numerous invitations to visit America shortly after the Nazi occupation of Denmark, Niels Bohr had nevertheless decided it was his duty to remain. He wanted to do what he could to preserve the scientific institutions which he had helped to build, and the scientists who worked within them. And, indeed, the work did continue. Bohr and his team had access to a cyclotron[1] and high-tension apparatus suitable for fission experiments. The lack of materials, especially metals, was alleviated somewhat by the Carlsberg Foundation, a generous sponsor of Denmark's greatest physicist, which loaned Bohr's institute a supply of metals from the Carlsberg brewery. Bohr probably thought he could sit out the war if not in comfort or free of concern, then at least in relative peace.

Eric Welsh thought rather differently. The veteran British SIS operative had figured that Bohr would be a valuable addition to Tube Alloys. Late in 1942 Tronstad had received a message indicating that Bohr would welcome the opportunity to see him again – interpreted as a hint that Bohr was

---

[1] Which the German physicists had not dared to touch.

ready to leave Denmark. Welsh talked to 'C',[2] Sir Stewart Menzies, the head of the SIS, and they agreed that an approach to Bohr should be made to sound him out about coming to Britain.

Shortly afterwards, in January 1943, Chadwick was approached by the SIS in Liverpool and asked if he would draft a letter of invitation to Bohr. Once the details of the proposed escape, or 'ex-filtration' plan had been explained, Chadwick agreed. His letter, dated 25 January, offered a warm welcome should Bohr decide to leave Denmark, freedom to work on any scientific problems of interest, and a veiled request for Bohr's support on the atomic programme. 'Indeed I have in mind a particular problem in which your assistance would be of the greatest help', he wrote.

The letter was reduced to a microdot and smuggled to Bohr hidden in the hollow handle of a key, stored on a ring alongside a number of other keys. A second key on the ring contained a duplicate microdot. Bohr was alerted to the imminent arrival of the message by Captain Volmer Gyth, an officer in the information division of the Danish general staff with connections to the Danish resistance. Gyth passed him a set of instructions to the effect that: 'Professor Bohr should gently file the keys at the point indicated until the hole appears. The message can then be syringed or floated out onto a microscope slide ... It should be handled very delicately.' Perhaps somewhat uncertain of his own abilities in the tradecraft of a spy, when Gyth offered to recover the microdot and provide him with a written version of the letter, Bohr gratefully accepted.

Bohr's judgement of the situation was, however, unchanged. His desire was to remain in Denmark and continue his work at the institute. As far as he understood, the possibility of extracting U-235 from natural uranium in sufficient quantities to make a bomb was completely impractical. He gave his reasons in a reply but he also left open the possibility of coming to Britain, recognising that his circumstances could easily change. 'However,' he wrote, 'there may, and perhaps in a near future, come a moment where

---

[2] This was a convention that dated back to Mansfield Cumming-Smith, the first head of the SIS, who would initial papers that he had read with a 'C', written in green ink. Ian Fleming would later adopt a similar convention for his fictional 'M'.

things look different and where I, if not in other ways, might be able modestly to assist in the restoration of international collaboration in human progress.' Gyth reduced Bohr's letter to millimetre dimensions, wrapped it in foil and arranged to have it inserted in the hollow tooth of a courier, hidden beneath a filling.

Further correspondence ensued, though the manner of transmission of subsequent messages was rather more conventional. Bohr explained in more detail why he thought a fission bomb was impossible.

Separate ways

After successfully completing their sabotage mission, the Norwegian commandos of Swallow and Gunnerside went separate ways, as Falkenhorst and Reichskommisar Josef Terboven ordered a massive search. Rønneberg led Idland, Kayser, Strømsheim and Storhaug north towards the Swedish border. They arrived on Swedish soil fifteen days later, exhausted from a 250-mile trek that had not been without incident but which had been relatively straightforward. On reaching London they were greeted warmly and given a nice cup of tea.

Poulsson and Helberg headed for Oslo, intending to lay low for a while before making contact with the Norwegian underground. From Oslo, Poulsson escaped into Sweden before returning to Britain for a short while. Helberg, who had done time in a Swedish prison and was therefore known to the authorities, planned to head back to the Hardanger Plateau when the dust had settled. Acting on incorrect advice, on 25 March 1943 he arrived back in an area that was still crawling with German troops. Realising he had been spotted, he set off on skis as three German soldiers gave chase. Two gave up after an hour. After two hours, Helberg turned and faced his pursuer. The German emptied his Luger, missing with every shot. Now it was Helberg's turn. He gave chase, bringing the German down with a single shot from his Colt .32.

More adventures were to follow. In darkness, Helberg fell over a precipice and broke his left shoulder. He reached his destination, a house he knew in the village of Rauland, only to find it full of German troops. He bluffed his way through the next two nights, drinking and playing cards with the troops, and even managed to get medical attention for his shoulder. He moved to a hotel in Dalen, where he was unfortunate to get caught up in an altercation between Terboven, who was staying in the next room, and a young, attractive Norwegian woman who had spurned Terboven's amorous advances. Helberg was rounded up with the other Norwegians in the hotel on the orders of a now incensed Terboven, and was told they were all to be sent to Grini concentration camp. Helberg jumped from the bus on the way to Oslo, avoiding grenades and pistol shots. He eventually managed to get to Sweden, avoided imprisonment, and boarded a plane bound for Britain on 2 June.

Haugland and Skinnarland moved their makeshift wireless operation to a location high in the mountains. They took cover under the snow and watched the German troops make a mess of the search on the Hardanger Plateau. Haugland completed Skinnarland's wireless training before joining his brother, whom he was surprised to find leading the resistance in Oslo. He provided the resistance with further SOE-style training in the use of explosives.

Haukelid and Kjelstrup headed west on the Hardanger Plateau, where they stayed for much of the summer of 1943. Kjelstrup's health began to suffer, and he returned to Britain to recuperate.

Somewhat improved apparatus

The loss of heavy water production from the Vemork plant was a major setback to the German programme. The loss was to prove temporary, however. Tronstad and Brun had believed that the destruction of the high concentration cells would halt production for a few years. But the damage was already repaired by 17 April 1943, and the plant was again producing small quantities of heavy water by the end of June.

By this time, the German War Office had ceased to take any interest in the programme. Diebner and his research team were transferred back to the broader Uranverein under the auspices of the Reich Research Council, although the team was allowed to continue working at the Army Ordnance

laboratory in Gottow. The two million Reichsmarks that had been promised by the War Office never materialised, and the Reich Research Council
was left with the task of finding the money for itself. Speer remained an
enthusiastic patron, however, and adequate funding was forthcoming.

Diebner may not have been a leading light in German theoretical physics, but he was an accomplished experimentalist. The reactor experiments
that had been carried out so far under Heisenberg's overall guidance had
relied on configurations in which uranium metal plates and quantities of
the heavy water moderator were organised in layers. Diebner had devised
an alternative configuration based on a three-dimensional lattice of equally
spaced cubes of uranium oxide or uranium metal immersed in a volume
of moderator. Ingeniously, he further figured that he could do without an
enveloping container of aluminium by freezing the heavy water moderator solid. In effect, the 'heavy ice' would function both as moderator and
support structure.

He set up such a configuration in the low-temperature laboratory of
the Reich Institute for Technical Chemistry. Reactor G-II consisted of
about 230 kilos of uranium in the form of cubes and 210 kilos of heavy ice,
arranged in a sphere about 75 centimetres in diameter. No self-sustaining
chain reaction was generated but there was clear evidence for neutron
multiplication, about one and a half times greater than the corresponding neutron multiplication in L-IV. Diebner was convinced that a self-
sustaining chain reaction would be achieved with sufficient uranium and
heavy water.

Heisenberg, however, made light of Diebner's achievements. In a conference held in Berlin on 6 May he acknowledged the results from Diebner's
group but declared that the latter's 'somewhat improved apparatus' had
'yielded the same result' as the previous year's L-IV design. Heisenberg was
planning a large-scale reactor experiment and had no intention of moving
away from the layer configuration.

Subsequent experiments at the Gottow laboratory bore out Diebner's
conviction. The team repeated the uranium–heavy ice experiment with the
same quantities of materials but this time with a lattice of uranium cubes
suspended on fine alloy wires in a volume of liquid heavy water at normal

laboratory temperatures. A further experiment with over 560 kilos of uranium and nearly 600 kilos of heavy water yielded even more promising results. It was clear that the lattice design was superior to anything that had yet been produced in Berlin or Leipzig.

Diebner started to draw up plans for an even larger reactor, but now ran into conflict with the demands of Heisenberg's experiments. Heisenberg preferred to continue with the layer configuration despite the evidence suggesting that the lattice arrangement might work better. At issue here was the very different experimental philosophies adopted by the two research groups. Heisenberg was content to build understanding of the physics through a series of reactor *experiments* designed to allow measurement of the values of fundamental nuclear constants. As Heisenberg later confided to Harteck, he preferred the layer configuration because the theory was much simpler.

Diebner was less concerned about the theory and wanted to build a working reactor as quickly as possible. When subsequent theoretical studies pointed to the superiority of Diebner's lattice configuration, Heisenberg remained stubborn. Professional pride may have been a factor, but the simple truth was that for Heisenberg the nuclear project was no longer his major preoccupation.

More ominously, perhaps, Heisenberg had so far perceived no need for cadmium control rods of the kind that had been used in the Chicago uranium–graphite pile, although he understood that these would be required in a working reactor. In truth, without control rods an experimental nuclear reactor reaching criticality would precipitate a major disaster.

A lot of experience in microfilm work

'Oh, I think that is true', Oppenheimer said, in answer to Pash's question concerning other groups interested in the work going on at the Rad Lab. 'But,' he went on to say, 'I have no firsthand knowledge. I think it is true that a man, whose name I never heard, who was attached to the Soviet consul, has indicated indirectly through intermediary people concerned in this project that he was in a position to transmit, without danger of leak, or scandal, or anything of that kind, information which they might supply.'

Oppenheimer explained that, speaking frankly, he was 'friendly' to the idea that the Russians – as allies of America in the war against Nazi Germany – be advised of the American work on the atomic bomb, but that he would not want this kind of information to get to the Soviets through the 'back door'.

Pash was all ears.

'Could you give me a little more specific information as to exactly what information you have?' Pash enquired. 'You can readily realise that phase would be, to me, as interesting, pretty near, as the whole project is to you.'

'Well, I might say,' replied Oppenheimer, 'that the approaches were always to other people, who were troubled by them, and sometimes came and discussed them with me.' He went on: '[T]o give more ... than one name would be to implicate people whose attitude was one of bewilderment rather than one of co-operation.'

In Oppenheimer's reply, the Chevalier incident had suddenly become one of several approaches, to several physicists working on the programme. Two of these physicists, Oppenheimer explained, were working with him at Los Alamos, and the other was a Rad Lab physicist who had departed, or was about to, for the Oak Ridge facility in Tennessee. It was, as he later admitted, a 'cock and bull story', designed – if Oppenheimer's flustered response could be called that – to throw Pash off the scent.

Oppenheimer had already named Eltenton, who, he now explained, was to arrange contact with someone from the Soviet consulate 'who had a lot of experience in microfilm work, or whatever the hell'. But Oppenheimer did not want to name Chevalier, who he believed had acted as an innocent messenger. When pressed by Pash to name his friend, Oppenheimer replied: 'I think it would be a mistake. That is, I think I have told you where the initiative came from and that the other things were almost purely accident ... The intermediary between Eltenton and the project thought it was the wrong idea, but said that this was the situation. I don't think he supported it. In fact, I know it.'

Pash pressed him further, but other than reveal the fact that the intermediary was a member of the Berkeley faculty, Oppenheimer refused to give a name. 'I want to again sort of explore the possibility of getting the name of the person on the faculty,' Pash cajoled, 'not for the purpose of taking him to task but to try to see Eltenton's method of approach.' Oppenheimer did not budge, and tried to downplay the significance of the incident. Surely, the transmission of information vital to America's allies was something that should in any case be happening through formal channels. The fact that this transmission was not happening meant that information passed through the 'back door' was obviously treason in substance, though, perhaps, not in spirit.

These were all sentiments that had been expressed by many in Oppenheimer's circle of 'leftwandering' friends and colleagues. They were not, however, the sentiments expected of the head of the Los Alamos laboratory, a leading contributor to one of America's most secret war programmes. Worse, Oppenheimer had started to spin a web of deceit, making the classic error of elaborating a lie in the mistaken belief that it would lend it authenticity. He had not yet been caught in this lie but, unknown to him, it had been caught on tape.

The meeting ended as it had begun – amicably. Pash arranged for a transcript of their conversation to be produced and sent it to Groves with a covering note. It made no difference.

By this time the FBI had received a rather extraordinary anonymous letter. The letter was dated 7 August 1943 and was written in Russian. It named Zarubin (Zubilin), Kheifets and Kvasnikov and many others as Soviet spies. It also accused Zarubin of involvement in the March 1940 massacre of nearly 15,000 Polish prisoners of war in the Katyn forest[3] and, rather bizarrely, of spying on the United States for the Japanese. The author clearly hated Zarubin, and urged the FBI to expose him to the Soviet authorities as a traitor, whereupon he would be summarily executed by Vasily Mironov, whom the anonymous author claimed was a Soviet diplomat and a loyal NKVD agent. Inevitably, the FBI was suspicious and didn't

---

[3] More recent estimates put the death toll at over 21,000.

know quite what to make of the letter, but there were enough independently verifiable references in it to make them pay attention.[4]

The FBI eagerly agreed to put Eltenton under surveillance. In early September a short note was intercepted from Weinberg to 'S' (presumed to be Steve Nelson), requesting that he, Weinberg, should not be contacted. A sure sign, Pash argued, that Oppenheimer had tipped him off. Peer de Silva added his own voice to the growing chorus. He wrote to Groves on 2 September: 'The writer wishes to go on record as saying J.R. Oppenheimer is playing a key part in the attempt of the Soviet Union to secure, by espionage, highly secret information which is vital to the United States.'

However, the intense surveillance of the radical young physicists at the Rad Lab had turned up no further evidence of espionage. The physicists were nevertheless removed from the programme and its proximity. Lomanitz had been drafted. Friedman was fired shortly after being given a position teaching physics to army recruits at Berkeley. Both Lomanitz and Friedman perceived their predicament to be a direct result of their union activity, and nothing more.

At Lomanitz's farewell party, Weinberg speculated that their troubles might be the result of something else, but held back from telling them that he might actually be the cause. In the meantime, Weinberg was left on the Berkeley campus under close surveillance in the hope that he would expose more of the Soviet intelligence network.

Oppenheimer had asked Bohm to join him at Los Alamos but Groves intervened, advising Oppenheimer that the transfer could not be sanctioned, giving the rather obscure reason that Bohm had relatives in Germany. Weinberg and Bohm took positions as teaching assistants at Berkeley, presenting the course on quantum theory that Oppenheimer had once taught.

---

[4] It seems the author of the letter was Mironov himself. He also sent a letter to Stalin denouncing Zarubin as a double agent, and both Zarubin and his wife were recalled from Washington in the middle of 1944. Zarubin was able to demonstrate that all his contacts with Americans were legitimate and both he and his wife were cleared. Mironov was then recalled to face charges of slander, but was discovered to suffer from schizophrenia and was subsequently discharged from the NKVD.

Lansdale interviewed Oppenheimer again on 12 September 1943, in Washington. Lansdale explained that in his position he could do little else but base his suspicions on past associations. What was he to make of:

> ... the case of Dr J.R. Oppenheimer, whose wife was at one time a member of the Party anyway, who himself knows many prominent Communists, associates with them, who belongs to a large number of so-called 'front' organisations, and may perhaps have contributed to the Party himself, who becomes aware of an espionage attempt by the Party six months ago and doesn't mention it, and who still won't make a complete disclosure.

But Lansdale also confessed that he believed Oppenheimer to be innocent of any wrongdoing: 'I've made up my mind that you, yourself, are OK,' he said, 'or otherwise I wouldn't be talking to you like this, see?'

'I'd better be – that's all I've got to say', Oppenheimer replied.

Thin Man and Fat Man

By the autumn of 1943 the road to the atomic bomb was clear to Oppenheimer and the team of physicists at Los Alamos, but no less fraught with difficulty.

Two huge facilities were now under construction at Oak Ridge for the large-scale separation of U-235. One of these, called Y-12, was an electromagnetic separation plant based on Lawrence's calutron design. Lawrence had estimated that to separate just 100 grams of U-235 per day would require about 2,000 calutron collector tanks, each set vertically between the pole faces of thousands upon thousands of tons of magnets. The tanks and magnets were organised in oval units – nicknamed 'racetracks' – with each racetrack consisting of 96 tanks. Groves believed 2,000 tanks – twenty racetracks – to be beyond the capabilities of the construction company, and cut the number back to 500, or five racetracks, anticipating that advances in the technology prior to completion would increase production rates and compensate for the difference.

The facility required a vacuum system and magnets that had never before been built on this, truly Lawrencian, scale. The magnets were 250 feet long, and weighed between 3,000 and 10,000 tons. Their construction had actually exhausted America's supply of copper, and the US Treasury had loaned the project 15,000 tons of silver to complete the windings. The magnets required as much power as a large city and were so strong that workers could feel the pull of magnetic force on the nails in their shoes. Women straying close to the magnets would occasionally lose their hairpins. Pipes were pulled from the walls. Thirteen thousand people were employed to run the plant. The first racetrack – Alpha I – began operation in November 1943. It promptly broke down.

Despite the enormous scale of Y-12, Groves had remained largely ambiguous about the prospects for electromagnetic separation. This was very new technology and therefore uncharted territory. About eight miles south-west of Y-12 a gaseous diffusion plant, called K-25, was being constructed. This plant was to be housed in a huge U-shaped building measuring half a mile long by 1,000 feet wide. At the time of its construction it was the largest building in the world. The plant would employ another 12,000 people. This was, at least, more familiar technology. But it was still all a gamble. The gaseous diffusion process itself was still the subject of intense research at Columbia University, and the problems with corrosion by uranium hexafluoride had yet to be solved.

The uncertainties over the separation of U-235 were to some extent compensated for by a growing degree of confidence that the gun method would work and, moreover, that a transportable weapon could be built.

An ordnance expert acting as adviser to the project had identified a flaw in the physicists' logic not long after the inaugural lectures at Los Alamos in April. The physicists had based their rather pessimistic estimates of the size of the gun that would be required on conventional gun designs. But these conventional designs had to take account of the need for the gun to be fired repeatedly. The gun firing the shy into the sub-critical mass of U-235 at the other end of a bomb would obviously have to fire only once, after which it would be reduced to atoms. This meant that the weight of the gun could be substantially reduced.

For U-235 the basic bomb mechanism was no longer the main problem. All they needed was enough fissile material.

But the Manhattan Project physicists were also backing another horse. Fermi's successful demonstration of a self-sustaining chain reaction in December 1942 had led directly to the creation of a much larger-scale reactor to produce plutonium at 'Site W' in Hanford, south-central Washington state. Construction had started in March 1943, with a 45,000-strong labour force. The first nuclear reactor, named the B-reactor or 105-B, began construction in August based on Fermi's uranium–graphite design. It would take a year to build the plant, and plutonium was not expected to be available in sufficient quantities for a bomb until early 1945.

However, unlike U-235, it was not at all clear that the gun method would be effective for a plutonium bomb. Too little was known at this stage about this new element's physical properties for any conclusions to be drawn, particularly in regard to spontaneous fission and problems of pre-detonation. If plutonium exhibited a greater tendency to pre-detonate, then the muzzle velocity from the largest gun would be insufficient. The plutonium shy would be fired too slowly to prevent the bomb from just fizzling out.

In contrast to the gun method, implosion offered the possibility of assembling a critical mass much faster and more reliably. Even better, Teller now realised, a very violent shockwave could compress a sub-critical mass of plutonium to super-criticality. Implosion would literally squeeze the mass to a super-critical density which would then explode, without the need to assemble a larger super-critical mass of normal density from a hollow sphere made up of separate components.

However, implosion would not be workable unless a spherical shockwave could be created with conventional explosives packed around the outside of the bomb core. The mathematical physicist John von Neumann had demonstrated that, to be effective, an implosive shockwave would need to be spherically symmetrical to within a tolerance of just 5 per cent. In early July, Neddermeyer had set to work with modest implosion experiments set up on a mesa south of the Los Alamos laboratory, across Los Alamos canyon. These involved detonating conventional explosives wrapped around short lengths of pipe, so that the pipes would close in on themselves to form flat metal bars. Early results looked distinctly unpromising, the pipes emerging bent and twisted, indicating that the shockwaves thus generated were far from symmetrical.

A uranium or plutonium bomb based on the gun method would be long and thin – about seventeen feet long with a diameter of about two feet. Serber named this design 'Thin Man', after the 1933 Dashiell Hammett detective story and the series of movies it had spawned. It was estimated that a plutonium implosion bomb, if implosion could be shown to work, would be a little over nine feet long and five feet in diameter. Serber named it 'Fat Man', for Kasper Gutman, the character played by Sidney Greenstreet in the movie *The Maltese Falcon*.

Experiments to investigate how bombs with such dimensions might be dropped from a B-29 bomber began in August 1943. The plane, which was just beginning large-scale production for the American war effort, would need to be modified to carry the bombs to their target, and the experiments were designed to discover precisely what modifications would be required. To preserve security, in their phone conversations air force personnel would refer to these modifications as though they were being made to the planes in order to carry Roosevelt (Thin Man) and Churchill (Fat Man).

Slept most of the way

By August 1943 the situation in Denmark had changed. The terms of Danish government co-operation with the German occupying forces had included protection for Denmark's 8,000 Jews. The growing boldness of the Danish resistance, and the increasing frequency of demonstrations, strikes and acts of sabotage, led German forces to declare martial law and re-occupy Copenhagen on 29 August. The Nazis started to arrest prominent Danish Jews.

On 28 September, Bohr received word from a sympathetic German woman working in the Gestapo offices in Copenhagen. She had seen the orders for Bohr's arrest. A cable from Chadwick and Cherwell advised him that his escape from Copenhagen had been granted priority by the British war cabinet. Bohr contacted members of the Danish resistance, and an escape route was prepared.

In early evening the next day, Bohr and his wife Margrethe walked to the Sydhavn quarter of Copenhagen close to the shores of the Øresund, the strait that separates Denmark from southern Sweden. They joined about a dozen others, including Niels' brother Harald and Harald's son Ole, in a small *Kolonihavehus*, not much more than a large garden shed, and waited for darkness to descend. At a pre-arranged time they crawled towards the beach, Bohr feeling rather self-conscious, and boarded a fishing boat that took them out across the Øresund. They then transferred to a large trawler and made their way to Linhamn, near Malmö in Sweden. They spent the rest of the night in the cells of the local police station in Malmö. Bohr travelled by train to Stockholm the next day, leaving Margrethe to await the arrival of their sons, who would shortly take the same escape route. Gyth was among those waiting at the station.

Gyth had alerted the SIS to Bohr's dramatic escape, and was told to advise Bohr that he should come to Britain as soon as possible. There were believed to be many Gestapo agents in Stockholm and Bohr was one of the most widely known scientists in Scandinavia. To evade watchful eyes, Bohr was escorted by Gyth in a taxi to a building used by the Swedish intelligence services. They climbed to the roof, crossed to the roof of a neighbouring building, descended and took another taxi. Once safely installed at the home of Oskar Klein, one of Bohr's colleagues from some years previously, Gyth passed on the message from the SIS and told Bohr that an unarmed Mosquito bomber was available at Stockholm's Bromma airport to transport him to England.

But Bohr was concerned for the fate of the 8,000 Jews he had left behind in Denmark. On the evening of Bohr's escape, two German freighters had arrived in Copenhagen harbour to transport the Jews to concentration camps in Germany. When he realised that the Swedish government had no plans to protest against the German intentions, Bohr made a personal plea to King Gustav V of Sweden.

In the meantime, a remarkable series of events had taken place. News of the impending deportation of Danish Jews had spread rapidly through the Jewish community. Within a few days virtually the entire Jewish population was hidden away, as offers of support flooded in from the general public. They hid in strangers' apartments or cottages, in their homes, in churches and in hospitals, among the patients. In the midst of tragedy, the people of Denmark had risen to the aid of their fellow citizens. Fewer than 300 Jews were caught by the German sweep, which began on the evening of 1 October.

The Swedish protest was broadcast on radio on 2 October, signalling to Danish Jews that there was safe haven to be found in Sweden. A mass evacuation followed, supported by the Danish resistance, local fishermen, the Swedish coastguard and even a German naval commander who declared that his fleet of coastal patrol vessels was in need of repair and could not put out to sea. Over the following two months, 7,220 Danish Jews escaped to Sweden.

With the crisis averted, Bohr left for Britain on 5 October on board a twin-engine Mosquito bomber. There was room in the empty, unpressurised, bomb bay for only a single passenger. Bohr had talked incessantly prior to take-off and had paid little attention to the instructions the pilot had given him. As the plane climbed to 20,000 feet to avoid the risk of anti-aircraft fire as it passed over the Norwegian coast, the pilot instructed Bohr to switch on his oxygen supply.

Unfortunately, the Nobel laureate's head was too large for the helmet he had been given. As the message was relayed via headphones in the helmet, Bohr didn't hear this instruction and promptly passed out through lack of oxygen. Sensing that something was wrong, the pilot descended steeply over the North Sea. By the time the plane landed, Bohr had recovered consciousness and appeared fine. He explained that he had slept most of the way.

Bohr was flown to Croydon airport near London, where he was met by Chadwick and an officer of the SIS. He was subsequently installed at the Savoy Hotel, where Chadwick informed him of the Frisch–Peierls memorandum, the MAUD report, Tube Alloys and the American bomb programme. Bohr was completely astounded. He may also at this time have been able to set the record straight regarding Lise Meitner's telegram. The reference to MAUD RAY KENT, to which Cockcroft had given such an ominous interpretation and which had led the MAUD Committee to be so named, was not a coded message at all. Maud Ray was a former governess to the Bohr children, now living in Kent.

Bohr dined with Anderson that evening. Anderson, who had been appointed Chancellor of the Exchequer in September following the unexpected death in office of Sir Kingsley Wood, explained that he would welcome Bohr as a member of Tube Alloys, part of a mission that was to be sent to join the American programme.

The impasse on Anglo-American co-operation on the atomic bomb had been resolved. A series of meetings through the summer months had helped to clear up misunderstandings about British post-war intentions. Britain's desire was for an independent atomic deterrent against an anticipated Soviet atomic arsenal of the future. It was not Britain's intention to acquire know-how at the American taxpayer's expense for commercial exploitation after the war. This appeared to do the trick: Stimson and Bush were mollified. At Churchill's request, Anderson had then drafted an agreement governing collaboration which Churchill subsequently amended.

In the meantime, Roosevelt himself had come to a decision, on the advice of Hopkins. He had decided that it was incumbent on him to honour the commitments regarding co-operation he had made to Churchill over a year before. Anderson and Bush talked Conant round. So, when Churchill tabled the draft agreement at the Quebec summit conference on 19 August, it was quickly (and, from the Americans' perspective, rather too hastily) signed. The first four articles were virtually those of the Anderson–Churchill draft. A fifth article set out the structure of a Washington-based Combined Policy Committee with representation from America, Britain and Canada. In the agreement Britain and America committed never to use the bomb against each other, never to use it against a third party without each other's consent, and never to communicate any information about atomic weapons to third parties without mutual consent. This agreement would cause considerable trouble later, but for now it meant the resumption of full collaboration.

The Americans accepted that the Manhattan Project should become the main focus of Anglo-American efforts to develop the atomic bomb, and that this should be supported by a British scientific delegation, or 'mission'. Anderson wanted Bohr to join the 30-strong British contingent that was now due to travel to America.

Bohr's son Aage, himself a promising young physicist, joined him in London a week later, and took the role of his father's personal assistant. The rest of the Bohr family stayed in Sweden.

Raid on Vemork (2)

Tronstad was greatly concerned by the news that the Germans had managed to get the Vemork heavy water plant working unexpectedly quickly after the successful SOE raid in February. Skinnarland, reporting from a makeshift radio station on the Hardanger Plateau, had estimated that the plant would achieve full production in mid-August. Tronstad was also concerned that a new 'combustion' method of separation could, if adopted at Vemork, lead to production at a much accelerated rate. The job that the SOE had set out to do with operations Freshman and Gunnerside was clearly not yet completed. Production was again delayed by small, limited acts of sabotage, in which vegetable oil was added to the distillation vats. But it was again obvious that this could not continue indefinitely. The Vemork plant had to be taken out of commission.

Tronstad tried to devise further large-scale sabotage operations, but the defences at the plant had been greatly strengthened. Vemork was now surrounded by barbed wire fences and minefields, and the garrisons at Vemork and Rjukan had been substantially increased. A further commando raid seemed out the question. The only alternative appeared to be a bombing raid. Tronstad and Wilson remained firmly opposed.

But Groves was insistent. He did not trust the British. He had not been informed of the failed Freshman raid until after it had ended in disaster. He had learned of the Gunnerside raid through a casual remark made by Akers in January. In the new spirit of Anglo-American collaboration embodied in the Quebec agreement, he now urged the British representatives of the Combined Policy Committee to agree appropriate action.

In fact, an SOE memorandum of 20 August 1943 had acknowledged that a bombing raid was the only viable option and should be given active consideration. The memo also advised that the Norwegian High Command and Norwegian government-in-exile should not be informed of such plans. By mid-October, a full-scale ground assault or a further sabotage raid had been firmly ruled out.

Groves was not prepared to take the risk that the German programme might be successful in producing, if not a bomb, then perhaps some sort of radiation weapon. Bohr's recollection of his discussion with Heisenberg in September 1941, and the diagram of a bomb that Bohr believed Heisenberg had drawn for him, merely compounded the situation. Groves ordered a bombing raid, his first combat decision after a lifetime in uniform.

A force of about 300 B-17 Flying Fortresses and B-24 Liberators of the American Eighth Air Force took off from airfields in East Anglia just before dawn on 16 November 1943, in poor weather conditions. Part of this force headed for targets near Stavanger and Oslo to divert German fighters away from the main force heading for Vemork. The raid had been carefully timed to coincide with the lunch break, between 11:30am and noon, when most of the plant's workforce would be off-site.

No fighters were encountered and the bombers arrived at the Norwegian coastline twenty minutes too early. The commander, Major John M. Bennett, ordered the fleet to circle back out over the sea and return for the bombing run at the right time. The decision traded civilian casualties for military: as the bombers returned to the coastline, one was shot down by anti-aircraft fire from the now fully alert coastal defences. The crew of a second parachuted into the sea as their plane spun out of the air with an engine on fire.

From their vantage point on the Hardanger Plateau, Haukelid and Skinnarland both watched as: 'Scores of American bombers were flying across Norway in broad daylight as if no German anti-aircraft defences existed. They began to circle over us and then proceed in an easterly direction, towards Rjukan.'

In the first wave of the attack about 145 bombers dropped more than 700 1,000-pound high-explosive bombs on the Vemork plant. Fifteen minutes later a second wave of about 40 bombers dropped 295 500-pound bombs on Rjukan. But in the Second World War so-called precision bombing was far from precise. The bombs fell everywhere. The plant itself received just two hits, damaging the top floors but leaving the electrolysis cells, in the basement, completely unharmed. The plant's power station was hit, as was the nitrate plant in Rjukan. Twenty-two civilians were killed.

The Norwegians were absolutely furious, and lodged formal protests with both the British and American governments. They argued that the attack 'seems out of all proportion to the objective sought'. Tronstad pointed out that he had given all the reasons why a bombing raid would not be successful four months before.

And yet the raid was successful, if not quite in the manner intended. The Germans had finally got the message that the Vemork plant was not safe, that the Allies would continue to attack it until it was utterly destroyed. Production of heavy water at Vemork was stopped and plans were laid to build a plant in Germany.

Oath of allegiance

Frisch was asked by Chadwick in November 1943 if he would like to work in America. 'I would like that very much', was his reply. 'But then you would have to become a British citizen', Chadwick warned. 'I would like that even more.' Within a few bewildering days he had pledged his oath of allegiance to the British Crown.

Chadwick had been eager to recruit the best scientists in Britain to join the delegation to America. To give the mission the best possible chance of success, he also sought to recruit nuclear physicists from outside Tube Alloys. Chadwick was well aware of Lise Meitner's discomfort in Stockholm, and asked if she was willing to join her nephew in America. Meitner was unambiguous: 'I will have nothing to do with a bomb', she replied.

British Nobel laureate Paul Dirac had acted occasionally as a consultant to the MAUD Committee on aspects of isotope separation and bomb physics. Keen to have such an eminent physicist in the delegation, Anderson telephoned Dirac in Cambridge and asked him to call into his office when next in London. Dirac agreed. As an afterthought, Anderson asked how often Dirac actually came to London. He replied: 'Oh, about once a year.' Dirac, too, refused to join the British mission.

All selections were subject to scrutiny by Groves, who had insisted that only British citizens would be accepted, and that they should arrive with full security clearance. Frisch's enthusiasm for British citizenship was not shared by Rotblat, however, who was adamant that he would remain a Polish citizen and return to Poland to rebuild physics there when the war was over. Rotblat judged this mission to be more important to him than working on the Anglo-American bomb programme. Chadwick made personal representations to Groves on Rotblat's behalf, giving assurances of Rotblat's integrity. By this time Chadwick had managed to build something of a rapport with the notoriously blunt, anti-British head of the Manhattan Project. Groves accepted, and Rotblat joined the delegation as a Polish citizen.

Chadwick, Frisch and Rotblat were to go to Los Alamos, with British physicists William Penney and James Tuck. Oliphant was to join Lawrence's team at the Rad Lab in Berkeley. Peierls and Fuchs were to head for New York to join the work on gaseous diffusion.

Fuchs had already pledged his oath and taken British citizenship a year earlier. British intelligence gave him a security clearance after a rudimentary background check. He advised his Soviet contact Sonja of his impending move. NKVD and GRU atomic intelligence activities had been consolidated with the NKVD First Chief Directorate under the umbrella of ENORMOZ just a few months earlier. Sonja moved quickly. At their next meeting she advised him that his controller in New York would have the cover name Raymond, and gave him a series of coded recognition signals he should use to make contact.

Barely a week after being approached by Chadwick, Frisch found himself at Liverpool docks, part of a group of about 30 scientists, some with their families. They were to board the *Andes*, a luxury liner that had been converted to bring American troops to Britain. Frisch had forgotten his ticket but Akers waved him through. He found he had an eight-berth cabin all to himself. 'Some of us got seasick,' Frisch wrote after the war about the delegation's journey to America, 'but otherwise the journey was uneventful, and the ship arrived safely, with perhaps the greatest single cargo of scientific brain-power ever to cross the ocean.'

And in their midst was a Soviet spy.

Niels and Aage Bohr headed for America on the *Aquitania*, early on the morning of 29 November, accompanied by an armed detective. When they arrived they were given the cover names Nicholas and James Baker.

Arlington Hall

In June 1942, the US Army Signals Security Agency had moved into new premises, a former private girls' school set in 100-acre grounds on Arlington Boulevard, Arlington, Virginia. The school had been named the Arlington Hall Junior College for Women, and the name Arlington Hall (or, sometimes, Arlington Hall Station) was retained. In many respects, Arlington Hall was the American equivalent of Britain's Bletchley Park.

By October 1943 the analysts at Arlington Hall had identified at least five different variants of the Soviet cipher system. One was used primarily for trade messages originating from Amtorg, the Soviet trade agency, and the Soviet Purchasing Commission which supervised Lend-Lease aid from America. A second was used by Soviet diplomats. The remaining three systems were being used by Soviet spies belonging to the NKVD, the GRU and the Naval GRU.

Lieutenant Richard Hallock had worked as an archaeologist and had studied the ancient languages of Babylonia before joining Arlington Hall. Now he was charged with studying a vast pile of paper consisting of about 10,000 coded Soviet trade and diplomatic messages – page upon page of apparently random and meaningless groups of five letters. But, of course, the groups of letters were not meaningless. Penetrate the cipher system and the coded messages would be revealed. Acquire a codebook or somehow break into the code and the messages themselves could be read. He confronted the stubborn question: where to begin?

He figured that at the beginning of each message there might be a reference to the subject matter that followed, a pattern that would, perhaps, be repeated for every message, much like addressing conventions in personal or business letters. He arranged for the clerks at Arlington Hall to produce punch cards containing the first five five-letter groups, converted to five-number groups, for all 10,000 messages. When these punch cards were run through a sorter, he noticed that seven messages conformed to a pattern. Although unrelated, the messages appeared to have been encrypted using the same cipher key.

For some unknown reason, some of the one-time pads had in fact been used more than once.

PART III

WAR

Chapter 11

UNCLE NICK

November 1943–May 1944

P ash and de Silva were convinced of Oppenheimer's guilt, but in the months of intensive surveillance following the taped con- versation in Berkeley they could uncover no further evidence of espionage. Lansdale was equally convinced that Oppenheimer was telling the truth, though not all the truth that could be told. Groves, who was himself convinced that Oppenheimer was already irreplaceable, began to weary of Pash's stubborn and fruitless pursuit. When the decision was taken in November 1943 to commission a secret intelligence mission in Europe, Groves supported the appointment of Pash to lead it. Pash left San Francisco for London.

Pash was gone but the surveillance continued. Oppenheimer was still under considerable suspicion. Groves and Lansdale had challenged him on a couple of occasions to volunteer the name of Eltenton's intermediary but Oppenheimer had refused. He would yield up a name only if ordered to do so, he said. Groves rationalised this as a childish desire on Oppenheimer's part not to 'rat' on a friend. His inclination was to let the matter rest.

But Groves could not let it rest for long. The investigation into Eltenton's involvement in suspected Soviet espionage against the Manhattan Project clearly could not progress much further without the name of his intermediary. In mid-December, Groves finally ordered Oppenheimer to co-operate. Reluctantly, Oppenheimer now gave up Chevalier's name, but insisted that his friend was merely an innocent messenger. Telegrams naming Chevalier were despatched to the Manhattan Project's chief security officers the next day. Oppenheimer had had little choice, but he was all too aware of the likely consequences of this revelation for his friend's career.

Chevalier had taken a year's sabbatical from Berkeley in July 1943, and in September moved to New York expecting to receive an assignment with the Office of War Information (OWI), which had been split off from the Office of Strategic Services (OSS)[1] the year before. He waited for news of his application for security clearance through the final months of 1943. In January 1944 he finally received word about his clearance. It was not good. His contact at the OWI had seen his FBI file in Washington. 'Someone obviously has it in for you', he said.

There remained the question of the identities of the three physicists whom, according to Oppenheimer, Chevalier was meant to have approached. Groves did not ask and Oppenheimer chose not to elaborate his lie any further. When the FBI pressed for more information, Groves ignored the requests.[2]

Pash had identified Jean Tatlock as a prime espionage suspect, most probably the all-important link in the intelligence network between Oppenheimer and the Soviet *rezidentura* in San Francisco. However, here again FBI surveillance turned up no further evidence.

In the first days of January 1944, Tatlock had descended into the blackness of depression. She visited her father on 3 January in a despondent mood and promised to call him the next day. She returned to her apartment and called a female friend, Mary Ellen Washburn, and invited her to come over. But Washburn could not come that night.

---

[1] The OSS was disbanded after the war, its functions split between the War and State Departments. The secret intelligence and counter-espionage branches were eventually to form the nucleus of the new Central Intelligence Agency in 1947.

[2] In a memorandum dated 5 March 1944, the FBI claimed that Oppenheimer had confessed that Chevalier had approached only one physicist: his brother, Frank Oppenheimer. This seems very unlikely, however. See Bird and Sherwin, p. 248.

After dining alone, Tatlock took a quantity of sleeping pills. She wrote a short note on the back of an envelope, wishing love and courage to all those who had loved her and helped her, declaring that she had wanted to live but 'got paralyzed somehow'. The note tailed off into illegibility, as the sleeping pills began to take effect. She partly filled the bathtub and may at this point have taken chloral hydrate – knock-out drops.[3] She passed out, slipped beneath the water, and drowned.

When she didn't call the next morning her father became concerned. After climbing through a window to gain access to her ominously silent apartment early the following afternoon, he discovered her body in the bath. He didn't call the police, however, and after carrying his daughter's body through to the living room and laying it on the sofa, he proceeded to search her apartment. Only when he had burned some of her private correspondence and photographs did he call a funeral parlour. Someone at the funeral parlour alerted the police.

Whatever it was that John Tatlock did not want discovered about his daughter, it is unlikely that this was anything to do with her activities as a Communist. Although the reasons for Jean Tatlock's suicide remained obscure,[4] her psychoanalysis had revealed latent homosexual tendencies. There were hints of lesbian affairs, including one with Washburn. It may have been denial of her homosexuality that had led her to take so many male lovers.

Whatever the reason, the woman whom Oppenheimer had loved – still loved – and had very nearly married, was now gone. Her death was a bitter blow and he was deeply grieved. On hearing of her death he took a long, quiet and contemplative walk in the woods surrounding Los Alamos.

---

[3] The autopsy revealed only a faint trace of chloral hydrate, uncorroborated by other evidence.

[4] The suspicious circumstances of her death have led some observers, including Jean's brother Hugh Tatlock, to speculate that she might have been assassinated.

Little Boy

The Italian physicist Emilio Segrè had found a little haven away from the bustle of the main Los Alamos compound. In December 1943 he had retreated to a small log cabin in secluded Pajarito Canyon, a few miles from the main laboratory. It was here that Segrè repeated experiments on the rate of spontaneous fission in natural uranium, experiments he had earlier performed in Berkeley. The results were much the same, but they hinted at a higher rate for U-235 than he had measured previously. Segrè wondered why.

His conclusion represented a significant discovery, one that was to bring a U-235 bomb much closer to realisation. It was simply a question of altitude, he reasoned. High up on the mesa, 7,300 feet above sea level, Segrè's samples were so much closer to the constant wash of neutrons that results from cosmic ray bombardment in the earth's upper atmosphere. The rate of spontaneous fission in U-235 was being affected by these stray neutrons. The closer the sample was to the upper atmosphere, the greater the number of stray neutrons and the higher the rate. At Berkeley, much closer to sea level, the density of stray neutrons was much lower.

This meant that the risk of pre-detonation in a U-235 bomb could be greatly reduced simply by shielding the bomb core from stray neutrons. The core material would no longer need to be quite as pure as had first been thought. And the muzzle velocity of the gun used to assemble the super-critical mass could also be reduced, shortening the length of the gun barrel and making the uranium bomb much more compact. Instead of Thin Man's seventeen feet, the bomb would now need to be no more than about six feet in length. This new design was codenamed 'Little Boy', Thin Man's smaller brother. The uranium bomb was now definitely a 'sure thing'.

But the fact remained that according to Lawrence's latest estimates, within the timescale envisaged for the Manhattan Project there was likely to be enough U-235 to build only one bomb. The threat of an atomic bomb would not be an empty one, but if the Allies used the bomb in the war against Nazi Germany sometime in early 1945 they would be unable to follow this with the threat of a second weapon, except through a very dangerous bluff. And if the Germans retaliated with a bomb of their own ...

Father confessor

Following their arrival at Newport News, Virginia in early December 1943, the physicists of the British Tube Alloys delegation had separated. Peierls, Frisch, Penney and Tuck went on to Los Alamos, although Peierls' stay was short. He and Fuchs had been assigned to join the work on gaseous diffusion and the problems associated with the operation of the large-scale diffusion plant at Oak Ridge. They worked from offices near Wall Street in New York rented by the British Ministry of Supply, and were officially listed as consultants to the Kellex Corporation, a subsidiary of the Kellogg engineering firm which was building the plant in Tennessee.

Chadwick and his wife Aileen arrived at Los Alamos in early 1944, after having taken a detour to see their daughters in Halifax, Canada, where they had lived since the summer of 1940, safe from German air raids. Rotblat arrived towards the end of February, and stayed in a spare room in the Chadwicks' relatively prestigious log cabin.

To the American physicists at Los Alamos, contact with colleagues who had so recently experienced life in war-torn Europe and who had lived for so long under the shadow of the Nazi threat brought home to them the real reason why they were all there, and what they were doing. The Britons' war-stories were sobering. Penney gave a colloquium on the destruction caused by the German aerial bombardment of London, in the detached, matter-of-fact style typical of a scientific discussion, a smile firmly affixed to his face. The smile hid much. Penney had lost his wife in the Blitz.

Niels and Aage Bohr had arrived in New York on 6 December, where they were hustled from the harbour to a hotel by intelligence agents concerned to keep secret their identity. Only when they were safely in their hotel room did they notice that their attempts to conceal their identities beneath pseudonyms had been rather frustrated by Bohr's luggage, which was boldly labelled NIELS BOHR.

Of course, not everyone whom Bohr met in his first days in America was aware of the pseudonym or the need for secrecy. In a Washington hotel, Bohr encountered a familiar face in the hotel elevator. It was Else von Halban, Hans von Halban's wife.

'Good evening, Mrs von Halban', he said.

'I'm not Mrs von Halban now,' she explained, 'I'm Mrs Placzek. Good evening Professor Bohr.'

'I'm not Professor Bohr now; I'm Mr Baker.'

Else had divorced Halban and married George Placzek in Montreal.

The Bohrs were met at the railway station in Lamy, New Mexico, and driven to a quiet stretch of road where they changed vehicles before driving on to Los Alamos. Their cover names Nicholas and James Baker were reinterpreted by the Los Alamos scientists as 'Uncle Nick' and 'Jim'. Oppenheimer organised a meeting on 31 December to give Bohr an opportunity to recount his visit from Heisenberg and to discuss the drawing he had carried with him from Denmark.

The physicists were puzzled by the crude drawing. Bohr referred to it as a drawing of a bomb, but it was clearly a drawing of a reactor. It showed alternating layers of uranium and heavy water moderator rather than the favoured lattice configuration that had been adopted in the construction of Fermi's first uranium–graphite pile. Bethe wondered if the Germans were so crazy that they wanted 'to throw a reactor down on London'. Bethe and Teller quickly estimated the explosive force of a uranium–heavy water pile, and concluded that it would be no greater than an equivalent mass of TNT.

But the general climate of fear fostered another possible conclusion. Perhaps Heisenberg had simply managed to keep the real intentions of the German programme secret, even from Bohr.

Bohr was apprised of the progress at Los Alamos. He quickly realised that 'they did not need my help in making the atom bomb'. His was to be a relatively modest, though vitally important, role on the Manhattan Project. Many of the physicists, now working long hours, six days a week to create the world's most dreadful weapon, had not themselves lived beneath the shadow of war. However, Bohr had directly experienced life as a Danish half-Jew under Nazi domination. He had been threatened with arrest by the Gestapo and had escaped in dramatic fashion with the help of the Danish resistance. His September 1941 conversation with Heisenberg had left him in no doubt that, under Heisenberg's leadership, everything was being done in Germany to develop atomic weapons. He provided a forceful and timely reminder of the threat of a Nazi weapon, and everything that this implied.

Bohr, revered by many gathered on the Hill as a father figure, became a father confessor, particularly for some of the younger physicists such as Feynman. Any scientist wrestling with his conscience over what he was helping to build would find the moral justification for it in Bohr's own experiences. 'He made the enterprise seem hopeful,' Oppenheimer observed after the war of Bohr's role, 'when many were not free of misgiving.' Bohr had a salutary effect on the morale of the Los Alamos physicists.

If Heisenberg's intention in September 1941 had really been to convince nuclear scientists not to engage in building weapons of mass destruction, then his failure was now complete, and utter.

Bohr paid frequent visits to the Chadwicks' cabin and got to know Rotblat, who had acquired a short-wave radio. Together they would listen to news of the war broadcast by the BBC World Service. Bohr maintained that: 'We must hear all the rumours before they are denied.'

A remark made by Groves over dinner one evening left a lasting impression on Rotblat. Groves volunteered the opinion that of course the real purpose of the atomic bomb was to subdue the Soviet Union. Rotblat had no illusions about the Soviet regime, but the Soviet Union was an ally in the war against Germany, and a nation whose people were making incredible sacrifices. Rotblat felt keenly 'the sense of betrayal of an ally'. His discussions with Bohr about the post-war implications of the bomb became more intense.

Complementarity of the bomb

If Bohr did not eventually contribute much of technical substance to the Manhattan Project, it was not only because he perceived the project to be in more capable hands, but also because his mind had already jumped a further step ahead. From what he could see and what he was told at Los Alamos, it was easy to conclude that atomic weapons were soon to be a hard fact of political life. He was astonished to learn that the British and American administrations had given little – if any – thought to the post-war challenges posed by the threat of the weapons that would soon be available. He had no doubt that there were scientists in the Soviet Union equally capable of building atomic weapons of their own.

He perceived parallels with *complementarity*, one of the central planks of his own philosophy of physics which he believed had implications for the way we can interpret aspects of our wider world. In the sub-atomic quantum world, complementarity attempts to rationalise the dual wave–particle behaviour of fundamental particles such as electrons. Under differ-ent, mutually exclusive circumstances, electrons will exhibit both wave-like behaviour and particle-like behaviour. It is impossible to rationalise this in terms of some kind of underlying reality because, according to Bohr's philosophy, such a reality is simply beyond the reach of our instruments, our observations and our understanding. We are therefore left with seem-ing contradictions. Here the electron is a wave. Here it is a particle. Bohr had argued in 1927 that although these behaviours are mutually exclusive, they are not actually contradictory. They are complementary behaviours of a deeper, forever unknowable, reality.

Atomic weapons have similar complementary properties, Bohr now realised. Under one set of circumstances, atomic weapons heralded an arms race leading, perhaps inevitably, to nothing less than the destruction of human civilisation. At the same time, under different circumstances, atomic weapons heralded the end of war, because in a war fought with atomic weapons there could be no victor. If political, cultural or religious differences were to be settled without an end-of-the-world, no-win sce-nario, then the advent of atomic weapons meant that recourse to war would no longer be thinkable. Differences would have to be settled in other, less violent, ways. The choice was plain. Arms race or international arms control?

An electron has no choice about how it is to behave, but a political organisation of society is in principle free to make choices, provided these are first recognised. Bohr believed that an arms race was not an inevitable consequence of the work now in hand at Los Alamos. Such a race could yet be avoided. He reasoned that the only way to achieve this was to adopt an 'open world' policy, taking potential enemies into confidence and building trust through dialogue. In early 1944 this meant gaining the confidence of the Soviet Union, now an ally in the war against Nazi Germany and Japan, but acknowledged by many as an enemy in waiting.

Sinking of the *Hydro*

The bombing raid on the Vemork plant convinced the Germans to rebuild the heavy water facility in Germany, and plans were laid to transport the last of the heavy water by sea, under armed guard. Rosbaud alerted the SOE to these plans early in January 1944, and Haukelid was asked to look at ways to prevent the heavy water from leaving Rjukan.

A one-man attack on the plant itself was impossible, so Haukelid turned his attention to the transport arrangements. The drums containing the heavy water, labelled 'Potash-lye', were to be moved by train from Vemork to the *Hydro*, a ferry that would cross Lake Tinnsjø, and thence by rail to a port and a waiting ship to Germany. The ferry itself was the least guarded and offered the prospect of minimising civilian casualties. On 9 February he sent a telegram back to London suggesting sabotage of the ferry. Although the saboteurs went on to express their misgivings that the operation was not worth the risk of reprisals that would certainly follow, the response from London was clear. The heavy water had to be destroyed.

Haukelid, Skinnarland and Rolf Sørlie laid plans to sink the ferry at the deepest point in the lake, sending the drums of heavy water to the bottom. The battle for heavy water was about to claim its last casualties.

Armed with sten guns, pistols and grenades, Haukelid and Sørlie smuggled themselves aboard the ferry in the early morning of Sunday, 20 February. On the inside of the hull they placed nineteen pounds of plastic explosive, high-speed fuses, detonators and timers hastily assembled from old alarm clocks. They set the clocks for 10:45am. They then made their escape.

The *Hydro* sailed on time at 10:00am carrying 53 passengers, crew and German guards, and 39 drums containing over 3,600 gallons of heavy water. The explosion ripped a hole eleven feet square out of the side of the 493-ton ferry. It sank within three minutes. Twenty-six drowned, including fourteen civilians, among them a couple and their three-year-old daughter. The other 27 managed to jump from the ferry as it sank, and were rescued from the icy water by local farmers and fishermen.

The heavy water sank to the bottom of Lake Tinnsjø, where it remains today.

Life on the Hill

Groves' belief in Oppenheimer's capabilities was genuine. The mismanagement that had characterised the first few months of Oppenheimer's tenure had given way to strong and effective leadership. Oppenheimer the scientific director of the Los Alamos laboratory emerged as a very different personality from Oppenheimer the academic physicist. He had adapted, chameleon-like, to the very different environment and the demands he now faced.

His was not a dictatorial style of leadership. For that he could always rely on Groves, more used to barking orders to army personnel than looking after the well-being of awkward, argumentative civilians. Instead, Oppenheimer would simply put forward persuasive and compelling arguments why someone on the project should do things his way rather than their way. He directed the work through politics and manipulation, carried out with considerable charisma and finesse. Most of the scientists on the project were aware they were being manipulated, but many accepted and welcomed it. 'I think that he really realised that the other person knew that this was going on,' explained one Los Alamos physicist, 'it was like a ballet, each one knowing the part and the role he's playing, and there wasn't any subterfuge to it.'

Subterfuge or not, Oppenheimer's concern for the well-being of the new citizens of Los Alamos transcended a strict interpretation of his duties as scientific director. A career in academic physics had hardly prepared him for the management problems he now faced in the closed and claustrophobic micro-society taking shape on the Hill. He was obliged to adopt roles ranging from mayor of a small boom-town to personnel manager to local priest.

He was daily confronted with the problems of squalid army buildings, poor facilities, a constant housing shortage, lack of suitable water supplies and intermittent electricity. He often found himself caught between his staff and Groves' almost criminal lack of diplomacy.[5] He had to deal with minor insurrections mounted by the scientists' wives, as they sought to improve their living conditions. Investigation of some minor infringements uncovered the 'WAC shack', a flourishing business conducted by a few young girls of the Women's Army Corps offering personal services to men stuck on the Hill with irrepressible desires, and money in their pockets. He got caught up in the often petty domestic troubles of his staff. As romances blossomed, Oppenheimer frequently attended the resulting wedding ceremonies, sometimes acting as a witness and occasionally giving away the bride.

Much of the trouble was, of course, created by the concentration-camp environment that civilian scientists and their wives now found themselves contending with. Aside from the fence around the entire facility, with its guarded gates, the Technical Area within the compound was also fenced and guarded. Only scientists and their assistants bearing white badges were allowed passage into the Tech Area. The oppressive atmosphere was fought with impressive quantities of alcohol, and humour. A new public address system was inaugurated with repeated paging for 'Werner Heisenberg, Werner Heisenberg'. This went on for two days before the operator was advised that she'd fallen victim to a prankster.

---

[5] Strolling over to a group of Italian physicists, which included Fermi and Segrè, Groves insisted that they should talk in English rather than their native 'Hungarian'.

Feynman fought his own small battles against the obsessive security. He and his wife Arline, in hospital in Albuquerque with tuberculosis, would write coded letters to each other in an attempt to defeat the censor. Even the mention of censorship was censored. As Feynman later explained:

> So finally they sent me a note that said: 'Please inform your wife not to mention censorship in her letters.' So I start my letter: 'I have been instructed to inform you not to mention censorship in your letters.' *Phoom, phoooom*, it comes right back! So I write, 'I have been instructed to inform my wife not to mention censorship. How the heck am I going to do it?'

Feynman learned to pick locks and crack safes. Physicists and mathematicians would tend to set the combinations of their safes using easily recalled mathematical constants such as the base of natural logarithms, e, or the value of π. He would open his colleagues' safes and leave them notes. On discovering that workmen had made a hole in the fence around Los Alamos to avoid having to walk around to the gate, Feynman used it to confuse the security guard by repeatedly coming in through the gate, without ever being seen going out. His exploits would be the subject of endless anecdotes told at the many parties hosted by the Oppenheimers, fuelled by Robert's legendary (and lethal) vodka martinis.

Apparently, nobody had thought to suggest that one consequence of the security-friendly isolation of Los Alamos and the relatively limited access to entertainment would be a baby boom. Oppenheimer could only plead to Groves that population control was not one of his duties. This was rather shamefaced pleading, however, as Kitty was now pregnant with their second child.

Teller had observed Oppenheimer's persuasive approach in action during the summer school at Berkeley the previous year, and had relished it. But a series of personal disappointments had by now dramatically changed Teller's perspective. He had grown to resent Oppenheimer's politicking. This resentment grew into confrontation.

Confrontation

Teller had gone to Los Alamos in April 1943 on the assumption that he would be responsible for leading a major stream of activity that would include work on the Super, the thermonuclear bomb. Oppenheimer's organisation chart had featured a Theoretical Division which Oppenheimer had initially thought to lead himself, until persuaded otherwise by Rabi. Teller had been involved in the American effort from its inception with Einstein's letter to Roosevelt in August 1939. In his own mind this made him one of the most senior physicists at Los Alamos. He might have therefore assumed he would be the logical choice to head the Theoretical Division but, at Rabi's suggestion, Oppenheimer had appointed Bethe.

In May 1943 a review of the priorities for the work programme at Los Alamos had concluded that, although work on the Super should continue at the laboratory, this was secondary to the work on fission weapons and should be restricted to theoretical study only. A further review in February 1944 concluded that the Super would need to be based on tritium, a substance that could be produced only in the large-scale nuclear reactors under construction at Hanford. But the priority at Hanford was to make plutonium, not tritium. Work on the Super could continue only so long as it did not interfere with the main programme. Teller sensed that he was being sidelined.

He tended to dismiss the theoretical physics of nuclear fission as yesterday's problem, already solved in principle. The technical difficulties associated with actually making a working bomb was not something that he felt should detain his intellect.

The Polish mathematician Stanislaw Ulam had been brought to Los Alamos from the University of Wisconsin to join Teller's group in the winter of 1943. Teller had immediately put him to work on the theory of the Super. Von Neumann subsequently directed Ulam to work on the hydrodynamics of implosion. Here the theoretical principles were relatively straightforward, but the shockwave calculations were extremely difficult. Ulam advocated a 'brute force' approach, crunching the numbers by machine computation and approaching a solution by trial and error.

When Bethe asked Teller to work on these calculations, Teller felt insulted. This was mind-numbing, hit-and-miss theory of the kind that Teller felt less qualified than others to carry out. He suspected that the calculations were so difficult that the work would not be completed in time to make any meaningful contribution to the construction of the fission bomb. Teller refused. 'Although Hans did not criticise me directly,' Teller later wrote, 'I knew he was angry.' Teller had, in effect, resigned from the division.

Oppenheimer moved quickly to placate him and thereby keep him at Los Alamos.[6] Despite the tremendous pressures on his own time, Oppenheimer offered to allow Teller and his group to continue working on theoretical aspects of the Super which they would review together once a week. Oppenheimer felt he had little choice: he valued Teller's input too much.

Although Bethe was an extremely competent theorist, he still needed the support of other able physicists to handle the volume of work. With Teller's attentions diverted away from priority work on fission weapons, Oppenheimer now scrambled to find a replacement.

Black comedy

Bohr took his concerns about the complementarity of the bomb to Felix Frankfurter, a friend from 1933, now a Supreme Court Justice and adviser to Roosevelt. Bohr did not speak openly of the Manhattan Project, but Frankfurter was aware that a large-scale project of some kind was under way. They found they could talk relatively freely about the implications of the programme without being explicit about its details, or even acknowledging that they were talking about a bomb. Bohr explained that with few

---

[6] Oppenheimer was selective about who he chose to defend, however. When Edward Condon fell into dispute with Groves over compartmentalisation, Oppenheimer chose not to support him and Condon left the programme. The theoretician Felix Bloch, who had been one of Oppenheimer's 'luminaries' at the summer school, became disenchanted with the regime and also left Los Alamos around this time. Oppenheimer let him go, and Teller saw him off.

exceptions, there was nobody he could talk to about his fears for the future. Frankfurter felt that Roosevelt would offer a sympathetic ear and agreed to raise the matter with the President.

Roosevelt's reaction to Frankfurter's intervention is clouded in controversy. According to Frankfurter, Roosevelt shared Bohr's concerns and agreed to a proposal that Bohr should discuss them further with Churchill. If true, it is hard to comprehend why Roosevelt would give Bohr, whom he had never met, a mandate for such an important mission, with potentially profound implications for post-war policy on atomic weapons. Roosevelt later denied he had given any such mandate. On the other hand, Roosevelt may have assumed that Bohr was already acting as an unofficial spokesman for the British.

Anderson, Bohr's principal link with the British administration, was also busily working behind the scenes to broach the subject of a post-war arms race with Churchill. Anderson was keen to pursue a policy of openness with the Soviet Union, advising them of the existence of the bomb and inviting them to collaborate on arms control. Churchill, however, was unyielding. Collaboration with the Soviet Union was, for him, simply out of the question.

Bohr and Cherwell met Churchill at Downing Street on 16 May 1944. C.P. Snow later called this meeting 'one of the blackest comedies of the war'. Churchill had no appetite for this discussion. He was in a bad mood, preoccupied with the plans for the impending Allied invasion of Normandy, set for the following month. He just failed to see the point: 'I cannot see what you are talking about', he said, scolding both Bohr and Cherwell as though they were schoolboys. 'After all this new bomb is just going to be bigger than our present bombs. It involves no difference in the principles of war. And as for any post-war problems there are none that cannot be amicably settled between me and my friend, President Roosevelt.'

Bohr foresaw what Churchill could not, or would not. The principles of war were indeed about to change. 'We are in a completely new situation that cannot be resolved by war', Bohr claimed, to anyone who would listen. As Bohr persisted in his efforts to address the issues, Churchill became more and more entrenched in his views.

Churchill had a misplaced faith in secrets. 'He was only too conscious,' wrote C.P. Snow, 'that British power, and his own, was now just a vestige. So long as the Americans and the British had the bomb in sole possession, he could feel that that power hadn't altogether slipped away. It is a sad story.'

The bomb was a secret that couldn't be kept. Klaus Fuchs was working to make sure of that.

Chapter 12

MORTAL CRIMES

February–December 1944

When Fuchs arrived in New York he checked first into the Taft Hotel near Times Square before relocating to the Barbizon Hotel. He spent Christmas 1943 in Cambridge, Massachusetts, with his sister Kristel, her husband Robert Heineman, and their two children. When many of the fifteen scientists of the British delegation assigned to work on gaseous diffusion returned to Britain after a few months, Fuchs moved once more, this time into a rented apartment on West 77th Street.

After war-threatened Birmingham, New York was a materially much more pleasant experience. Food was rationed in America but this was not as restrictive as in Britain. Restaurants were packed with native New Yorkers, tourists and servicemen. Theatres and concert halls experienced brisk business. Not that Fuchs had much time for recreation.

On a cold Saturday afternoon in early February 1944 he walked along Henry Street, in Manhattan's lower East Side. If anyone had paid him any attention, they might have noted some fairly odd behaviour. Following the instructions he had received from Sonja, he walked along the street carrying a tennis ball. Outside the Henry Street Settlement House he was approached by a short, dark, rather portly American with thinning hair and thick eyeglasses. The man was wearing gloves and, again somewhat incongruously, carrying a second pair of gloves in his hand. This was Fuchs' recognition signal.

'Can you tell me the way to Grand Central Station?' the man asked.

They exchanged a couple of meaningless remarks. The proper codes had been given and received. 'Raymond?' Fuchs asked.

Raymond had heard nothing of the Allied atomic bomb programme. As they walked on down the street, Fuchs filled in the background about atomic fission, isotope separation and the work on gaseous diffusion. Fuchs did not yet have any documents to pass on. They made arrangements for their next meeting.

Fuchs would never know his contact by any name other than Raymond. He was Harry Gold, a Jewish, Swiss-born chemistry graduate who had been brought to America when he was just two years old. He had started spying for the Soviet Union in 1936. He passed on information about industrial chemical processes he obtained from Pennsylvania Sugar, where he worked, via the Soviet foreign trade agency Amtorg.

Although Gold was sympathetic to the Soviet vision of a just society (or, at least, the version of this vision promulgated in Communist propaganda), his was espionage that was not born from some deeply-held conviction about Communism or the Soviet Union. Gold never became a party member. Nor did he seek compensation for the information he acquired. Gold was a social inadequate. It seems he began his career in espionage largely because he was asked to do this by someone who had helped him get a job. He became a spy because he was grateful.

Gold would weave elaborate fantasies about himself and his life. He lived in Philadelphia, single and alone. Yet he would talk tearfully of his wife, of her betrayal and how the break-up of his marriage had denied him access to his children. He talked about his younger brother Joe, who had been killed in action in the Pacific. In truth, his younger brother – who really was called Joe – was not only alive but had been decorated for bravery.

His access to industrial secrets was rather limited, however, and he quickly exhausted his usefulness as a spy. Gold was instead put to work as a courier before being given his latest assignment. Fuchs met with Gold several times at various public places. As his work on the Manhattan Project got under way, Fuchs once again passed on reports that he had prepared himself. These were mostly technical papers concerned with the operation of K-25, the large-scale gaseous diffusion plant at Oak Ridge, and the barrier materials that had now been adopted. Despite Groves' obsession with secrecy, the scientists working on the project would frequently take classified documents home with them, under strict instructions not to let them out of their possession. Fuchs would pass his written reports to a secretary for typing. All typed reports and any copies were strictly controlled, but there appeared to be no controls on the handwritten originals. He would pass these to Gold, holding on to the reports until the very end of their meetings, so that if they were discovered it would be Fuchs who would be found with classified documents in his possession.

In his turn, Gold would pass the papers, together with his own reports of his conversations with Fuchs, to his Soviet contact. This was New York vice consul and NKGB[1] agent Anatoly Yatskov.

Fuchs had warned Gold that the various projects running within the Manhattan Project were quite strictly compartmentalised, and that he anticipated that he would be transferred to work somewhere in the southwestern US later in the year or early in 1945. Fuchs believed he explicitly mentioned New Mexico, but Gold heard 'somewhere in Mexico'. Fuchs gave him Kristel's address in Cambridge and explained that if contact was broken for any reason, he would leave a message for Gold with his sister. It was still nevertheless quite a shock for Gold when, on 5 August, Fuchs failed to appear at a pre-arranged meeting outside the Bell movie theatre in Brooklyn. Fuchs missed the next meeting as well.

Yatskov acquired Fuchs' New York address and advised Gold to go to the apartment and investigate. Bearing of copy of a novel by Thomas Mann in which he had inscribed Fuchs' name and address, Gold went to the apartment with the cover story that he was returning a book that Fuchs had lent him. Getting no answer from the apartment, Gold made enquiries of the

---

[1] The Soviet security services underwent a number of organisational changes during the war. In July 1943 the foreign intelligence and security services of the NKVD was split out as a separate commissariat, called the People's Commissariat for State Security (Narodny Kommisariat Gosudarstvennoye Bezopasnosti, or NKGB).

janitor, who told him that Fuchs had moved away. Yatskov advised Gold to sit tight.

Oppenheimer's solution to Teller's departure from the mainstream activity of the Theoretical Division was to bring Peierls to Los Alamos. Bethe and Peierls had known each other for many years (Bethe had stayed with Peierls and his wife in England in 1934). Both Bethe and Oppenheimer considered Peierls to be the best physicist available to work on the theory of implosion. By the summer of 1944 the Oak Ridge plant was up and running and there was little need for the Tube Alloys physicists to remain in New York to work on diffusion theory. It was agreed that Peierls would move to Los Alamos, and that Fuchs would join him.

Fuchs had been unable to attend his meetings with Gold because he was already at Los Alamos.

The plutonium-240 crisis

The small pilot nuclear reactor at Oak Ridge, called X-10, had gone critical for the first time in November 1943. Its purpose was to generate larger quantities of plutonium for testing ahead of regular supply of reactor-produced materials from the large-scale reactors that were being built by Du Pont at Hanford. But what the Los Alamos physicists now discovered about reactor-produced plutonium threatened to rule out completely the idea of a plutonium bomb. And, without a plutonium bomb, all the Manhattan Project could hope to produce was just one U-235 bomb.

The physicists discovered that plutonium produced in a uranium reactor behaves rather differently from the tiny quantities of plutonium produced by neutron bombardment in a cyclotron. A year before, Glenn Seaborg had warned of the risk that plutonium produced in a reactor might be contaminated with quantities of the plutonium isotope Pu-240, formed from Pu-239 by capture of another neutron. In July 1944 Segrè showed in his makeshift cabin-laboratory that Seaborg had been right. The longer the plutonium was left to accumulate in a nuclear reactor – and hence the more plutonium was produced – the higher was the proportion of Pu-240. When it finally arrived, the plutonium from the Hanford reactors would likely contain a high proportion of this isotope.

The problem was that Pu-240 is quite unstable, emitting alpha particles and acting as a source of background neutrons. Segrè found that the rate of spontaneous fission in the X-10 plutonium was significantly higher than in the cyclotron samples, behaviour that could be ascribed to a very high spontaneous fission rate in Pu-240. With the gun method it was believed to be possible to assemble the sub-critical components within about a ten-thousandth of a second. Calculations quickly showed that the high spontaneous fission rate in Pu-240 would release a flood of neutrons into the assembling mass before it had reached its optimum configuration. Any attempt to assemble a super-critical mass of plutonium using the gun method would result only in pre-detonation. The bomb would fizzle but not explode, because the gun method simply could not create a super-critical mass fast enough. An assembly mechanism at least 100 times faster was required. Implosion was the only candidate.

This was a double blow. The gun method was ruled out because of the contamination by Pu-240. Purifying the plutonium meant separating Pu-240 from Pu-239, isotopes differing by only one neutron in their nuclei, a much harder task even than separating U-235 from U-238. The promise of plutonium – a promise of access to fissile materials without the pain of a difficult physical separation process – was cruelly snatched away.

Oppenheimer met with Conant, Compton, Fermi, Groves and Nichols to discuss the problem in Chicago on 17 July. There was no real prospect of purifying the plutonium. Without purification, the gun method could not be used for a plutonium bomb. Conant suggested an alternative approach based on mixtures of uranium and plutonium. But this would be a low-yield weapon, generating an explosive force equivalent to a little more than a few hundred tons of TNT. Conant wondered if building such a weapon would help build confidence in the push for a larger weapon, but Oppenheimer argued that taking this approach would result in unacceptable delays.

'It appears reasonable,' Oppenheimer wrote in conclusion the next day, 'to discontinue the intensive effort to achieve higher purity for plutonium and to concentrate attention on methods of assembly which do not require a low neutron background for their success. At the present time the method to which an over-riding priority must be assigned is the method of implosion.'

It was Oppenheimer's worst nightmare. As the physicists debated what to do next, the question of German progress was inevitably raised. Had the Germans encountered, and already solved, this same problem? 'We finally arrived at the conclusion that they could be exactly up to us, or perhaps further', Rabi noted. 'We were very solemn. One didn't know what the enemy had. One didn't want to lose a single day, a single week. And certainly, a month would be a calamity.'

Neddermeyer and his team studying implosion in the Ordnance Division had taken a rather plodding academic approach to the problem. 'I think [Oppenheimer] felt very badly because I seemed not to push things as for war research but acted as though it were just a normal research situation', Neddermeyer later admitted. In an attempt to get more traction, in January 1944 Oppenheimer had persuaded George Kistiakowsky to join Los Alamos full-time. Kistiakowsky had been travelling to and from the Hill in his role as consultant to the Manhattan Project. This was merely one part of a whirlwind commute between Pittsburgh, Florida, Washington and New Mexico for the NDRC's explosives group. He had an interesting overseas appointment in the pipeline, and Oppenheimer had had to pull out all the stops and apply his legendary charm.

When Kistiakowsky arrived on the Hill for good he announced: 'I am old, I am tired, and I am disgusted.' Oppenheimer had made sure he got good accommodation, and sold the former Cossack one of his own saddle horses for a knock-down price. The horse was called, appropriately, 'Crisis'.

Kistiakowsky had worked in the Ordnance Division on implosion through the spring and early summer of 1944, trying desperately to referee the interminable disputes between Neddermeyer and the division head, Captain William 'Deke' Parsons. It was a clash between military-style management and academic sensibility. 'The two never agreed about anything and they certainly didn't want me interfering', Kistiakowsky said. He made little progress, and threatened to leave.

Explosive lenses

It was around this time when the ultimate solution to the implosion problem was outlined by James Tuck, a Manchester-born and -educated physicist. Tuck had worked with Szilard in Oxford in 1937 and at the outbreak of the war he had been appointed as a scientific assistant to Cherwell, working on the development of armour-piercing anti-tank weapons. It was his expertise in the use of shaped charges that had led to his inclusion in the British delegation.

Neddermeyer had been wrestling with the problem of creating a perfectly spherical shockwave by varying the shape of the explosion, the type of explosive and the number and positioning of the detonators. But the shockwave produced by a point detonator expands spherically outwards through the explosive material, just as the ripples on the surface of a pond expand outwards from the point where a pebble has been thrown. Put several detonators in close proximity and the result is an unpredictable combination of diverging and interfering shockwaves, much like the turbulence on the surface of a pond when many pebbles are thrown in at once.

This, Tuck argued, was not a new problem. American and British efforts to develop armour-piercing shells in which all of the force of the explosive charge is directed towards the armour had resulted in the development of explosive *lenses*. These work according to the same kinds of principles by which ordinary lenses direct and focus light waves. The medium of an optical lens affects the velocity of light passing through it, exerting different effects in different places in the medium such that the light is 'gathered' and focused to a point. An explosive lens consists of a series of charges with different rates of detonation, such that the explosive shockwave is similarly gathered, and focused. Surround the spherical core of plutonium with explosive lenses which are then detonated simultaneously and, Tuck suggested, the result would be a perfectly spherical shockwave directed inwards towards the core.

It was not immediately identified as *the* solution. Developing explosive lenses appeared far more complex than simply trying to get a uniform spherical shockwave from conventional explosives. However, initial experiments, run separately to Neddermeyer's implosion work, appeared promising. Geoffrey Taylor, a leading British expert in hydrodynamics, arrived at Los Alamos in May 1944 and lent the weight of his opinions. The brute force hydrodynamic calculations seemed to tell against the simple solution and, gradually, the Los Alamos physicists began to acknowledge that explosive lenses might present the only solution. But it also became clear that success with explosive lenses would require considerable trial-and-error experimentation.

Oppenheimer now took a huge gamble. The problem of spontaneous fission in reactor-bred plutonium meant that if a plutonium bomb was going to be at all possible, it would have to be an implosion bomb. Somehow, they were going to have to make implosion work. They now had no choice but to throw the book at it.

Oppenheimer finally lost patience with Neddermeyer. 'Oppenheimer lit into me. A lot of people looked up to him as a source of wisdom and inspiration. I respected him as a scientist, but I just didn't look up to him that way. I didn't look up to him. From my point of view, he was an intellectual snob. He could cut you cold and humiliate you right down to the ground.' In desperation, Oppenheimer now moved to place his bet. He elected to reorganise the laboratory. In August 1944 he created two new divisions out of the old Ordnance Division. These were G Division (for 'Gadget'), charged with developing the physics of implosion and the Fat Man bomb design, to be led by Robert Bacher, and X Division (for 'eXplosives') which would focus on the development of explosive lenses. He removed a bitterly disappointed Neddermeyer and persuaded a reluctant Kistiakowsky to take charge of X Division.

Parsons was busy with the development of the uranium gun, but was angry with the way he had been out-manoeuvred by Oppenheimer and Kistiakowsky. 'Parsons was furious,' said Kistiakowsky; 'he felt that I had by-passed him and that was outrageous. I can understand how he felt but I was a civilian, so was Oppie, and I didn't have to go through him.'

Kistiakowsky took charge of a division consisting of about a dozen or so scientists, many of whom had been close colleagues of Neddermeyer's. Within a few months the division had grown to 600, including 400 army physicists and engineers recruited to a Special Engineering Detachment (SED). The SEDs were enlisted personnel, many college-educated and some with Ph.D.s, ordered by the army to work at Los Alamos.

Among them was David Greenglass, a machinist assigned to Kistiakowsky's division.

A problem underestimated

Igor Kurchatov was by nature a patient man, with a temperament well suited to managing a fundamentally important, large-scale scientific programme. However, by September 1944 his patience had worn thin.

Kurchatov had gathered around him a group of talented scientists to support the Soviet bomb programme. This group included Yuli Khariton, who had gained his Ph.D. in Cambridge in 1928 and who had studied explosives at the Institute for Chemical Physics in Leningrad. Before the war, Khariton had worked with Yakov Zeldovich on the theory of nuclear chain reactions. Also included in the group were Flerov, Issak Kikoin, Abram Alikhanov and Aleksandr Leipunskii, all graduates of the Leningrad Polytechnic Institute. Kurchatov's brother Boris joined the group in the middle of 1943.

A new, secret laboratory had been established at the Seismological Institute in Moscow under the overall control of Mikhail Pervukhin, People's Commissar of the Chemical Industry. To disguise its purpose, it was called simply Laboratory No. 2. As more and more scientists joined the programme, the laboratory expanded into adjacent premises before relocating in April 1944 to new buildings in the north-west of Moscow, near the Moscow River.

Kurchatov now established his programme along familiar lines. He took for himself the responsibility of designing and building the Soviet

Union's first nuclear reactor, eventually fixing on a uranium–graphite pile with a lattice configuration.[2] He was well aware that his biggest headache was going to be the lack of sources of uranium for the pile. He estimated that he needed about 60 tons, compared to the one to two tons available to him in 1943. Surveys of potential uranium deposits in Central Asia had suggested that a little over ten tons could be available by early 1944, implying that it would take five to ten years to acquire sufficient uranium for a nuclear reactor. The People's Commissariat of Non-ferrous Metallurgy had been asked to find over 100 tons of uranium 'as soon as possible', which, against the urgent demands of the war against Germany, implied the lowest priority.

Without a reactor there could be no plutonium. Kurchatov asked that his team establish a cyclotron at Laboratory No. 2 as a matter of urgency. This could be assembled from the component parts of the Fiztekh cyclotron, which had become somewhat scattered and some of which were now located dangerously close to the front line. The cyclotron was needed to generate minute quantities of plutonium with which to perform critical physical measurements.

Kikoin took charge of the effort to identify methods for the large-scale separation of U-235. This, again, was work that was obviously contingent on Kurchatov's ability to acquire sufficient quantities of natural uranium. Kikoin first examined centrifuge techniques, before moving on to gaseous diffusion. Thermal diffusion was also studied. In 1944, Lev Artsimovich joined Laboratory No. 2 to work on electromagnetic separation.

There remained the matter of bomb design. Kurchatov had known Khariton for nearly twenty years and, as the Soviet programme got under way, he asked Khariton to lead the design effort. Khariton was initially reluctant to abandon his work on anti-tank weapons, but Kurchatov was ultimately persuasive. Without quantities of materials sufficient for

---

[2] Kurchatov did not learn about the successful Chicago pile until the end of July 1943. His decision to focus efforts on a uranium–graphite lattice configuration was based on the assessments of his own scientists.

experimental study, however, Khariton could make little progress. He began some basic experiments to investigate the gun method.

The cyclotron became operational on 25 September 1944. Kurchatov and the cyclotron team celebrated this success with champagne. But any feeling of elation that Kurchatov felt was quickly overtaken by a strong sense of frustration. From the intelligence materials on ENORMOZ that he had seen, it was obvious that the Americans had embarked on a major project. The Soviet programme, by contrast, was severely hampered by the lack of materials. Without sufficient uranium, all they could do was theorise and carry out experiments that nagged at the very edges of the problems they needed to solve. A few days later, Kurchatov poured his exasperation into a letter to Beria:

> But in our country, in spite of great progress in developing this work on uranium in 1943–1944, the state of affairs remains completely unsatisfactory. The situation with raw materials and questions of separation is particularly bad. The research at Laboratory No. 2 lacks an adequate material-technical base. Research at many organisations that are co-operating with us is not developing as it should because of the lack of unified leadership, and because the significance of the problem is underestimated in these organisations.

Near the edge of mortal crimes

Bohr's attempts at atomic diplomacy had broken on the rocks of Churchill's misplaced faith in secrets. But his arguments had persuaded many in Churchill's inner circle, including Cherwell and Anderson. Bohr returned to Washington on 16 June 1944 and reported on his failed meeting with Churchill to Frankfurter a few days later. Frankfurter advised Roosevelt. Despite Bohr's lack of success, Roosevelt continued to make encouraging noises and invited Bohr to discuss the issues directly with him.

Bohr first summarised his views in a short memorandum, which spoke of 'forestalling a fateful competition about the formidable weapon'. Father and son worked on the memorandum as Washington steamed in a summer heatwave. Aage typed the various drafts as his father wrestled with the precision of his language, while darning socks and sewing buttons. Bohr met with Roosevelt on 26 August.

This was a very different meeting. It lasted for more than an hour. Unlike Churchill, who was grumpy and appeared to have taken an instant dislike to Bohr, Roosevelt was cordial and found him interesting. Most importantly, Roosevelt found Bohr's arguments persuasive. According to Aage Bohr:

> Roosevelt agreed that an approach to the Soviet Union of the kind suggested must be tried, and said that he had the best hopes that such a step would achieve a favourable result. In his opinion Stalin was enough of a realist to understand the revolutionary importance of this scientific and technical advance and the consequences it implied.

The meeting concluded on a positive note. Bohr summarised the main points of their discussion in a letter to Roosevelt which reached him the day before he was due to depart for another meeting with Churchill in Quebec.

Bohr had grounds for optimism, but had not counted on Churchill's wilfulness and the extent of his influence over the American President. At the end of the Quebec conference, Roosevelt and Churchill adjourned to Roosevelt's private estate in the Hudson Valley at Hyde Park, New York. At this meeting Churchill moved to stomp on any attempt to reveal the existence of the bomb programme to the Soviet Union, or indeed to anyone – friend or foe – outside the terms of the Quebec agreement. In a secret aide-mémoire drafted, it would seem, largely by Churchill at the end of their meeting, the two leaders agreed to maintain the utmost secrecy. They acknowledged that: 'When a "bomb" is finally available, it might perhaps, after mature consideration, be used against the Japanese, who should be warned that this bombardment will be repeated until they surrender.'

Churchill reserved particular ire for Bohr: 'Enquiries should be made regarding the activities of Professor Bohr and steps taken to ensure that he is responsible for no leakage of information particularly to the Russians.'

At issue was a letter Bohr had received from the Soviet physicist Peter Kapitza via the Soviet embassy in London in April that year. In this letter, dated 28 October 1943, Kapitza had invited Bohr to the Soviet Union, using veiled language that could be interpreted as an invitation to collaborate on the development of atomic weapons.[3] Bohr had sent a non-committal reply, which he had been sure to send first to the SIS for approval. He had handled the approach carefully and correctly.

But Churchill sensed treachery. To his mind, Bohr was in close correspondence with a Russian professor, an old friend to whom he had written and might be writing still. In a memo to Cherwell written the day after his Hyde Park meeting with Roosevelt, Churchill went even further: 'It seems to me Bohr ought to be confined or at any rate made to see that he is very near the edge of mortal crimes.'

Churchill could not understand the complementarity of the atomic bomb. Roosevelt may have had his own reasons for acquiescing to Churchill's views. Both may have chosen to fix on doubts about Bohr's trustworthiness as the world's first atomic diplomat as a way of ducking the issues that Bohr was trying to raise. In any event, despite Bohr's best endeavours, the prospects for a post-war nuclear arms race had now been greatly enhanced.

Cherwell and Anderson leapt to Bohr's defence, and Bush and Conant did likewise across the Atlantic. Bohr was not arrested.

Breaking the American monopoly

It was a bit of a puzzle. If you had just decided that your concern for what your country would do with a monopoly on atomic weapons was sufficiently strong to overcome your moral qualms about betraying your country's military secrets, what did you do next? Where did you go? You couldn't just look up a list of Soviet spies in the phone book.

---

[3] Kapitza had sought permission from Molotov before writing to Bohr, though it is likely that the purpose of his letter was simply to encourage Bohr to come to the Soviet Union, for the good of Soviet science generally.

Theodore Hall was just nineteen years old. Despite his youth, he had been identified by Los Alamos talent-spotters and had joined the still-growing ranks of physicists on the Hill in January 1944, together with his friend and fellow Harvard graduate student Roy Glauber. They had arrived at Lamy railway station on 27 January, having journeyed with an older physicist who had simply introduced himself as Mr Newman. They later realised he was John von Neumann, held in awe by the teenage physicists as the man who had transformed the mathematical basis of quantum theory in the 1930s.

The new arrivals received their briefing from Bacher and devoured Serber's 'Los Alamos Primer'. Hall was put to work in the experimental physics division (P Division). He quickly made a positive impression. In April he was assigned to work on implosion, studying the radioactivity of samples of radium-lanthanum that had been imploded using high explosive packed around the outside. These 'Ra-La' experiments, as they were called, were set up at a small laboratory in Bayo Canyon, south-east of the main Los Alamos site. They provided a measure of the uniformity of the explosive shockwave that had been created. As the effects of the Pu-240 crisis rippled through the laboratory, the Ra-La experiments took on considerably greater significance.

Hall was a Jewish radical. He had once been a member of the American Student Union, several chapters of which were associated with the Young Communist League, but he had resisted being part of what he came to think of as a Communist front. He never joined the Communist Party. At Harvard, he had joined the Marxist John Reed Society, but did so under the influence of his eloquent room-mates at Leverett House rather than from political conviction. Nevertheless, as his first summer at Los Alamos wore on, concern that America was building a monopoly in atomic weapons gnawed at him. America was a democratic state but, Hall now pondered, could another depression drive it towards fascism, as had happened in Germany? And what would a fascist America do with its atomic monopoly?

After all, there was some sympathy and much talk among some of the Los Alamos scientists about the need to inform the Soviet Union of their work. Surely, Hall reasoned to himself, the post-war world would be a safer place if the secrets of the atomic bomb were shared with the Soviets. Years later he recalled what was going through his mind:

> Thinking back to the rather arrogant 19-year-old I then was, I can recall quite well what was on my mind at the time. My decision about contacting the Soviets was a gradual one, and it was entirely my own. It was entirely voluntary, not influenced by any other individual or by any organisation ... I was never 'recruited' by anyone ... As I worked at Los Alamos and understood the destructive power of the atomic bomb, I asked myself what might happen if World War II was followed by a depression in the United States while it had an atomic monopoly ... It seemed to me that an American monopoly was dangerous and should be prevented.

Having decided to share with the Soviet Union those secrets to which he had access, he then had to face the challenge of finding an appropriate channel.

In October 1944 he took two weeks' leave from Los Alamos and returned to New York to celebrate his twentieth birthday. He confided his intentions to Saville Sax, another Harvard student friend, and together they tried to figure out how to make contact with the Soviets. Through a series of rather comical trial-and-error approaches to Artkino Pictures, the Soviet film distributor, and Amtorg, they were both directed towards the Soviet journalist Sergei Kurnakov, who contributed to the Communist Party's newspaper, the *Daily Worker*. But Kurnakov seemed rather cagey and promised little.

'Do you understand what you are doing?' he asked Hall. 'Why do you think it is necessary to disclose US secrets for the sake of the Soviet Union?'

'There is no country except for the Soviet Union which could be entrusted with such a terrible thing', was Hall's reply.

Hall handed over a report he had written about Los Alamos which listed the scientists working there, including Oppenheimer, Bethe, Bohr, Fermi, von Neumann, Kistiakowsky, Segrè, Penney, Compton, Lawrence, Urey and Teller. In fact, Kurnakov was a Soviet agent, though a rather low-level one. He did not have the authority to make any commitments to the young would-be spies.

Concerned that Hall would return to Los Alamos without establishing the needed contacts, towards the end of October Sax blundered into the Soviet consulate, where he met Yatskov. Sax gave him another copy of Hall's report on Los Alamos. Yatskov promised little more, but subsequently checked the students' story with Kurnakov. There had been rumours of a secret 'Laboratory V', involved in research on the properties of U-235 and plutonium for the purposes of building an atomic bomb. Hall's report appeared genuine. Yatskov's most important source inside the Manhattan Project had gone missing in August, and he sorely needed a replacement.

Kvasnikov was unavailable, so Yatskov consulted with the NKBG station chief in New York, Stepan Apresyan. They agreed that Hall and Sax were worth a risk. Yatskov instructed Kurnakov to make contact with Hall before he returned to Los Alamos. Rather embarrassingly for Hall, this brief meeting took place at Penn Station as Hall waited with his parents for a train to Chicago. Kurnakov advised him that he and Sax had been accepted into 'the club'. It was agreed that Sax would be Hall's go-between.

Triumvirate

Harry Gold went to see Fuchs' sister Kristel in Cambridge, Massachusetts, in early November 1944. He explained that he was a good friend of Fuchs who had lost contact and happened to be in the Boston area on business. Fuchs had given him her address and he had decided to try to find out what had happened to him. In fact, Fuchs had contacted his sister from Chicago, either en route to Los Alamos or during a subsequent visit to the Met Lab. She told Gold that he had gone to some unknown location in the south-west, but that he intended to return to Cambridge to be with them for Christmas. Gold left a sealed message containing a contact telephone number.

Gold reported back to Yatskov, who no doubt breathed a sigh of relief. The unknown location in the south-west was obviously Los Alamos. On 16 November he sent a cable to Moscow Centre announcing that Fuchs had been found and should be in a position within months, possibly weeks, to provide information from 'Camp No. 2', one of the codenames for Los Alamos.

A few weeks later, on 29 November, David Greenglass celebrated his second wedding anniversary with his wife, Ruth, in Albuquerque. He had started at Los Alamos on 5 August and had been assigned to Group E-5 in Kistiakowsky's X Division, initially working on high-speed cameras before moving on to explosive lenses. Husband and wife were missing each other terribly, and decided to get together in New Mexico rather than wait until Greenglass could take some leave towards the end of the year. Ruth's trip to New Mexico had been paid for by David's brother-in-law, Julius Rosenberg.

Both Julius and his wife Ethel Rosenberg, Greenglass's elder sister, were dedicated Communists. As a graduate student in electrical engineering at City College in New York in 1935, Julius Rosenberg had joined the Steinmetz Club, the CCNY campus branch of the Young Communist League. A year later he founded the chapter of the American Student Union that Hall would subsequently join in 1938. Hall never met Rosenberg, but his suspicion that the ASU was a Communist front was justified.

Both the Rosenbergs had encouraged the teenage David Greenglass to join the Young Communist League. He became a committed convert. On joining the army shortly after his wedding in late 1942, he would write letters to Ruth which mixed declarations of love with declarations to the Communist cause: 'Victory shall be ours and the future is Socialism's.' Both would sign their letters to each other 'comrade'.

David and Ruth Greenglass were aware that the Rosenbergs were somehow involved in industrial espionage for the Soviet Union, although they were vague on the details. Julius had tried to 'soften up' Greenglass with the suggestion of going into business together after the war. When Greenglass discovered that he was to be assigned to work at a secret location, he alerted Rosenberg via his wife. 'I have been very reticent in my writing about what

I am doing or going to do because it is a classified top secrecy project and as such I can't say anything', he wrote before arriving on the Hill.

The close surveillance of correspondence into and out of Los Alamos meant that any reference to the real subject of the dialogue between husband and wife had to be disguised, but on 4 November 1944 Greenglass passed the message back that he 'most certainly will be glad to be part of the community project that Julius and his friends have in mind'.

During their anniversary celebrations later that month, Ruth told her husband that, according to Rosenberg's information, he was working on the atomic bomb. He was very surprised, as this was information that had not been shared with him through official channels. Ruth then put Rosenberg's by now familiar proposition to him. America and the Soviet Union were allies in the war against Nazi Germany and Japan. In the face of their common enemies, the Soviets were therefore entitled to information about American military projects.

Ruth had misgivings, and David's own mind was initially clouded with doubts. But he worshipped his brother-in-law as a hero and, by next morning, his conviction had overcome his doubts. Greenglass agreed to spy for the Soviet Union. He began by providing to Ruth some general information about the layout of the laboratory and the number of scientists working there. He mentioned Kistiakowsky's name, and those of Oppenheimer and Bohr.

Julius Rosenberg was assigned the principal liaison role, though he felt technically unqualified and asked his Soviet controller – Alexander Feklisov – for support in debriefing Greenglass during their meetings. Greenglass returned to Los Alamos, eyes and ears now fully alert for any information that might be of value.

Fuchs had arrived at Los Alamos just over a week after Greenglass. He was given a room in the bachelors' quarters next to Feynman. They became good friends. Feynman would tease him about his personality, and suggested that he find a girlfriend. Joking about which of them would more likely be a Nazi spy, Fuchs argued that it must be Feynman, because of his frequent visits off site to see his wife in hospital in Albuquerque. Feynman agreed.

Fuchs quickly made a very strong, positive impression on his peers and superiors. Bethe regarded him as one of the most valued physicists working in the Theoretical Division. He impressed Oppenheimer. Though neither a group leader nor division head, he was invited to attend meetings of the Co-ordinating Council, allowing him to become intimately familiar with the work of every division at Los Alamos.

Pressure of work meant that Fuchs could not make the trip to Cambridge to be with his sister and her family at Christmas. The visit was postponed to February 1945.

Of the triumvirate of Soviet spies now working at the very heart of the Manhattan Project, Fuchs was clearly the most important. He had access to a wide range of secrets centred on the single most pressing problem that the physicists now faced – implosion. Hall was close to these same secrets, though as a junior scientist his access was more limited. Greenglass was a machinist, busy making moulds in which high explosive lenses would be cast. None of the spies knew about the others, but together their information provided independent corroboration of the details of the work in hand.

On 25 December, Kurchatov received another pile of espionage materials. Among the documents was a review of activity at an unnamed laboratory – most likely Los Alamos – from March 1943 to June 1944. Fuchs had not by this time had an opportunity to re-establish contact with Gold or pass any information about the work at Los Alamos. The review describes details of instrumentation that Greenglass would not likely have been familiar with. Of the trio, Hall is therefore the most likely source, the review possibly being part of the document that he had handed to Kurnakov and Yatskov in New York.

Kurchatov remained impressed with the materials. 'On the basis of theoretical data,' he wrote, 'there are no grounds to believe that in this respect there will be any difficulties in implementing the bomb.'

Too many sixes

Cecil Phillips had been just nineteen years old when he joined the cryptanalysts at Arlington Hall in the summer of 1943. He was transferred to work on the 'Russian diplomatic problem' in May 1944.

After six months' intensive study, he began to notice new patterns among the strings of seemingly random numbers. At the beginning of some of the coded messages there appeared a number group with a higher than average number of sixes. Richard Hallock had looked for references to the content at the beginning of each message and had stumbled on the fact that the Soviets had been using duplicate one-time pads. Now Phillips had found another vulnerability. The number group with more than its fair share of sixes was unencrypted. It was the first number group on the one-time pad that had been used to encrypt the message once it had been coded. By including it at the beginning of the message, the clerk in the receiving station could identify which one-time pad had been used and so decrypt and then decode the message.[4]

By sorting the messages using the number group with too many sixes, Phillips could identify which messages had been encrypted using duplicate one-time pads. And from any such pair of messages, the cryptanalysts could start to unpick the one-time pad encryption.

---

[4] This appears to have been a response to an April 1944 request to all stations from Moscow. By this time the Soviets had discovered the existence of an American counter-espionage project and were concerned for the security of their coded and enciphered messages. The NKVD requested that cipher clerks change the system by adopting a new message starting point or key-pad indicator. The change was discovered by Phillips and actually made the messages more vulnerable.

Chapter 13

ALSOS AND AZUSA

January–December 1944

The Alsos mission was John Lansdale's idea.

Largely for want of any other way to engage the enemy, the Allied forces that had secured victory in North Africa had launched an invasion of Sicily on 10 July 1943, a prelude to the invasion of the Italian mainland, Churchill's 'soft under-belly' of Europe. Lansdale had suggested that a scientific intelligence mission should follow hard on the heels of the invading forces, its main purpose being to discover precisely how far the German atomic bomb programme had progressed.

Groves approved, and Lansdale spent the summer garnering support for the mission, which was codenamed Alsos.[1] It was formally approved in September 1943 and involved army intelligence, the US Navy, representatives of the Manhattan Project and the OSRD. Pash was appointed as the mission's military leader, which gave Groves some relief from Pash's relentless pursuit of the Los Alamos laboratory's scientific director.

But by this time the British believed they already knew what was happening with the German programme. Eric Welsh of the SIS and Michael Perrin of the Directorate of Tube Alloys argued in November that the Alsos

---

[1] Groves learned too late that *alsos* is the Greek word for grove.

mission was unnecessary. The SIS had gathered information from scientists who had encountered various members of the Uranverein in Norway, Sweden and Switzerland. Earlier in the year, Hahn had told Lise Meitner in Stockholm that there would be 'no practical utilisation of fission chain reactions in uranium for many years to come'. The physicist Paul Scherrer in Zurich reported a conversation he had had with the German chemist Klaus Clusius, who said that efforts to separate U-235 had been abandoned. The SIS spy Paul Rosbaud no doubt provided confirmation that the German bomb programme was all but stalled.[2]

There was also circumstantial evidence. While the German rocket programme was mentioned occasionally in coded radio messages that had been deciphered by the cryptanalysts at Bletchley Park, they could find no mention of uranium, or atomic bombs. The German physicists had resumed publishing their research results in the German scientific literature. The results so reported appeared to be genuine, rather than fabrications designed to mislead Allied scientists. In January 1944 the British assessed all the evidence to hand: they concluded that the Germans were not in fact carrying out any large-scale work on an atomic bomb.

However, the British did not see fit to share the sources of their intelligence with the Manhattan Project, and Groves was not persuaded. While he acknowledged that a German atomic bomb was perhaps unlikely, he felt that the risk was still too great. There was always the possibility that British intelligence agents were being fed carefully-constructed disinformation.[3] Oppenheimer interpreted the now openly published German research

---

[2] The full extent of Rosbaud's intelligence activities remain buried in classified SIS files. In 2006 campaigners secured the services of Cherie Booth QC, wife of former UK Prime Minister Tony Blair, in an attempt to get Rosbaud's files declassified. However, this is not going to happen anytime soon. In April 2008, the Investigatory Powers Tribunal concluded that it was legal to withhold records beyond 30 years and, therefore, to neither confirm nor deny the records that they hold. See http://www.cabinetoffice.gov.uk/about_the_cabinet_office/080425meeting.aspx

[3] This was not as far-fetched as it might now seem with the benefit of hindsight. The Allies frequently used carefully staged events to feed disinformation to the Germans and disguise real intentions. For example, the Allied invasion of Sicily was, in part, successful because plans recovered from the body of a major in the Royal Marines found floating in the sea off the Spanish coast indicated the imminent Allied invasion not of Sicily, but of Greece.

results as deliberate, feigned or enforced ignorance of scientific facts which were by now recognised to be crucial to the Allied bomb programme.

Groves was in no mood to take chances. Fourteen civilians soon to perish aboard the *Hydro* would be part of the price to be paid for surety.

Tinker, tailor, catcher, spy

Yet even as Lansdale gained support for the Alsos mission, Groves and another of his intelligence aides, Army Corps of Engineers Major Robert Furman, set in train a parallel mission through the Office of Strategic Services. Created by Roosevelt a few months before America's entry into the war, the OSS was intended to be America's equivalent to the British SIS. Under William 'Wild Bill' Donovan, a New York lawyer and war hero, the new intelligence organisation was taking its first faltering steps on its way to becoming the Central Intelligence Agency. But its reputation in these early years was poor. Many dismissed the organisation as a bunch of hopeless amateurs.

Groves was hedging his bets. He was now so desperate for information about the German programme that he was willing to back two independent missions to Italy to find out whatever they could. And, in typical Groves style, compartmentalisation meant that the two missions knew nothing of each other.

The OSS mission, codenamed AZUSA, was disguised beneath Project Larson, designed by the OSS to target and 'liberate' Italian rocket scientists. Its real target was the German atomic programme. In November 1943 Furman briefed one of the OSS operatives assigned to the mission – former Boston Red Sox catcher Morris (Moe) Berg.

It says something about the OSS recruitment policy in these first few years of its existence that a moderately famous major-league baseball player could find employment as a spy. However, Berg was no ordinary major-league baseball player. He had studied languages at Princeton and the Sorbonne, and was fluent in many. It was at Princeton that he

---

The plans, together with personal letters which helped to authenticate the identity of the body, were all fabrications produced by British naval intelligence.

had developed his talent for baseball. Despite his modest fame, he also possessed an unerring ability for disappearing into the background, to come and go with hardly anyone noticing.

Furman told him very little about the Allied bomb programme, but he was also pragmatic: 'We told people generally what to look for without telling them why', he later admitted. 'A guy like Berg could learn more than you wanted him to. He was their hot rod, one of their best.'

Berg figured it out. The newspapers were in any case full of reports about the threat of Nazi super-weapons, based on the principles of energy release from split atoms, which according to speculation could blow up half the globe. While waiting for his travel orders, Berg buried himself in physics textbooks. He studied atomic theory and quantum mechanics, taking particular interest in Heisenberg's famous uncertainty principle. When William Fowler, a nuclear astrophysicist at the California Institute of Technology, met Berg shortly after the war he concluded that Berg understood the uncertainty principle at least as well as he did himself. Fowler went on to win the 1983 Nobel prize for physics.

But Groves was still not done. There was yet another way that the Allies could prevent any further German progress. Bohr had confirmed what many Allied physicists had suspected: Heisenberg had taken a leading role in the German programme. Oppenheimer and Chadwick both acknowledged that Heisenberg was Germany's leading theoretical physicist and the single most powerful mind involved with the German programme. An idea that had been mooted among some of the Los Alamos physicists now resurfaced. Why not simply deny the German programme its most valuable intellectual resource? Why not prevent the Germans from making any further headway by kidnapping the programme's scientific leader?

Unlike many of the Manhattan Project physicists who had met or worked with Heisenberg, Groves was not saddled with a deep intellectual respect for a former colleague. For Groves, assassination of Heisenberg was also therefore a possibility to be considered. The German Nobel laureate was known to make occasional visits to neutral Switzerland to give lectures on academic subjects. Groves now pondered the merits of taking Heisenberg out of the equation by kidnapping or assassinating him. He confronted Lansdale with this proposal, which he hinted had come from the OSS.

To Lansdale, this was certainly no recommendation. He was appalled. He robustly rejected the suggestion, citing as reasons political fall-out with the Swiss government and the simple fact that kidnapping or killing Heisenberg would betray the existence of an Allied atomic weapons programme. He thought the proposal uncharacteristic of Groves, and when he heard nothing further of it he presumed the idea had been quietly dropped.

Groves simply passed it to Furman, who continued discussions with the OSS. In February 1944 Colonel Carl Eifler accepted the mission. Six feet tall and a muscular 280 pounds, Eifler had graduated from the Los Angeles Police Academy and had served in the US Border Patrol before being called to active service. His jeep had been strafed by Zeros during the Japanese attack on Pearl Harbor. He had commanded the OSS's Detachment 101, which fought a particularly dirty war against the Japanese in the jungles of Burma, before injuries forced a return to Washington.

In Washington he had been tasked with the assassination of Chiang Kai-Shek. Now plans were drawn up to kidnap Heisenberg in Germany, transport him to Switzerland, board an American military plane, parachute into the Mediterranean and rendezvous with a submarine which would bring Heisenberg to America. Even by the standards of the OSS, it was a crazy plan.

If anything went wrong (which, given the plan, was very highly likely), Eifler, a crack shot who would frequently demonstrate his prowess with a pistol after several whiskies, was instructed to 'deny the enemy his brain'. He left for London in late March.

However, a significant problem had now arisen. Heisenberg had disappeared. Neither American nor British intelligence knew where he was.

Heisenberg's 'ambition'

Abraham Esau, the head of the Reich Research Council's physics section and administrative head of the Uranverein, had made himself unpopular with many Uranverein scientists. By October 1943 he had also lost the support of Albert Speer. Towards the end of the year he was replaced by Walther Gerlach, professor of physics at Munich University. Göring had approved the change, which formally came into effect on 1 January 1944.

Gerlach was urged to accept the post by both Heisenberg and Hahn. Unlike Esau, Gerlach possessed a physicist's appreciation for the work in hand and was passionate about 'pure' research. No less committed than his colleagues to the German cause, he nevertheless saw his appointment as an opportunity to help salvage German physics from the 'Schwindel' of war priorities, starting with the Uranverein.

This, however, would be no easy task, and Gerlach was quickly over-whelmed. He scurried between Munich and Berlin, hurried between meetings with the Uranverein physicists and I.G. Farben, which was building a new heavy water facility next to its chemical plant at Leuna, near Merseberg in Saxony-Anhalt, eastern Germany. He failed to produce timely reports to the Reich Research Council, which grew impatient. As deadlines slipped, Gerlach would unload his troubles to his good friend Paul Rosbaud, with whom he would take lunch two or three times a week.

Despite having sought Heisenberg's advice about taking the role, Gerlach was wary of Heisenberg's 'ambition'. This perception may have stemmed from the politicking that had surrounded Heisenberg's appointment as director at the Kaiser Wilhelm Institute for Physics. Or it might have been born from a general inability to fathom Heisenberg's personal agenda.

Whatever it was, this agenda had taken Heisenberg on many trips to parts of occupied Europe at the invitation of the Reich Education Ministry, or prominent Nazis. As ever, the impression he left with his foreign colleagues was ambiguous at best, odious at worst. In Holland he had lectured Dutch physicist Hendrik Casimir on history and world politics, explaining that it was the historic mission of Germany to defend Western culture against the 'eastern hordes'.

Although many Germans were in denial over the existence of Hitler's 'final solution',[4] Heisenberg was at least familiar with aspects of it. In December 1943 he had accepted an invitation to visit his old school friend Hans Frank, now General Governor of Poland in Krakow, and to give a lecture at the cultural propaganda institute. Frank had overseen the ruthless annihilation of the Jewish ghettoes in Krakow and Warsaw.[5] Of course, familiarity with mass murder does not imply acceptance or complicity, but Heisenberg's willingness to embrace the demands of his position, and his insensitive, discomfiting political views often gave the impression that he was very much part of the Nazi establishment.

On 24 January 1944 Heisenberg found himself on the way back to Bohr's institute in Copenhagen, this time for a very different kind of mission. Having discovered that Bohr had escaped, on 6 December the German military police had occupied Bohr's institute and arrested the only two people who had been found inside it at the time, a physicist and a laboratory technician. Heisenberg and Diebner spent three days in Copenhagen trying to sort out what should happen next.

Heisenberg's proposals all carried conditions that he believed were likely to be acceptable to the occupying authorities. Either the Germans could take over the institute for war research, the institute could be stripped of its equipment, including the cyclotron, and returned to Danish control, or the institute could be returned provided no war research was conducted there. Weizsäcker had already advised Heisenberg that he had no wish to run the institute as a German laboratory on behalf of the Uranverein. It hardly mattered. All the options were roundly rejected by the Danish physicists.

---

[4] In the autobiography of her life with Heisenberg, his wife Elisabeth wrote of the reaction of her father to reports of mass executions of Polish Jews that Heisenberg had shown him: 'So this is what it has come to, you believe things like this! This is what you get from listening to foreign broadcasts all the time. Germans cannot do things like this. It is impossible!': Heisenberg, Elisabeth, p. 49.

[5] Frank was found guilty of war crimes at the Nuremberg trials and executed on 16 October 1946. In his testimony to his defence lawyer, Frank said: 'A thousand years will pass and still Germany's guilt will not have been erased.' See: http://www.law.umkc. edu/faculty/projects/ftrials/nuremberg/ franktest.html

Among them was Christian Møller, who no doubt recalled with bitterness Heisenberg's behaviour during his last visit in September 1941.

An impasse seemed likely. In the event, it was Ernst von Weizsäcker's Foreign Office which insisted that Bohr's institute be returned to the Danes without condition and the imprisoned physicists be released.

If the Danish physicists felt any gratitude towards Heisenberg for his role in getting the matter resolved, this must surely have evaporated a few months later, when Heisenberg returned to Copenhagen to lecture at the German Cultural Institute. During this last visit, Heisenberg dined publicly with SS-Oberstürmbannführer Werner Best, the German plenipotentiary in Denmark and a former deputy of Reinhard Heydrich. But for the intervention of the Danish people and the Swedish authorities, Best would have consigned 8,000 Danish Jews to their fate in concentration camps in Germany.

The Allied bombing of Berlin was now making it increasingly difficult to continue research. The Kaiser Wilhelm Institute for Chemistry received a direct hit in what Gerlach referred to as a 'catastrophic' air raid on 15 February 1944. Hahn's laboratories were destroyed, together with many of the papers he had collected in a lifetime's work at the institute. The Institute for Physics had only been superficially damaged, and Heisenberg joined other staff in an attempt to save books from Hahn's library.

The quiet Dahlem suburb of Berlin was targeted not because of its strategic military or industrial significance. Groves had asked for the area to be bombed specifically to kill the German scientists or at least drive them 'out of their comfortable quarters'. It worked. Plans to evacuate both the institutes of chemistry and physics were accelerated.

You are looking at the Alsos mission

By the time the Alsos mission had been approved, Allied forces had established bridgeheads in Calabria, the toe of Italy, and in Salerno. The force in Calabria met with little resistance. Salerno was a different matter, the Allies taking heavy casualties in five days of fierce fighting. Nevertheless, within a few weeks the American Fifth Army, commanded by Lieutenant

General Mark Clark, and the British Eighth Army under General Bernard Montgomery had advanced and secured control of much of the southern part of the country. Mussolini had already been deposed and imprisoned. The Italians had surrendered on 3 September.

The Germans, however, had not. In a daring raid, Mussolini was sprung from prison and re-installed as a 'puppet' dictator. The Italians were largely powerless as the Germans and the Allies now fought for possession of their country.

Although declared a 'scientific' mission, the real purpose of Alsos was atomic intelligence. The mission had a long list of potential scientific targets, but its principal objective in Italy was to interrogate Edoardo Amaldi and Gian Carlo Wick, leading members of the nuclear physics group at the University of Rome whose names had been supplied by Fermi. The Alsos team included a number of scientific advisers drawn from MIT, Cornell University and Bell Laboratories.

The team set off for Naples on 16 December 1943 and arrived after a long and tortuous journey. By this time the Allied advance north up the length of Italy's mainland had stalled. Field Marshal Albert Kesselring had made good use of Italy's mountains and rivers to form a strong defensive line across the width of the country. Behind this defensive line lay an even more formidable obstacle – Monte Cassino, an imposing sixteenth-century Benedictine monastery set atop a mountain above the junction of the Rapido and Liri valleys, about 80 miles south of Rome. Securing Highway 6, the road to Rome, would mean first securing Monte Cassino.

Rather than try to fight their way through Kesselring's defences, Allied commanders thought to circumvent them by landing forces at Anzio, a few miles south of Rome and 60 miles behind the enemy line. The British First and American Third Divisions landed at Anzio on 22 January 1944. Kesselring launched a counter-attack on 30 January, as soon as the Allies thought the bridgehead to be secure. A further counter-attack on 16 February nearly forced an Allied retreat. Far from helping the drive towards Rome, the force at Anzio was now effectively neutralised.

In frustration, Pash returned to Washington in February with little to show for his efforts. Berg hadn't budged far from Washington, still waiting for his travel orders.

Kesselring had advised the Allied forces that, because of the historical significance of the monastery on Monte Cassino, he had ordered his units not to occupy it. However, Allied reconnaissance reported German troops inside the monastery and it was believed that it would make too good an artillery observation post for the Germans not to use it. Clark's infantry had mounted a series of attacks through January and early February, and had sustained heavy losses. Lieutenant General Bernard Freyberg, commander of the New Zealand Second Division of the Eighth Army, which replaced the bloodied American units, judged that men were more important than monasteries, and requested that it be bombed. The monastery was pounded to rubble on 15 February, but the bombing left the cellars intact, and German troops now moved quickly to occupy what remained and fortify their positions.

Further attacks were repelled, again with heavy casualties. The final assault was launched on 11 May 1944. Units of Moroccan troops versed in the art of mountain warfare penetrated a weak point in the German defensive line; though the point was weak only because the Moroccans were able to scale the 'unscaleable' Petrella Peak, which had been left undefended. On 18 May a Polish reconnaissance group found the monastery abandoned. On 25 May Kesselring's fall-back defensive line was finally breached at Piedimonte San Germano. Two days earlier, the Allied forces trapped at Anzio had broken out and advanced from their beachhead.

Kesselring requested permission from Hitler to retreat from Rome, and Hitler agreed: Rome was a place of culture that had to be spared. Clark's Fifth Army entered the city on 4 June. The occupation of Rome had been dearly bought. Over 54,000 Allied troops and 20,000 German troops had perished in the battle for Monte Cassino.

Berg's travel orders finally came through and he departed from Washington on 4 May, heading for London. Towards the end of May he travelled to Algiers, and he was in Italy in early June. He checked into the Hotel Excelsior just two days after Rome had been liberated. Pash had got there a little before him, following closely behind Clark's jeep as it swept through the city early on the morning of 5 June.

Pash headed immediately for Amaldi's home on Via Parioli. Amaldi was friendly and co-operative, and agreed to Pash's request to remain in the city. He was no doubt surprised when, shortly after Pash's departure, Berg turned up on his doorstep asking that he prepare to leave immediately for America.

When an agitated Amaldi appeared at Pash's hotel to explain that he was now being asked to travel to Naples, on the orders of President Roosevelt, Pash smelled a rat. He had sat fuming through the winter months waiting on an OSS mission to spring Amaldi and Wick from Rome by submarine, a mission that not only never happened but, as he had discovered himself directly from Amaldi, never really existed at all. Amaldi explained that his visitor was waiting for him in the hotel lobby. Pash hastened to confront him, now full of rage at what he suspected was yet another example of OSS incompetence.

Berg remained slouched in an easy chair as Pash introduced himself.

'Colonel, looks like you and I are going to have to reach an understanding', Berg began.

Something in Pash snapped, as the Alsos and AZUSA missions collided. 'Attention!', he yelled. Berg scrambled to his feet, explaining that he needed to escort Amaldi to Naples because the Alsos mission was waiting for him.

Pash vented his frustration and hurled a torrent of abuse. 'Captain,' he said, 'you are looking at the Alsos mission. No doubt you're from OSS … You have no business in Rome. If I run across you again, I'll bring charges, and I can think of plenty. Now get out …'

Rather nonplussed but undaunted, Berg returned to Amaldi's home the next day. Pash did not understand the physics and had no choice but to wait for scientific members of the Alsos mission to catch up with him in Rome. But Berg understood enough of the physics and started his questioning immediately. He continued to talk to Amaldi, Wick and other Italian physicists as June wore on.

It turned out that the Italian physicists knew next to nothing about the German bomb programme. Although they had worked on nuclear fission before and during the early years of the war, they had not been approached by the German physicists or asked to get involved. Hahn had visited Rome in 1941 but had avoided discussion of nuclear fission research. Amaldi was sure that the Germans were working on fission but felt that nothing could come of this for ten years at least. He was convinced that Heisenberg was not working on a fission project, as Heisenberg was a theoretician, not an experimentalist.

Wick knew a little more about Heisenberg's activities. Heisenberg had taught him physics in Leipzig and Wick had corresponded with his former teacher throughout the war. Heisenberg had written to him of the damage caused by the Allied bombing of Berlin and Leipzig. When asked where Heisenberg was now, Wick was cagey. He would say only that Heisenberg had moved south, to a 'woody region' of Germany.

Haigerloch

Gerlach realised that the Uranverein would make no progress with reactor research in heavily bombed Berlin. Earlier in the year, Bagge had overseen the evacuation of about a third of the Kaiser Wilhelm Institute for Physics to laboratory facilities established in Hechingen, the ancestral home of the House of Hohenzollern set against the backdrop of the Black Forest in south-western Germany. By the end of July 1944 many of the Uranverein physicists had relocated there. Hahn had moved with the Institute for Chemistry to Tailfingen, a village about ten miles south of Hechingen.

Heisenberg's home in Leipzig had been destroyed by Allied bombing in December 1943. He had spent the first half of 1944 commuting between the laboratories in Berlin and Hechingen and the village of Urfeld, about 50 miles south of Munich in the Bavarian Alps, where he had moved his family to escape the worst of the bombing. He settled in Hechingen in the summer of 1944, leaving Wirtz in Berlin to oversee the reactor experiments.

Gerlach now scoured the neighbouring countryside for a new location suitable for reactor research, one in a narrow valley that could not be easily targeted by enemy planes. He found what he was looking for in Haigerloch, a picturesque medieval village about ten miles from Hechingen. The village is perched in a steep limestone valley formed by loops of the Eyach river, a tributary of the Neckar.

The Reich Research Council requisitioned an old wine cellar that had been hewn from solid rock at the foot of an imposing cliff next to the Swan Inn. It was in this cellar that equipment for the reactor experiments was installed. The Haigerloch laboratory was codenamed the 'Speleological Research Institute'.

The biggest problem remained lack of sufficient quantities of heavy water. Although four drums of heavy water had been recovered from the sinking *Hydro*, these were heavily contaminated. More drums from the wrecked plant had followed, bringing the total amount of pure heavy water available to the Uranverein to about two and a half tons. Heisenberg had used more than half this precious supply for reactor experiments in a specially constructed bunker laboratory in Berlin. However, the reactor had been built from uranium plates, 'for the sake of method', despite the demonstrated superiority of lattice configurations using cubes of uranium. These experiments were largely unsuccessful and had taken the Uranverein no further forward.

More heavy water was needed for further experiments at both the Berlin laboratory and the reactor now being built at Haigerloch. However, the I.G. Farben works at Leuna were destroyed by Allied bombing on 28 July 1944, and with them went the prospects for a new heavy water plant. The heavy water concentration cells at the Vemork plant were dismantled in preparation for the journey to Germany in August. Half the cells would go to Berlin and the remainder would come to Haigerloch.

Some valuable assets, some liabilities

Just two days after the liberation of Rome, the Allies launched the largest combined amphibious and airborne assault in history against Hitler's

Atlantic Wall. Operation Overlord was launched on 6 June 1944, with the objective of landing five divisions on the beaches of Normandy. It was a marvellously orchestrated attack. Elaborate deception played its part. Dummy planes and landing craft installed along the Kent coast – the only part of Britain that the Luftwaffe could now reconnoitre from the air – and German agents who had been 'turned' by MI5, all led Field Marshal Gerd von Rundstedt to expect an attack in the Pas de Calais. Had it not been for Field Marshal Erwin Rommel, sent by Hitler to take command of Army Group B in northern France in January 1944, the Normandy beaches would have been lightly defended.

But the defences that had been erected on Rommel's orders had been probed secretly by British commandos aboard midget submarines. Tanks had been ingeniously adapted to drive through water and detonate land mines ahead of the invading forces. Even as the Allied forces embarked, German radar was tricked into thinking they would strike elsewhere along the coast. Bad weather also played a part. The forecast for 4 and 5 June had led Rommel to believe that an invasion was not imminent, and he had left for Germany for a weekend with his family. What German forecasters did not anticipate was that the weather would be fine again a day later.

The south shore of the Bay of the Seine had been divided into five targets: Utah, Omaha, Gold, Juno and Sword. Only at Omaha beach did the attack very nearly fail. Landed at low tide to avoid defensive obstacles, the troops of General Omar Bradley's First Army found themselves exposed on Omaha's steep rise, trapped in a 'killing zone' in the face of heavy defences manned by the veteran 352nd Infantry Division which, unknown to the Allies, Rommel had brought forward to defend the beach some months before. For a time the headlong collision of hot metal, flesh and blood evoked nightmarish scenes from the First World War. Elaborate timetables for achieving objectives sank into chaos and confusion. Establishing a foothold on Normandy soil had always been about throwing sufficient numbers of Allied troops onto its beaches, but it was the heroism of individuals that turned near-defeat into victory that day.

As the Allied forces took their first tentative steps into northern France, attention turned once again to the German atomic programme. The Alsos mission to Italy had produced very little new intelligence, but was declared by all its senior sponsors as a success. A second mission was planned, following the front line troops into France. Pash was to continue as military leader, but this time there would be a scientific head of the mission, someone who knew the European physicists well and who could distinguish important scientific intelligence from irrelevancies or disinformation. After a brief search, Samuel Goudsmit was appointed, for reasons that he was unable to fathom. When a few months later he read the dossier on him that had been carelessly misfiled among his own papers, he learned that he had 'some valuable assets, some liabilities'.

His assets clearly outweighed his liabilities, whatever they were. Goudsmit had studied physics at the Universities of Amsterdam and Leiden, and had carried out research with Friedrich Paschen in Tübingen in Germany and with Bohr in Copenhagen. Together with fellow Dutchman George Uhlenbeck he had in 1925 discovered the phenomenon of electron self-rotation, what later came to be called electron *spin*.[6] Both Goudsmit and Uhlenbeck had moved to the University of Michigan in 1927.

Goudsmit knew all the leading European physicists and their specialities, and spoke their languages, literally and figuratively. His aptitude for scientific puzzle-solving was well suited to the kind of detection that would be required for the Alsos mission. As a student in Amsterdam he had taken a course in police detective work.

He also had personal reasons for accepting a leading role in the second Alsos mission. His parents had been arrested in Nazi-occupied Holland in late 1942 or early 1943, and aside from a farewell letter from the Theresienstadt concentration camp he had heard nothing from them since. Goudsmit had appealed to Heisenberg for help, but as far as he knew nothing had come of it. He left for Britain in June 1944.

Pash insisted on some ground rules between Alsos II and the OSS in order to avoid a repeat of the embarrassing encounter in Rome. It was

---

[6] Originally thought to be visualisable literally as a bit of negatively-charged matter spinning on its axis as it orbits the positively-charged nucleus, it was subsequently accepted that electron spin is an entirely relativistic quantum-mechanical phenomenon with no counterpart in classical physics.

agreed that the OSS would continue its own atomic intelligence-gathering in neutral countries only. Goudsmit subsequently received copies of OSS reports on the German programme. These were largely circumstantial, attributing stories of explosions and fires to uranium research. The reports even included the rumour that a uranium bomb had exploded in Leipzig, killing several scientists.

On 25 August, the President of the provisional government of the French Republic, Charles de Gaulle, re-entered Paris in triumph. Pash was in among the French army units that entered Paris from the south, and was therefore among the first American army personnel in the newly-liberated city.[7] His target was the Collège de France, and Frédéric Joliot-Curie. Pash managed to dodge the sniper fire, found Joliot-Curie in his office and, very gently, took him prisoner. Together they celebrated the liberation of Paris with champagne, drunk from laboratory beakers. Goudsmit joined them on 28 August.

Despite the fact that Joliot-Curie had met a succession of Uranverein physicists passing through his laboratory, there was little he could add to the body of intelligence on the German atomic programme. He explained that he had received assurances from Schumann in 1940 that the Paris cyclotron would not be used for war research. In telling of his encounters, one name was new to the Alsos team: Kurt Diebner. If anyone was organising atomic weapons research in Germany, Joliot-Curie said, it would be him. Diebner's name was placed close to the top of the Alsos watch list.

Further evidence of German atomic research remained stubbornly elusive until the liberation of Strasbourg towards the end of November 1944. Weizsäcker was known to have a nuclear physics laboratory in the city and an office at the university. But he was not among the physicists who had been found at the laboratory, pretending to be doctors (the laboratory was in a separate building on the grounds of Strasbourg hospital). For the first time in the mission, Goudsmit found himself interrogating enemy physicists. It was a grim business. 'Thank God I didn't know them personally,'

---

[7] Roosevelt had agreed that de Gaulle's Free French would enter the city first. The American forces had halted just outside the city limits.

1. When Austrian physicist Lise Meitner (pictured right) was forced to flee Nazi Germany, her long-term collaborator, German chemist Otto Hahn, continued to write to her from Berlin about his experiments on uranium. The results described in Hahn's letters were 'startling'. (AIP Emilio Segrè Visual Archives, Brittle Books Collection)

2. Austrian physicist Otto Frisch joined his aunt, Lise Meitner, in the Swedish seaside village of Kungälv for Christmas 1938. It was to prove the most momentous visit of his whole life. When he returned to Copenhagen, he and Meitner had discovered nuclear fission. (AIP Emilio Segrè Visual Archives, Segrè Collection)

3. Both Danish Nobel laureate Niels Bohr (pictured right) and German Nobel laureate Werner Heisenberg visited America in 1939. Working with American physicist John Wheeler at Princeton, Bohr identified the role of U-235 in nuclear fission. Heisenberg faced intense interrogation from his colleagues about his reasons for returning to do physics in Nazi Germany. (Photograph by Paul Ehrenfest, Jr, courtesy AIP Emilio Segrè Visual Archives, Weisskopf Collection)

4. Persuaded by the 'Hungarian conspiracy' – Leo Szilard, Edward Teller and Eugene Wigner – Albert Einstein agreed to sign a letter to US President Franklin Roosevelt. The letter warned of 'extremely powerful bombs of a new type'. In this picture, taken in the late 1940s, Einstein and Szilard recreate the moment. (Time & Life Pictures/ Getty Images)

5. The faculty of the Berkeley Radiation Laboratory, gathered beneath the magnet of the unfinished 60-inch cyclotron. J. Robert Oppenheimer is in the back row, fifth from the left. Robert Serber and Ernest Lawrence are in the front row, second and fifth from the left, respectively. Edwin McMillan is in the second row, fourth from the left. Philip Abelson is on the extreme right of the front row. McMillan and Abelson published early work on neptunium, work that was eventually to lead to the discovery of plutonium. (Ernest Orlando Lawrence, Berkeley National Laboratory, courtesy AIP Emilio Segrè Visual Archives, Fermi Film Collection)

6 (above). Working with Frisch in Birmingham in early 1940, German émigré physicist Rudolf Peierls, pictured here with his Russian wife Genia, realised that an atomic bomb was feasible. The Frisch–Peierls memorandum summarised their reasoning and led to the formation of the MAUD Committee. (AIP Emilio Segrè Visual Archives, photograph by Francis Simon)

7 (left). James Chadwick won the Nobel prize in 1935 for his discovery of the neutron. Together with Polish physicist Joseph Rotblat, he had independently concluded that a uranium bomb might be feasible. He strongly supported the Frisch–Peierls memorandum. (Photograph by Bortzells Esselte, Nobel Foundation, courtesy AIP Emilio Segrè Visual Archives, Weber and Fermi Film Collections)

8. The Frisch–Peierls memorandum encouraged American physicists to greater efforts and, with the publication of the report of the third National Academy review group, the S-1 project was established. Pictured here, from left to right, are: Ernest Lawrence, Arthur Compton, Vannevar Bush, James Bryant Conant, Karl Compton and Alfred Loomis. (Ernest Orlando Lawrence, Berkeley National Laboratory, courtesy AIP Emilio Segrè Visual Archives)

9. American physicist J. Robert Oppenheimer was asked to lead work on the physics of fast-neutron chain reactions in May 1942. He was subsequently to become the scientific director of the project's weapons laboratory at Los Alamos, and the 'father of the atom bomb'. (Los Alamos Scientific Laboratory, courtesy AIP Emilio Segrè Visual Archives)

10. Edward Teller (pictured left) and Enrico Fermi joined the Manhattan Project's Metallurgical Laboratory in Chicago. Fermi worked to build the world's first nuclear reactor. Initially unsure what to do, Teller worked with Emil Konopinski on the physics of a thermonuclear bomb. (AIP Emilio Segrè Visual Archives)

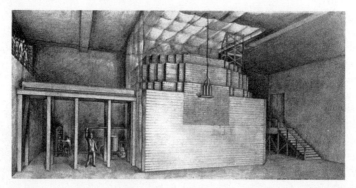

11. The world's first nuclear reactor, a uranium–graphite 'pile', was successfully tested in a squash court at the University of Chicago on 2 December 1942. There are no photographs of the completed reactor – this is an artist's impression. (Archival Photofiles (apf2-00503), Special Collections Research Center, University of Chicago Library)

12. West Point graduate Leslie Groves was 'probably the angriest officer in the United States Army' when he was appointed to lead the Manhattan Project in September 1942. Groves moved quickly to establish the facilities necessary for the production of the first atomic bombs. (Los Alamos National Laboratory)

13. Norwegian chemist Leif Tronstad escaped from Nazi-occupied Norway in September 1941. He became head of Section IV of the Norwegian High Command, based in Britain. Through his colleague Jomar Brun, he encouraged sabotage of the heavy water plant at Vemork, which he had helped to design, and worked with the SOE in the planning of sabotage raids. (Hydro)

14. Explosive charges laid by Norwegian commandos during Operation Gunnerside wrecked the heavy water concentration cells. Nikolaus von Falkenhorst, commander-in-chief of the German forces in Norway, declared the operation 'the finest coup I have seen in this war'. However, the plant was operational again within a few months. (Hydro)

15. The heavy water plant was part of the larger Norsk Hydro Vemork facility designed to produce fertilisers, perched high in the fjords near the Norwegian town of Rjukan. The main building can be seen here on the right. The plant can be reached only via a narrow suspension bridge, visible to the left of the picture, which spans a deep ravine. (Hydro)

16. Inaugurated in April 1943, the Los Alamos laboratory was an isolated collection of shoddy army buildings with poor facilities, inadequate housing and intermittent electricity. The scientists and their wives fought the oppressive, concentration-camp atmosphere with humour and impressive quantities of alcohol. (Los Alamos National Laboratory)

17. German émigré physicist Klaus Fuchs joined in the work of the MAUD Committee in May 1941. He became a Soviet spy towards the end of that year. Fuchs was part of the British Tube Alloys delegation and joined the Manhattan Project in December 1943 to work on gaseous diffusion. He relocated to Los Alamos in August 1944. This picture is the photograph from Fuchs' Los Alamos staff badge. (Los Alamos National Laboratory)

18. American physicist Theodore Hall was just nineteen years old when he was recruited by Los Alamos talent-spotters in January 1944. He became a Soviet spy in October of that year, passing secrets initially with the help of his student friend Saville Sax. (Los Alamos National Laboratory)

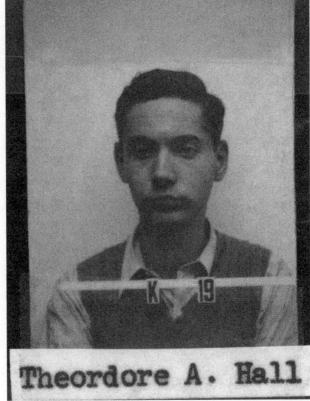

Theordore A. Hall

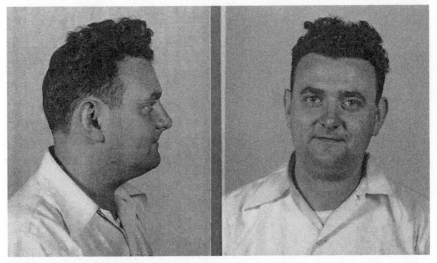

19. David Greenglass worked as a machinist in X Division, part of a Special Engineering Detachment assigned to Los Alamos in August 1944. He was recruited as a Soviet spy by his brother-in-law, Julius Rosenberg, in November. (US National Archives and Record Administration)

20. Soviet physicist Igor Kurchatov was appointed by the State Defence Committee to lead its atomic programme in February 1943. Following the German invasion of the Soviet Union in June 1941, Kurchatov had vowed not to shave his beard until the enemy was defeated. He was the principal scientific recipient of the espionage materials supplied by Fuchs, Hall and Greenglass. (VNIIEF Museum, courtesy AIP Emilio Segrè Visual Archives)

21. The second Alsos mission followed Allied forces into Germany with the aim of tracking down German atomic physicists and materials. This 'council of war' in Hechingen features SIS operative Eric Welsh (back row, second from left), Charles Hambro (sitting, second from left), John Lansdale, head of security for the Manhattan Project (sitting, fourth from right), Michael Perrin, deputy director of Tube Alloys (sitting, third from right), and Boris Pash (sitting, extreme right). (Courtesy of Brookhaven National Laboratory)

22 *(right)*. The Alsos mission captured leading German physicists involved in atomic research, together with their experimental apparatus such as this spherical reactor. (AIP Emilio Segrè Visual Archives, Goudsmit Collection)

23 *(below)*. The last German experimental nuclear reactor, B-VIII, was reassembled in a cave laboratory in Haigerloch. The experiment failed. The reactor was discovered by the Alsos mission in April 1945. Michael Perrin realised that the reactor was too small to go critical. (AIP Emilio Segrè Visual Archives, Goudsmit Collection)

24. The first 'Fat Man' plutonium bomb was assembled for the Trinity test atop a 110-foot tower. The bomb consisted of about five kilos of plutonium, formed into a sphere about the size of a small orange. The plutonium core was surrounded by a uranium tamper and 32 explosive lenses. (Los Alamos National Laboratory)

25. The Fat Man plutonium bomb exploded with a yield equivalent to 18,600 tons of TNT. Oppenheimer, wearing his trademark porkpie hat, and Groves inspect the twisted remains of the tower structure following the Trinity test explosion. (Digital Photo Archive, Department of Energy (DOE), courtesy AIP Emilio Segrè Visual Archives)

26. The atomic bombing of Hiroshima and Nagasaki was just one more catastrophe on top of a long list of utter catastrophes, one that brought the war to an end with such a powerful exclamation that it burned into the consciousness of all who lived then, and all who have lived since. This picture shows the mushroom cloud developing above the ruins of Nagasaki. (US National Archives)

27 *(left)*. British physicist and Soviet spy Alan Nunn May was exposed by the defection of cipher clerk Igor Gouzenko in Ottawa in September 1945. May was arrested on 4 March 1946. He was tried at the Old Bailey on 1 May and sentenced to ten years in prison. (Getty Images)

28 *(below)*. A new generation of Soviet physicists, 'possessed by a true war psychology', embraced the challenge of creating weapons capable of ever-increasing devastation. Andrei Sakharov (pictured here on the left) regarded himself as a soldier in a new scientific war. (AIP Emilio Segrè Visual Archives, Physics Today Collection)

he wrote to his wife, 'and I kept my own identity hidden until the very end, when I had them put on a truck and taken to a camp.'

The captured physicists were unco-operative, but a more detailed picture of the German bomb programme emerged from files taken from Weizsäcker's office which Goudsmit read eagerly, by candlelight. Among the papers was a letter that Weizsäcker had written to Heisenberg, dated 15 August 1944, criticising some of the latter's calculations. The letter had been torn up and never sent (a more moderate version of the same letter was subsequently recovered from Heisenberg's files). To Goudsmit's eyes, the subject matter was astonishingly basic at so late a stage in the German programme:

We found references to 'special metal', which was obviously uranium, memos about the difficulty of obtaining the 'special metal' in slabs instead of in powdered form, letters confirming our information that it was the Auer Company that produced the metal for the German experiments. We learned that 'large scale' experiments were being performed at an Army proving ground near Berlin. We found parts of computations which clearly applied to the theory of a uranium pile.

A pile, not a bomb. 'We've got it!' Goudsmit declared.

'I know we have it, but do they?' Pash misunderstood.

'No, no!' Goudsmit said. 'That's it. They don't!'

These documents, Goudsmit believed, accurately reflected the state of the German programme towards the end of 1944. Hitler had been advised of the possibility of a 'super-weapon' more than two years previously, but it seemed that experiments carried out as recently as August had not advanced much beyond a preliminary phase. The general lack of secrecy supported this conclusion – the documents quite clearly revealed the whereabouts of the Uranverein scientists, in Tailfingen, Hechingen and Haigerloch.

The documents recovered from Strasbourg were studied by scientific advisers to the Manhattan Project and the OSRD. Groves wondered if the information had come rather too easily. Reconnaissance photographs of the Hechingen area had revealed what looked like an Oak Ridge-sized isotope separation plant.[8] But, as far as the Alsos team was concerned, the evidence was indisputable.

'Isn't it wonderful that the Germans have no atom bomb?' Goudsmit enquired of Furman.

Furman's answer surprised him. 'Of course you understand Sam,' Furman said, 'if we have such a weapon, we are going to use it.'

The objectives of the Alsos mission now shifted to accommodate what had been discovered. As Soviet forces advanced on eastern Germany, the Alsos mission became a race to capture the German physicists and whatever materials could be recovered to keep them out of Soviet hands.

Assassination plan

Donovan met Eifler at OSS headquarters in Algiers in June 1944. Eifler was preparing to proceed with the mission to kidnap Heisenberg, but Donovan now advised him that the mission had been scrubbed. Donovan broke the news gently, but the truth was that Eifler had lost the confidence of Groves, Furman and Donovan himself. There was no way Eifler's crazy plan could be executed without leaving behind a considerable mess.

Eifler was off the mission, but despite what Donovan had told him, the mission was still very firmly on.

In November, Berg was briefed about Paul Scherrer, director of the Physics Institute at the Eidgenossische Technische Hochschule (the ETH) in Zurich. Scherrer was well connected with the European physics community and had a reputation for organising well-attended lectures by visiting scientists, including German nuclear physicists. He also gathered intelligence for the SIS and the OSS. The OSS bureau chief in Bern, Allen Dulles, valued Scherrer highly as a source of intelligence.

Word reached the OSS office in Bern that Heisenberg was to give a lecture at the ETH sometime around 15 December. Berg headed for Paris, arriving on 10 December. It is not clear precisely who briefed Berg about

---

[8] It was actually a plant designed to squeeze desperately-needed oil from a seam of shale.

his next mission. Furman was in Washington, but Goudsmit was in Paris and spent several days with Berg. Goudsmit was aware of Heisenberg's scheduled lecture, and gave Berg a small container of heavy water as a gift for Scherrer. No official documentation of a combined Alsos-AZUSA mission exists, but Berg wrote notes during his briefing. 'Gun in my pocket', he wrote. 'Nothing spelled out, but – Heisenberg must be rendered *hors de combat*.'

Some years later Berg related sketchy details of the mission to an OSS colleague, Earl Brodie. 'If anything Heisenberg said convinced him the Germans were close to a bomb,' Brodie later explained, 'then his job was to shoot him – right there in the auditorium. It would probably have cost Berg his life – there would have been no way to escape.'

Berg arrived at the ETH in Zurich on 18 December 1944 together with another OSS agent, Leo Martinuzzi. He used several cover stories. In one he was a Swiss physics student (barely plausible, as this was his 42nd birthday). In another he was an Arab businessman. In a third he was a French merchant from Dijon. He carried in his pocket a small pistol and a deadly cyanide pill. The cyanide pill was for his own suicide, in the event that he was compromised, or as his only means of 'escape' after he had killed Heisenberg.

Berg passed himself off convincingly. He followed Heisenberg's lecture intently, understanding little of the S-matrix theory that was being discussed. He used the time to make copious notes, including a detailed description of the man he might yet have to kill. He spotted Weizsäcker in the audience, and on a seating plan of the auditorium that he had sketched he wrote 'Nazi' next to Weizsäcker's name.

He might not have understood the physics, but he understood enough to realise that there appeared to be no immediate threat. This left him in an uncertain position. He wrote: 'As I listen, I am uncertain – see: Heisenberg's uncertainty principle – what to do to H.'

The lecture ended without incident. Berg introduced himself to Scherrer and presented him with the gift from Goudsmit. Scherrer was aware that Berg was an OSS agent but was unaware of his mission, though he knew of the Allies' interest in Heisenberg. Scherrer had been ambiguous about his

German colleague, but having spent several days in his company prior to the lecture he had come to the conclusion that Heisenberg was anti-Nazi. Heisenberg was deeply distressed that, in the fall-out from the failed plot to assassinate Hitler on 20 July, Max Planck's son Erwin had been sentenced to death.[9] Scherrer now passed these conclusions on to Berg, who wondered if Heisenberg should be invited to come to America. Scherrer thought this a good idea, and in turn invited Berg to join the group at his home for dinner later that week.

But even as Scherrer painted his sympathetic portrait for Berg, Heisenberg was once again busy betraying a distinct lack of tact and diplomacy. Over dinner with his Swiss colleagues after the lecture, he eagerly consumed news reports of Rundstedt's offensive against Bastogne in Belgium, in what would come to be known as the Battle of the Bulge, declaring triumphantly that 'they're coming on!'

Heisenberg accepted Scherrer's invitation to dinner on the understanding that he would not discuss politics. But his fellow guests had no such understanding, and he was soon bombarded on all sides by questions and challenges. When asked to admit that Germany had all but lost the war, Heisenberg replied in familiar style: 'Yes, but it would have been so good if we had won.'

Berg overheard the remark. For him, this was the final confirmation he needed that the Germans were nowhere near to developing an atomic 'super-weapon'. If they were, why would the programme's leading physicist openly declare that the war was lost?

Berg arranged to leave dinner at Scherrer's home at the same time as Heisenberg, and together they made their way through the city's dimly-lit streets. This would have been a perfect opportunity for murder. Instead, as they walked, Berg continued to question Heisenberg about his views of the German regime. Berg's Swiss-accented German aroused no suspicions.

Eventually, they parted company. Berg's pistol had remained in his pocket. Heisenberg had no idea he had come so close to death.

---

[9] Erwin Planck was hanged at Plötzensee Prison in Berlin on 23 January 1945.

Chapter 14

THE FINAL PUSH

January–June 1945

Frisch had settled into accommodation in one of the original Los Alamos school buildings, a large blockhouse constructed from huge tree trunks known as the 'Big House'. He had been greeted cordially on arrival by Oppenheimer, wearing jeans, open-necked shirt with sleeves rolled up, and his trademark pork-pie hat. Oppenheimer tended to greet all the members of the newly-arrived British delegation the same way: 'Welcome to Los Alamos, and who the devil are you?' Frisch was amazed at the scientific talent gathered on the Hill. He had the impression that if he struck out on any evening in any direction and knocked on the first door he came to, he would find interesting people inside.

Working in the Critical Assemblies Group in G Division, Frisch busied himself with various small projects mostly involving the development of experimental instruments. However, he also continued to nag at the problem of critical mass in uranium. Despite the progress that had been made, the physicists were still largely ignorant of the precise experimental conditions under which uranium enriched with U-235 or pure U-235 would form an explosive super-critical mass. Frisch was surrounded by theoretical physicists of the first order. Some, like Ulam, had confessed to Frisch that they had now sunk so low as to include actual numbers in their calculations instead of just abstract mathematical symbols. But still, to a seasoned experimentalist like Frisch, theory was one thing, practice quite another.

Enriched uranium began to arrive at Los Alamos from Oak Ridge in early 1945, largely thanks to a revelation that Oppenheimer had experienced about eight months previously. The scientists had backed each method of isotope separation like horses in a race, without realising that connecting the methods in series – using the output from one method as the feedstock for another – might offer a more efficient route to enriched uranium and weapons-grade U-235.

As early as January 1943, Philip Abelson had proposed to use a liquid thermal diffusion technique to enrich uranium for reactor research at the Met Lab. Abelson was working for the US Navy, and compartmentalisation meant that Oppenheimer did not learn of this work until over a year later. In April 1944 he had realised that the horse-race analogy was 'a terrible scientific blunder'. He now saw that using even slightly enriched uranium as a feedstock for the calutrons at Oak Ridge would greatly increase their efficiency. Abelson's thermal diffusion plant could provide a temporary alternative to the gaseous diffusion plant K-25, still delayed by problems with the porous barrier materials. Groves authorised the construction of a thermal diffusion plant at Oak Ridge in June 1944. He gave the contractors a 90-day deadline.

In fact, satisfactory barrier materials for K-25 were delivered to Oak Ridge and the first charge of uranium hexafluoride passed through the plant on 20 January 1945. The thermal diffusion plant, S-50, was enriching uranium by March.

With regular supplies of enriched uranium now arriving on the Hill, Frisch devised an ingenious way to test their theories and, at the same time, determine precisely how much fissile material would be required to make a bomb. He submitted a proposal for a series of experiments to the Co-ordinating Council, which oversaw the various projects at Los Alamos. Much to his surprise, his proposal was approved.

The group had plenty of experience with assemblies formed from stacked blocks of uranium hydride, and had advanced towards a critical assembly by reducing the hydrogen content as the proportion of U-235 was increased. Such a 'naked' assembly, which Frisch called a 'Lady Godiva', was quite dangerous. Frisch himself almost suffered what would have been a fatal dose of radiation when he leaned too closely to one such sub-critical assembly. His own body reflected back some of the neutrons that would otherwise have escaped harmlessly, and these reflected neutrons caused the assembly to go critical. He noticed that the small red lamps monitoring the neutron intensity had stopped flickering. Instead, they were glowing continuously, the neutron counters by now overwhelmed. He hastily shut the experiment down.

The challenge was to figure out how to work with critical and super-critical assemblies in relative safety. Frisch's idea was to assemble blocks of enriched uranium hydride into a near-critical configuration, but with a hole running through the centre. Through this hole would then be dropped another block of enriched uranium hydride – called the core – sufficient to make the whole assembly go critical just for an instant as the core passed through the hole and dropped out the bottom.

Feynman, sitting in judgement on the Co-ordinating Council, found the experiment intuitively appealing. He said it was 'like tickling the tail of a sleeping dragon'. The experiment was henceforth known as the Dragon experiment. 'It was as near as we could possibly go towards starting an atomic explosion without actually being blown up.' There were obvious dangers. If the core were to get stuck on its way through, the assembly would go critical and irradiate the physicists with potentially lethal doses of radiation. Frisch was confident that the experiments could be done safely, however, and stipulated that nobody should ever work on the assembly alone.

Now working as a group leader, Frisch built the first of a series of such assemblies at a small laboratory in Omega Canyon, some distance from the main Los Alamos laboratory facilities, during the winter months of early 1945. He worked around the clock to make the first precise measurements of critical mass in U-235. The experiments were very successful. In the split second during which the core passed through the assembly, a large burst of neutrons was produced and the temperature of the apparatus increased by several degrees. The largest measured energy production was twenty million watts, for just three thousandths of a second, raising the temperature of the assembly by six degrees. This was the first time a super-critical mass of enriched uranium had been studied in the laboratory. By early April 1945, sufficient U-235 was available to form assemblies from blocks of pure metal.

Following the script

The experimental X-10 reactor at Oak Ridge represented an intermediate scale between the first Chicago pile and the large reactors constructed at Hanford. The Chicago pile had operated at a very modest output – barely one watt. X-10 had achieved one million watts. The three Hanford reactors, designated B, D and F, were built and operated by Du Pont and designed to run at 250 million watts. Each reactor consisted of a cylinder of graphite measuring 28 feet by 36 feet, weighing about 1,200 tons, containing 2,004 equally spaced aluminium tubes drilled along its length. Uranium slugs the size of rolls of quarters were sealed in aluminium cans and inserted into the tubes. Cooling water was pumped through the tubes and around the uranium slugs at the rate of 75,000 gallons per minute. The reactors had one purpose only: to produce plutonium. No attempt was made to capture the heat energy released by the reactor and convert it to electricity.

Fermi oversaw the loading of the uranium fuel slugs into the Hanford B reactor on 15 September 1944 and brought the reactor to 'dry' criticality, low-power operation close to the threshold of criticality which did not require cooling water. The physicists at Hanford then added more fuel slugs and carried out more experiments. Everything was in order. The reactor behaved precisely as it should. It was pushed towards full power on 26 September.

John Wheeler had joined Wigner's group at the Met Lab in early 1942 and moved to the Du Pont offices in Wilmington, Delaware in March 1943. A year later he relocated once again to Hanford. Due some late nights 'babysitting' the reactor later in the week, he had decided to go home and get some sleep on the night of the 26th. But when he got to his office the next morning he found that not all had gone to plan: 'the reactor was not exactly following the script.'

The reactor had started up as expected and had reached a record output of nine million watts. Then it began to lose reactivity, and as the output declined the operators had tried to maintain it by withdrawing the cadmium control rods. 'It was as if the engine of your car got sick as you were driving along a level road,' Wheeler later wrote, 'and you had to push farther and farther down on the accelerator pedal to maintain speed; eventually the pedal would be all the way to the floor and the car would start to slow down.' By mid-afternoon on 27 September, the control rods had been pulled virtually all the way out to maintain the reactor output. By early evening the control rods were all the way out, yet still the reactor shut down.

Fermi wondered if water had leaked into the reactor, but Wheeler suspected something else. Only a few weeks after joining the Met Lab he had written a report on the possibility of 'self-poisoning' of the reactor by its own reaction products.

When in 1938 Hahn and Strassman found barium among the products resulting from the bombardment of uranium with neutrons, they had discovered the stable end result of a long and complex series of nuclear reactions. When U-235 absorbs a neutron, the unstable U-236 nucleus fissions. One of the possible fission reactions produces zirconium, Zr-98, and tellurium, Te-135, and three neutrons. The zirconium isotope is radioactive, and goes on to produce niobium and then molybdenum. Likewise, the radioactive tellurium isotope decays first into iodine, then xenon, then caesium, and finally barium.

If just one of these decay products has a high affinity for neutrons, Wheeler had reasoned, then it would tend to inhibit the chain reaction, soaking up free neutrons until there were insufficient numbers to keep the chain reaction going. As more and more of the 'poison' was produced, it would become more and more difficult to maintain the reactor output. Eventually, the poison would overwhelm the reaction and the reactor would shut down. Wheeler had made some further estimates in April 1942 and concluded that self-poisoning would be a significant problem only if one of the intermediate reactor products had an appetite for slow, 'thermal' neutrons about 150 times that of U-235 itself.

When the B reactor was examined, no water contamination could be found. Self-poisoning became the most obvious conclusion. In the early hours of 28 September the reactor recovered, building again to nine million watts by the afternoon before declining again. This further suggested that the poison was itself radioactive, with a half-life of about eleven hours, roughly the time it took the reactor to come back to life. Wheeler checked a table of measured half-lives and identified the culprit. It was the isotope of xenon, Xe-135, later found to have an appetite for neutrons more than 4,000 times greater than that of U-235.

Fixing the problem was relatively straightforward once it was understood. Of course, nothing could be done about the physics of the nuclear reactions. The reactor would always poison itself with Xe-135. The solution was to play a numbers game, adding more uranium fuel to the reactor to ensure that more neutrons would be produced than could be absorbed by Xe-135 at its steady-state concentration. Fortunately, the reactor design had allowed for such contingencies: extra tubes had been drilled at substantial cost and delay to the reactor construction. Wheeler's prudence now paid off. The extra uranium fuel that was required could be added without a major redesign and rebuild.

The Hanford D reactor went critical on 17 December 1944, and the repaired B reactor went critical once more eleven days later. The F reactor went critical in February 1945. On 4 February the reactors hit the designed operating output of 250 million watts. Plutonium production was now in full swing, with a theoretical yield of 21 kilos of plutonium per month. A jubilant Groves estimated that they would have enough plutonium for eighteen atomic bombs in the second half of 1945.

Solid core compression

But the plutonium would serve no purpose unless a way could be found to detonate it. Kistiakowsky's X Division had laboured hard through the winter of 1944–45. The woods surrounding Los Alamos reverberated to an endless series of explosions that increased in intensity as the scientists scaled up their experiments. The group consumed about a ton of high-performance explosives each day, cast into moulds to produce shaped charges weighing about 50 pounds each and machined with elaborate precision.

To support the push for implosion, G Division scientists had developed a series of diagnostic tests that could be used to tell just how symmetrical the implosive shockwave was. In addition to the Ra-La experiments that Hall was involved with, the scientists were also using various types of X-ray photography, high-speed photography and magnetic field measurements. Von Neumann had designed an arrangement of explosive lenses consisting of a fast-burning outer layer and a slow-burning inner component which worked together like a magnifying glass, shaping and directing the shockwave towards the bomb core. Each lens transformed the initial explosive burst from a spherical shockwave expanding outwards to a spherical shockwave converging to a central point. A second layer of fast-burning explosive accelerated and strengthened the implosive wave.

On 7 February a Ra-La test showed encouraging improvement, although spherical compression of a solid core could not yet be demonstrated. It was progress, but lens development was still behind schedule. At a meeting on 28 February involving, among others, Oppenheimer, Groves, Conant, Bethe and Kistiakowsky, the chemical composition for the explosive lenses was finalised and the overall design approach for a plutonium bomb agreed. On 1 March Oppenheimer established the 'Cowpuncher' committee, headed by physicist Samuel Allison – recently freed from his responsibilities at the Met Lab – and including Bacher and Kistiakowsky. The committee's job was to 'ride herd' on the final stages of plutonium bomb development. A few days later Oppenheimer called a halt to any further refinement of the explosive lens design.

Kistiakowsky was wary of Allison, who he suspected had been tasked by Oppenheimer to watch over his shoulder. As pressure mounted, nerves began to fray. Although Oak Ridge was now reliably producing weapons-grade U-235, in the likely timeframe of the war, the production rate was sufficient only for a single bomb. Hanford was now reliably producing plutonium sufficient for several bombs. But a simple gun-type detonation scheme could not be used successfully with reactor-bred plutonium. Everything now hinged on the development of implosion. Kistiakowsky, a chemist in an elite community of physicists, was now confronted with battle lines drawn up based on scientific discipline. 'On one occasion I was forced to say to Oppie in this top level council where I was the only chemist, "You're all ganging up on me because I'm not a physicist." To which Oppie replied smilingly, "George, you're an outstanding third rate physicist."'

Although their confidence had grown as the work had progressed, the many uncertainties of plutonium bomb design were still painfully apparent. No matter how valuable the quantities of plutonium that were now beginning to arrive from Hanford, the Los Alamos scientists could not be sure that the Fat Man design would work without a full-scale test.

Plans for such a test had been laid the previous year. A site had been identified at one corner of the US Air Force's Alamogordo bombing range in the New Mexico desert. It measured 24 miles long and eighteen miles wide. Inspired by a John Donne sonnet – 'Batter my heart, three person'd God ... Your force, to break, blow, burn, and make me new' – Oppenheimer codenamed the site Trinity. Oppenheimer appointed Harvard physicist Kenneth Bainbridge to plan and direct the test.

By mid-March, experimental evidence had been obtained for solid core compression in an implosive shockwave so smoothly symmetrical that the results agreed closely with theoretical predictions. The news must have been accompanied by considerable sighs of relief all round. On 11 April, Oppenheimer wrote to Groves to tell him the good news. The production rate of U-235 at Oak Ridge had suggested that a uranium bomb would be ready by 1 July. Oppenheimer now advised Groves that a plutonium bomb would be available by 1 August.

Death of a president

The next day, 12 April, Roosevelt died in Warm Springs, Georgia. He had been sitting for a portrait when he suffered a massive cerebral haemorrhage.

The Los Alamos community was shocked. Many mourned the departure of a much-loved national leader, a father-figure who had presided over America for thirteen years. Some wondered if the Manhattan Project would continue. At a memorial service convened on the Hill the following Sunday, 15 April, Oppenheimer delivered a eulogy. He quoted from the Hindu scripture, the Bhagavad-Gita:

> 'Man is a creature whose substance is faith. What his faith is, he is.' The faith of Roosevelt is one that is shared by millions of men and women in every country of the world. For this reason it is possible to maintain the hope, for this reason it is right that we should dedicate ourselves to the hope, that his good works will not have ended with his death.

Roosevelt's successor, Harry S. Truman, had been Vice President for only a few months. He asked of Roosevelt's widow, Eleanor: 'Is there anything I can do for you?' She replied: 'Is there anything we can do for you? For you are the one in trouble now.'

As a Democrat senator, Truman had gained some fame and considerable respect with a campaign to reduce wastefulness in military expenditure, through the US Senate Special Committee to Investigate the National Defense Program, popularly known as the 'Truman Committee'. He had nagged away at what seemed to be major accounting discrepancies in the War Department budget. At the time, Secretary of War Stimson had thought him an untrustworthy nuisance. As Truman was not a member of the trusted inner circle of the American administration, he had not been made aware of the existence of the Manhattan Project. He never did discover the origin of the discrepancies. Until now.

It fell to Stimson to reveal where the money had gone.

B-VIII

The Alsos mission had not found Weizsäcker in Strasbourg because towards the end of August 1944 he had already moved to Hechingen, ostensibly for a six-week visit to the Kaiser Wilhelm Institute for Physics; but in fact he had no intention of returning to war-torn France. Shortly after returning with Heisenberg from Switzerland came news of the failure of Rundstedt's offensive in the Ardennes. The Americans had suffered more than 80,000 casualties, the Germans slightly more. The last of the German army reserves were spent and the Luftwaffe had been largely destroyed. By early February 1945 the battle lines were redrawn more or less where they had been the previous December.

The war was lost but, Gerlach now reasoned, the peace could still be won. If German physicists could bring a nuclear reactor to criticality in one final push, then this would surely be a major achievement. One which could, perhaps, be used to bargain for better conditions in post-war Germany. Besides, as the war's final stages dealt death even more indiscriminately, Gerlach was all too aware of the 'safe haven' provided by the nuclear programme. 'Again and again the officials overseeing research and development have pressured us to [carry out a large-scale experiment]', he had written in a letter to Heisenberg the previous October. 'And I have to admit that they are right in the sense that many people have been exempted from front-line service in order to carry out such experiments.'

The signs were very promising. In Berlin, Wirtz reported that the reactor experiments in the bunker laboratory had yielded a neutron multiplication of 3.37 using graphite as a shield and neutron reflector. The Uranverein physicists scrambled for enough heavy water to carry out their final experiments. By 29 January, Wirtz had overseen assembly of B-VIII, their largest heavy water pile so far, consisting of hundreds of uranium cubes and a ton and a half of heavy water. The layer configuration that Heisenberg had stubbornly refused to abandon had finally given way to the much more promising lattice structure.

But time was fast running out. Soviet forces were advancing on Berlin at an alarming rate. Gerlach had no choice but to evacuate with the precious materials. Later that same day he told his friend Rosbaud that they were leaving Berlin 'with the heavy stuff'.

Gerlach, Wirtz and Diebner travelled first to Kummersdorf, then south-west to a laboratory in Stadtilm, about 200 miles from Berlin, where Diebner had made preparations to complete his experiments. When

Heisenberg found out he was incensed. Even now he was not prepared to allow materials assigned to his research group to be used by his rival Diebner. Heisenberg and Weizsäcker made the perilous journey from Hechingen to Stadtilm to protest that the materials should now come to the cave laboratory in Haigerloch. Heisenberg prevailed. By the end of February, B-VIII was being reassembled in Haigerloch.

It was not to be. The B-VIII reactor showed a ten-fold increase in neutron intensity as the last of the heavy water was pumped in. This was a result better than anything the Uranverein had officially achieved thus far, but the physicists estimated that to achieve criticality they would need a reactor at least half as large again.[1]

Allied forces crossed the Rhine in February and by 6 March they had taken Cologne. Heisenberg planned to abandon theory and use the last of the uranium and heavy water from Stadtilm in a last-ditch effort to coax the reactor towards criticality. By the end of March American troops were converging on Stadtilm. On arriving back in Munich in early April, Gerlach found that he could no longer contact the physicists in Stadtilm. A few days later he was no longer able to make contact with Berlin.

Special task force

It seems likely that Kurchatov first learned of the implosion principle from Theodore Hall. The young spy had used a simple cipher based on Walt Whitman's *Leaves of Grass* to arrange a rendezvous in Albuquerque with his Harvard friend Sax in December 1944. At that meeting Hall had passed to Sax a couple of pages of handwritten notes describing the first results of the Ra-La experiments and a theoretical summary of the implosion principle.

---

[1] There is tantalising evidence that Diebner's last Gottow experiment, G-IV, may have actually gone critical for a short time and ended in an accident. There is also evidence that Diebner's group may have been working on crude atomic weapons, using enriched uranium and lithium deuteride, in an experimental project known to Gerlach but kept secret from Heisenberg, Weizsäcker, and other Uranverein physicists. See Karlsch and Walker, p. 18.

Kurchatov declared the material to be of great interest. In a report dated 16 March 1945, he wrote:

> It is difficult to give such a conclusion a final assessment, but the implosion method is undoubtedly of immense interest, is fundamentally correct, and should be subjected to close scrutiny both theoretically and experimentally.[2]

Sax had not long delivered this information to his Soviet contact when Hall was drafted into the US Army, a victim of the pressure to recruit as many young men as possible for active service. Qualified scientists were exempted from the draft, but Hall had not completed his doctorate, and was not therefore sufficiently qualified. In Santa Fe, Hall protested to the recruiting officer, who discovered that Hall was needed back in Los Alamos. A compromise was reached: Hall was sent to Fort Bliss in Texas for his army induction and then returned to Los Alamos, in uniform, as a private in the army's Special Engineering Detachment. Hall later put his experience down to the result of horse-trading between Oppenheimer and Groves. 'Oppenheimer was agitating to get sidewalks', Hall said. 'And Groves was agitating to have people drafted. They reached a consensus: I was drafted and the sidewalks appeared.'

Fuchs arrived at his sister's home in Cambridge in February 1945. She immediately informed him of Gold's visit the previous November, and gave him the contact telephone number that he had left in her possession. Fuchs thus called Yatskov, who dropped everything to get to Philadelphia and ask Gold to go at once to Massachusetts. Yatskov gave Gold an envelope containing $1,500, payment to Fuchs for services rendered thus far, but insisted that Gold should not push Fuchs to accept the money.

After more than six rather nervous months, Gold met once more with Fuchs in the Heinemans' spare bedroom. Fuchs briefly summarised the

---

[2] As there is no single word in Russian meaning 'implosion', Kurchatov used the cumbersome term 'explosion towards the inside'. It seems likely that this report would also have been partly based on espionage materials derived from Greenglass.

activities at Los Alamos and gave Gold a street map of Santa Fe. He suggested a future rendezvous in Alameda Street, on 2 June. In the meantime, he proposed to write a short report on everything he knew and pass this to Gold in a couple of days' time at a pre-arranged meeting in Boston. Gold offered him the money, sealed in an envelope, but when he understood what it was he refused to accept it. Gold returned the envelope, unopened, to Yatskov.

Gold and Fuchs met again a few days later. In his report Fuchs had summarised information on the high spontaneous fission rate in reactor-bred plutonium, various aspects of implosion bomb design, including multipoint detonation and explosive lenses, the critical mass of plutonium compared with that of U-235, and current thinking on the design of the initiator. The report reached Kurchatov on 6 April.

By this time, the Soviet physicists had already become convinced of the relative advantages of implosion over the gun method for a plutonium bomb, but Fuchs' report provided considerable additional detail. 'All these are very valuable data,' Kurchatov wrote, 'but of particular substance are indications with regard to conditions conducive to achieving symmetry of the implosion effect, which is crucial to the very essence of the method.'

Thanks to espionage by Fuchs, Hall and Greenglass, the Soviet physicists were no more than a few months behind the latest developments at Los Alamos. However, Kurchatov was well aware that knowledge of fission physics and the principles of atomic bomb design were no substitute for practical experience with the materials involved. And, until the Soviet Union found a solution for its uranium supply problems, practical experience was going to be hard to gain.

But there was some progress on this front, too. By early 1945 Soviet forces had liberated Czechoslovakia with help from both Czech and Slovak resistance groups. A provisional Czechoslovak government was installed in Kosice on 4 April 1945. Edvard Beneš, President of Czechoslovakia's government-in-exile in London, had maintained friendly relations with the Soviet Union throughout the war in an attempt to avoid a Communist coup at the war's end. In March 1945 he appointed Czechoslovak Communist exiles in Moscow to key positions in his cabinet. He also agreed a secret deal to allow the Soviets to source uranium ore from his country's Joachimsthal mines.

On 23 March, Beria had suggested to Stalin that a special task force be established to 'grope in Germany and search there for novelties of German atomic technology and for its creators'. In essence, the force was to operate in a manner similar to the 'trophy brigades' that had requisitioned (or looted) anything of value left behind by the fleeing German population as the Red Army advanced westwards towards Berlin. Beria chose one of his deputies, Colonel General Avram Zavenyagin, head of the NKVD 9th Chief Directorate, to lead the mission. Kurchatov was asked to submit suggestions for several search teams, and Khariton and Artsimovich were asked to provide scientific guidance.

Each unaware of the other, the Soviet mission now joined Alsos; scavengers of the Old World and New World wheeling and circling as they sought atomic technology, materials and personnel from the ruddy carcass of a defeated Germany.

Capture

Towards the end of March an Alsos team led by Pash captured Bothe and the Uranverein's only cyclotron in the university town of Heidelberg. This was the first captured physicist known to Goudsmit personally, and he was unsure how to proceed. Bothe greeted him warmly, and they shook hands, which was specifically prohibited by regulations established to prevent fraternisation with the enemy. 'I am glad to have someone here to talk physics with', Bothe declared. 'Some of your officers have asked me questions, but it is evident they are no experts on these subjects. It is so much easier to talk with a fellow physicist.'

Bothe treated the encounter as though this were a routine visit of a scientific colleague, and proudly gave Goudsmit a tour of his laboratory as he talked about the research they had done. Only when Goudsmit raised the question of 'war problems' did Bothe stop short, saying that he could reveal nothing that he had promised to keep secret.

At the Yalta conference in February 1945, Roosevelt, Churchill and Stalin had ratified the proposed post-war occupation zones for Germany. Initially there were to be three: American, British and Soviet; but Stalin agreed that France might have a fourth zone of occupation provided this was carved from territory within the American and British zones. In the event, the American and British administrations agreed to cede two non-contiguous areas to the French.

This left something of a problem for the Alsos mission. The southernmost of these areas included the Württemberg-Hohenzollern region, and the small towns of Tailfingen, Hechingen and Haigerloch. Groves trusted the French little more than the Soviets, and various plans were now devised to destroy or capture the last remnants of the German atomic programme before they fell into French hands.

Goudsmit argued persuasively that everything they had learned of the German programme indicated that there was little justification for a bombing raid or missions which might put Allied lives at risk (Pash was preparing nervously for his first parachute drop, on Hechingen). Events forced their hand. As French Moroccan troops closed in on the area, Pash hastily organised an ad hoc force of army engineers to get there first, by road. On 21 April, he drove into Haigerloch without incident.

The cave laboratory was quickly found. Eager to get in on the act, Perrin, Welsh and Charles Hambro, representing British interests in the mission, arrived shortly afterwards together with Lansdale and Furman to inspect the facilities. Perrin was the only one of the group to have seen Fermi's original Chicago pile and he quickly concluded that the Haigerloch reactor would have been too small to go critical.

Uranverein physicists Bagge, Wirtz, Weizsäcker and Horst Korsching were captured in Hechingen a few days later. Max von Laue, though not part of the Uranverein, was also detained. Faced with this evidence of the sheer scale of the Alsos mission, and after lengthy interrogation, Wirtz and Weizsäcker yielded the uranium, heavy water and, eventually, documents that they had carefully hidden just a few days earlier.

Hahn was captured in Tailfingen on 25 April. He was waiting for them, his suitcase already packed. From Hahn, Goudsmit learned devastating news. His parents had perished in an Auschwitz gas chamber. Heisenberg had written a personal letter pleading for their release, but they had been murdered five days before it was sent. Heisenberg could, perhaps, have done little more to intervene on their behalf, but Goudsmit would never forgive him for not trying harder.

Heisenberg himself was nowhere to be found, but his disappearance was not a mystery. The German physicists made no secret of the fact that he had left Hechingen a few days before, in an attempt to get back to his family in Urfeld, 120 miles away. His journey, largely by bicycle through the detritus of war, was nerve-wracking and exhausting. Aside from the constant threat of Allied fighters roaming the skies searching for targets, there was also the risk of instant, roadside justice meted out by SS units carrying out Hitler's order to fight to the end. The SS officers would simply pronounce sentence and shoot or hang 'deserters' on the spot.

It took Heisenberg three days to get to Urfeld, arriving on 23 April. 'And finally, suddenly and unexpectedly, I saw him coming up the mountain, dirty, dead tired and happy', his wife Elisabeth wrote years later. They celebrated the announcement of Hitler's suicide with their last bottle of wine, and with tears of relief. On 3 May he was visiting his mother, who had also moved to Urfeld and was staying at an apartment in the village, when he received a call from Elisabeth. He returned home promptly to find Pash and a small detachment of men waiting to take him into custody. For Heisenberg his capture was a mixed blessing: 'I felt like an utterly exhausted swimmer setting foot on firm land.'

The Alsos mission caught up with Gerlach and Diebner in Munich on 1 May, and with Harteck in Hamburg.

A photograph that had been recovered from Heisenberg's office in Hechingen showed Heisenberg and Goudsmit together, taken at the dockside in Ann Arbor during their last meeting in the summer of 1939. At this meeting, Goudsmit had challenged Heisenberg about his decision to stay in Germany and had urged him to come to America. Meeting again in Heidelberg under rather different circumstances, Goudsmit now repeated the suggestion, but Heisenberg told him: 'No, I don't want to leave. Germany needs me.' When Heisenberg asked if there was a similar atomic programme in America, Goudsmit lied and said no. Heisenberg offered help: 'If American colleagues wish to learn about the uranium problem, I shall be glad to show them the results of our researches.'

Goudsmit found this all rather 'sad and ironic'. The Tube Alloys and Manhattan Project physicists had always thought themselves to be competing in a race. Yet the German atomic programme hadn't even succeeded in building a working reactor. Heisenberg had lost the race long ago. But he still did not know this yet.

In the ruins of Berlin

Berlin was encircled by the Soviet First Belorussian Front and First Ukrainian Front, and on 20 April – Hitler's birthday – Soviet artillery began pounding the city. On 29 April the Soviet Third Shock Army crossed the Moltke Bridge, a short distance from the Königsplatz and the Reichstag. Hitler signed his last will and testament and married Eva Braun. Both committed suicide the next day.

Zavenyagin headed the first of the Soviet search missions to Berlin, arriving on 3 May and occupying a building in Berlin-Friedrichschagen. The search teams variously included Artsimovich, Kikoin, Flerov and Khariton, the latter dressed in the uniform of an NKVD lieutenant-colonel. Kurchatov did not join, perhaps concerned that the NKVD might in future lean towards German, rather than Soviet, science to drive the Soviet atomic programme.

On 4 May the search team visited the Kaiser Wilhelm Institute for Physics in Dahlem. The equipment had long since been moved to Hechingen, but the search team discovered detailed documentation giving a complete description of the German atomic programme. The Soviet physicists were not greatly impressed. Some of Heisenberg's calculations for the critical size of a nuclear reactor had already been surpassed by Soviet scientists. It was obvious from these documents that the German programme had not advanced very far.

German forces surrendered unconditionally on 7 May. Three days later a Soviet search team headed for Manfred von Ardenne's laboratory in

Berlin-Lichterfelde. The team found Ardenne waiting, and was this time impressed with what had been achieved at his private laboratory. Along with a prototype electron microscope, the search team found a prototype calutron. Ardenne and Zavenyagin discussed plans for the future, and on 21 May Ardenne flew to Moscow, ostensibly to sign an agreement on the creation of a new physical-technical institute in the Soviet Union.

Ardenne probably figured that he would not be returning to Germany any time soon, when on his arrival in Moscow an interpreter asked: 'Haven't you brought your children with you?' Within minutes of his departing Berlin for the airport, about 100 Soviet soldiers had suddenly appeared and started packing everything that could be moved. A few weeks later, the rest of Ardenne's laboratory staff arrived in Moscow by train.

Anticipating that the Soviets would capture the Auer uranium production facilities at Oranienburg, Groves had sought to ensure that all they would find was rubble. The Auer plant had been heavily bombed by B17s of the American Eighth Air Force on 15 March. Auer's chief scientist, Nikolaus Riehl, had moved with some of his staff to a small village west of Berlin in the hope that they would meet British or American forces before the invading Soviet forces arrived. However, he was tracked down by Artsimovich and Flerov, who suggested that he join them for a few days of discussion. That few days was to become ten years.

Riehl had been born in St Petersburg, the son of a Siemens Company engineer, and spoke fluent Russian. A highly competent chemist and mineralogist, he was not allowed to return home. Instead he accompanied the Soviets to what remained of the Auer production facilities, where 'demounting and loading of everything that was not nailed down or riveted proceeded at full speed'. On 9 June, Riehl departed for Moscow with his family and some of his staff.

Khariton and Kikoin now began a systematic search for uranium. They left Oranienburg just before 100 tons of fairly pure uranium oxide was discovered in the rubble. And a chance encounter, some detective work and interrogations by agents of Soviet army counter-intelligence – Smersh – led them to a leather-tanning plant in Neustadt am Glewe, where more than 100 tons of uranium had been hidden. With the uranium from

Oranienburg, the Soviet mission was able to send back about 300 tons of uranium oxide and other compounds.

This would be enough to support development of the Soviet Union's first nuclear reactor.

Chapter 15

TRINITY

April–July 1945

Groves had taken a dislike to Leo Szilard the very first time they had met in October 1942. Ever since then, Szilard had lodged like an increasingly painful thorn in Groves' backside. Growing ever more annoyed at Szilard's erratic behaviour, his outspoken criticism of the Manhattan Project's administration, unauthorised travels and intransigence over his chain-reaction patent claims, Groves had the FBI put him under close surveillance. His contract with the Met Lab had expired at the end of 1942 and was not immediately renewed. He had been obliged to sign over his patents as a condition of his reinstatement.

He rejoined just as the work programme at the Met Lab was winding down, with many physicists heading for Hanford or Los Alamos. The mood at the Met Lab became somewhat restless and depressed. With some time on his hands, he turned his attention to post-war atomic politics.

The problem that he and other like-minded Manhattan Project physicists now confronted was simply stated. The reason for the very existence of the project had been fear of German atomic weapons. Yet, by early 1945, the defeat of Germany seemed certain and fear of a Nazi super-weapon groundless. Rotblat had come to similar conclusions, and had taken the unilateral decision to leave the project and Los Alamos in December 1944.[1]

This restlessness was not confined to the Met Lab. As work on the Hill accelerated, so too did the intensity of the discussion among many scientists, their sense of moral responsibility growing along with their concern for the likely impact of the 'gadget' on future civilisation. Oppenheimer attended many of the discussion meetings, gently but persuasively defusing the situation. If atomic weapons were to foreshadow the end of all conventional war, as Bohr had argued, then Oppenheimer believed the world needed first to see the effects of these new weapons if they were to be properly respected.

To Szilard and many other scientists, it was equally certain that the shock of the first use of such weapons by America would spark an arms race with the Soviet Union. A system of international controls was needed if such a race was to be avoided. In March 1945 he decided to write once more to Roosevelt, and again enlisted Einstein's support. Unable to share the full details of his memorandum with Einstein, who had never become involved in the Manhattan Project, he asked only that Einstein write him a letter of introduction. Szilard also thought to write to Eleanor Roosevelt, and was delighted to receive an invitation to meet her in Manhattan on 8 May.

The meeting never happened. Roosevelt's death on 12 April forced Szilard to start again.

Szilard now had to find a channel to Truman. He enlisted the support of a young Met Lab mathematician, Albert Cahn. Cahn hailed from Kansas City, which sat at the heart of Truman's political base. Born in Missouri, Truman had served as a judge in the Jackson County Court before defeating the Republican incumbent for Missouri, Roscoe C. Patterson, in the 1934 Senate elections. Through Cahn's contacts, Szilard was able to secure an appointment with Truman at the White House on 25 May.

---

[1] By this time, army intelligence had built up a thick dossier on Rotblat, filled largely with fabrications. It was nevertheless agreed that he would leave without revealing his real reasons: he was to say that he was worried about his wife in Poland. See Brown, p. 284.

But Szilard did not get to meet the new president. On the day they were to meet, Truman's appointments secretary instead pointed Szilard in the direction of James Byrnes, in Spartanburg, South Carolina. Somewhat puzzled, Szilard agreed and made the necessary travel arrangements.

At the time of Szilard's visit, Byrnes was retired and did not appear to have any role in the new administration. A close confidant of Roosevelt, he had headed the government's Economic Stabilisation Office and War Mobilisation Board, and had served informally as a kind of 'assistant' president, running the country while Roosevelt ran the war. He had been the favourite for selection as Roosevelt's running mate for the US presidential election in 1944 in place of Henry Wallace but, much to his disappointment, had lost out to Truman. At Roosevelt's request, Byrnes had attended the Yalta conference, where he had listened intently and taken copious notes in shorthand.

But Byrnes was not retired. Although the Yalta conference was the limit of his experience of foreign affairs, within days of assuming the presidency, Truman had decided that he should be his new Secretary of State.

The meeting between the erratic physicist and the astute politician from South Carolina went predictably. Szilard made an unfavourable impression on Byrnes, and Byrnes' seeming inability to grasp the significance of atomic energy left Szilard fearful that a US–Soviet arms race was inevitable. Byrnes suggested that the Soviets might be suitably impressed and therefore more 'manageable' if they were shown an example of American military supremacy. 'I was rarely as depressed as when we left Byrnes' house and walked to the station', Szilard wrote.

Targets

As Szilard walked disconsolately back to the station, Groves' Target Committee was meeting for the last time at the Pentagon. The committee had been tasked with finalising the precise details of the first use of atomic weapons against Japan. Oppenheimer served as an adviser. It had previously identified four potential targets for further study – Kyoto, Hiroshima, Yokohama and Kokura Arsenal – and it was now agreed that three targets would be reserved: Kyoto, Hiroshima and Niigata. The committee had agreed the height at which the bomb, in this first instance a uranium bomb using the Little Boy design, should be detonated. It had examined the bomb's likely radiological effects.

Colonel Paul Tibbets joined the Target Committee meeting in Washington. Tibbets had served as a personal pilot to Dwight D. Eisenhower and had a reputation as the best pilot in the US Army Air Force. He had been selected to command a special air combat group, the 509th Composite Group, assembled specifically to provide a combat delivery system for the atomic bomb.[2] He was scheduled to fly the converted B-29 that would deliver Little Boy to its target. His crews had been training in Cuba, gaining experience flying with heavy loads over water. Movement orders for the 509th had been issued in April, and an advance party had arrived at North Field, on Tinian Island in the northern Pacific, just ten days before on 18 May.

The uneasiness that had settled on the Los Alamos and Met Lab physicists was to a large extent shared by Stimson. He, too, was brooding over what the advent of atomic weapons might mean for the future security of mankind.

Stimson was 77 years old, and had been appointed a US Attorney in 1909 by Theodore Roosevelt, Franklin's fifth cousin. He possessed a strong sense of morality, of faith in humanity and international law. He was horrified by the way that war on such an unprecedented scale had blunted this sense of morality among the Western democracies. The Allied firebombings of Dresden, Hamburg, and then of selected Japanese cities, represented a kind of total war that he disliked intensely. He was determined, somehow, to use the atomic bomb in a way that would minimise civilian casualties. 'The reputation of the United States for fair play and humanitarianism,' he said to Truman, 'is the world's biggest asset for peace in the coming decades. I believe the same rule of sparing the civilian population should be applied, as far as possible, to the use of any new weapon.'

---

[2] The 509th was termed a 'Composite' Group as it included both transport aircraft and bombers.

At the very least, the former ancient capital city of Kyoto, a major cultural centre, should be spared. Stimson had spent his honeymoon there. Groves, for once, was out-manoeuvred. Kyoto was dropped from the target list.

Interim Committee

To address some of the post-war issues surrounding the use of atomic weapons, with Truman's blessing Stimson had set up an Interim Committee, so called because it represented a temporary stop-gap for a permanent policy organisation to be established by the US Congress or international treaties. Its task was to develop the American position on the use of atomic weapons in wartime, and debate some aspects of post-war atomic policy, including international control. The idea had come originally from Bush and Conant. Stimson was chairman, and Bush, Conant, MIT president Karl Compton (Arthur Compton's brother), Assistant Secretary of State William Clayton, Undersecretary of the Navy Ralph Bard and Stimson aide George L. Harrison were appointed as members. Byrnes was a late addition. At the Interim Committee's first meeting it was recognised that the Manhattan Project scientists might themselves have an important contribution to make, and Arthur Compton, Lawrence, Fermi and Oppenheimer were recommended for appointment to a scientific panel.

Separate briefings by Stimson and Groves and the first few meetings of the Interim Committee gave Byrnes the background he needed. The first full meeting of the committee and its scientific panel was held at the Pentagon on 31 May. Army Chief of Staff General George C. Marshall, Groves and a couple of aides were also invited to attend.

Individual committee members quickly established their own positions. On the subject of control and inspection, Oppenheimer argued for free exchange of atomic information, with emphasis on the peace-time uses of atomic power, preferably before the bomb was actually used. This, he suggested, would greatly strengthen America's moral position. His unspoken desire was, no doubt, to see that the ultimate legacy of the Manhattan Project, and his own legacy as scientific director of Los Alamos, should be the enlargement of human welfare rather than weapons of mass destruction.

But free exchange of information raised questions concerning the nature of an international control organisation required to police the uses to which such information might be put. Stimson was not sure: 'The Secretary asked what kind of inspection might be effective and what would be the position of democratic governments as against totalitarian regimes under such a program of international control coupled with scientific freedom.' Totalitarian regimes were not, after all, noted for their commitment to the strict terms of international agreements. At issue, of course, was the anticipated behaviour of the Soviet Union.

Oppenheimer had faith in Russia's scientists, and pointed out that Russia had always been very friendly to science. Perhaps the Soviet Union could be approached in a tentative fashion and be advised of the Manhattan Project in the most general terms in the hope of securing their co-operation. His statement dimly echoed his August 1943 conversation with Pash and Johnson. Back then he had said that he would not want classified atomic information to get to the Soviets through the 'back door'. Now he was looking for a front door, for a step towards Bohr's 'open world'.

He found an unexpected ally in Marshall, who argued that the intransigence of the Soviets was more often imagined than real. When real, their lack of co-operation derived from their understandable concern for their own security. 'General Marshall was certain that we need have no fear that the Russians, if they had knowledge of our project, would disclose this information to the Japanese. He raised the question whether it might be desirable to invite two prominent Russian scientists to witness the [Trinity] test.'

Had Churchill been present at this discussion, it is not difficult to imagine his reaction to Marshall's proposal. Churchill was not present, but Byrnes was. 'Mr Byrnes expressed a fear that if information were given to the Russians, even in general terms, Stalin would ask to be brought into the partnership. He felt this to be particularly likely in view of our commitments and pledges of cooperation with the British ... [He] expressed the view, which was generally agreed to by all present, that the most desirable program would be to push ahead as fast as possible in production and research to make certain that we stay ahead and at the same time make every effort to better our political relations with Russia.'

It was a familiar formula, one which effectively bolted the door that Oppenheimer was trying to prise open. Byrnes thought he understood enough about the atomic bomb to interpret the American lead in atomic weapons technology as a powerful bargaining chip in post-war US–Soviet relations. In other words, he had not understood the bomb at all. Pushing ahead in production and research could have only one consequence. And this was not the one he intended.

The meeting adjourned for lunch and resumed again at 2:15pm. Of the morning participants, only Marshall was now absent. The discussion over lunch had switched to the impending use of the bomb against Japan, and this discussion continued as the meeting resumed.

If growing moral qualms affected any of the members of the Interim Committee, these are not reflected in the notes of the meeting. The only issue debated at length concerned the psychological impact of the bomb. The opinion was expressed that the impact of one bomb might not be much different than the impact of current firebombing raids on Japanese cities.

As there was still considerable uncertainty over the yield of an atomic bomb (Compton had earlier estimated an explosive force anywhere between 2,000 and 20,000 tons of TNT), Oppenheimer did not argue with the observation. 'However, Dr Oppenheimer stated that the visual effect of an atomic bombing would be tremendous. It would be accompanied by a brilliant luminescence which would rise to a height of 10,000 to 20,000 feet. The neutron effect of the explosion would be dangerous to life for a radius of at least two-thirds of a mile.'[3]

After further discussion of various types of targets, Stimson summarised the committee's interim conclusions:

---

[3] The military doctrine of rapid dominance, also known as 'shock and awe', is based on the use of overwhelming force and spectacular displays of power to undermine an enemy's will to fight. The doctrine was developed at the US National Defense University in 1996, its authors citing the atomic bombing of Japan as an early example.

The Secretary expressed the conclusion, on which there was general agreement, that we could not give the Japanese any warning; that we could not concentrate on a civilian area; but that we should seek to make a profound psychological impression on as many of the inhabitants as possible. At the suggestion of Dr Conant the Secretary agreed that the most desirable target would be a vital war plant employing a large number of workers and closely surrounded by workers' houses.

It was not much of a compromise. Stimson wanted to avoid unnecessary civilian casualties, but the entire population of Japan had in effect been drafted in support of the war effort. In Japan, bombing a war plant closely surrounded by workers' houses meant bombing men, women and children. Without warning.

Oppenheimer had argued at Los Alamos that the war should not end without the world knowing about the bomb and its effects. The Manhattan Project had to be at least driven to the point where the bomb could be tested. His scientists had been persuaded, even pacified. But Oppenheimer had not explained that this would mean bombing a large number of workers and their families without warning. Yet this is what he was now accepting.

It seems that nobody at the meeting thought to raise the question about possible psychological impacts on the Soviets.

Groves had been fairly quiet throughout, at least according to the formal meeting notes. As the meeting wound to a close, Groves took the opportunity to raise the issue of certain troublesome scientists, 'of doubtful discretion and uncertain loyalty', who had plagued the atomic programme since its inception. Groves was aware of Szilard's meeting with Byrnes and his attempts to influence the conclusions of the Interim Committee, and there can be no doubt about the identity of the scientist he had in mind. The committee agreed that nothing could be done until after the bomb had been tested or used, at which point 'steps should be taken to sever these scientists from the program'.

Oppenheimer remained silent.

The Franck report

The four members of the scientific panel agreed to meet again at Los Alamos on 16 June to finalise their formal recommendations on the use of the bomb. They were forbidden from revealing details of the membership of the Interim Committee or the conclusions that had been reached thus far, but on his return to Chicago, Compton told the Met Lab scientists that their views would be welcome. If the scientists could prepare submissions in the next two weeks, he would convey them to the scientific panel at its next meeting.

Compton's invitation now sparked a flurry of activity. The Met Lab scientists formed themselves into groups to review future atomic organisations, research, education and security, production of atomic materials, and the political and social problems posed by the bomb. Szilard was asked to chair the group on production, but declined. He preferred to focus on his contribution as a member of the political and social problems group under the chairmanship of James Franck, with whom Oppenheimer had once worked in Göttingen. The group also included Glenn Seaborg and Russian-born biophysicist Eugene Rabinowitch.

It was Rabinowitch who drafted the group's report, but Szilard's ghost lies beneath many of its words. The report is forthright in establishing the most logical consequence of American attempts to maintain supremacy in nuclear armaments:

[A]ll that these advantages can give us, is the accumulation of a larger number of bigger and better atomic bombs – and this only if we produce those bombs at the maximum of our capacity in peace time, and do not rely on conversion of a peace time nucleonics industry to military production after the beginning of hostilities. However, such a quantitative advantage in reserves of bottled destructive power will not make us safe from sudden attack. Just because a potential enemy will be afraid of being 'outnumbered and outgunned,' the temptation for him may be overwhelming to attempt a sudden unprovoked blow

– particularly if he would suspect us of harboring aggressive intentions against his security or 'sphere of influence.'

To avoid an arms race it would be necessary to exercise remarkable self-restraint:

> **From this point of view a demonstration of the new weapon may best be made before the eyes of representatives of all United Nations, on the desert or a barren island.** The best possible atmosphere for the achievement of an international agreement could be achieved if America would be able to say to the world, 'You see what weapon we had but did not use. We are ready to renounce its use in the future and to join other nations in working out adequate supervision of the use of this nuclear weapon.' ... If the United States would be the first to release this new means of indiscriminate destruction upon mankind, she would sacrifice public support throughout the world, precipitate the race of armaments, and prejudice the possibility of reaching an international agreement on the future control of such weapons.

By now wary of the formal channels of communication with senior politicians, Franck himself sought to bring a copy of his report directly to the attention of Secretary Stimson. Compton met him in Washington, but they were told that Stimson was out of the city (he was not). Stimson's aide, Harrison, assured them that he would pass the report on. Compton hastily drafted a covering note. 'In this note it was necessary for me to point out that the report, while it called attention to difficulties that might result from the use of the bomb, did not mention the probable net saving of many lives, nor that if the bomb were not used in the present war the world would have no adequate warning as to what was to be expected if war should break out again.' Compton thus undermined the report even as he helped Franck seek attention for it. Franck's report was immediately classified.

A few days later, Compton carried the report to the meeting of the scientific panel at Los Alamos and passed copies to Oppenheimer, Lawrence and Fermi. To men more accustomed to drawing testable conclusions from their evaluation of established scientific facts, the political conclusions they now had to draw from complex political and military uncertainties must have seemed daunting.

It is likely that Compton himself argued that the unannounced use of the bomb against Japan would swiftly end the war and save countless American lives. Japan was defeated but prized its honour as a nation above all else. 'In spite of such disastrous damage,' Compton later wrote, 'the Japanese militarists seemed ... unwavering in their determination to fight to the finish.' The Japanese military culture promoted suicide as preferable to the shame of defeat, as witnessed by the kamikaze attacks on Allied destroyers, which peaked in the battle of Okinawa in April. Compton's brother Karl, a director of the Office of Field Service of the OSRD, had recently departed for Manila to oversee the provision of radar support ahead of the planned American invasion and occupation of Japan. Compton hoped that use of the bomb might render invasion unnecessary.

'We didn't know beans about the military situation in Japan', Oppenheimer later admitted. 'We didn't know whether they could be caused to surrender by other means or whether the invasion [of Japan] was really inevitable. But in the backs of our minds was the notion that the invasion was inevitable because we had been told that.'

With use of the bomb perceived as a *necessity* to end the war, the argument about saving American lives prevailed. The scientific panel issued a short report to the Interim Committee on the immediate use of nuclear weapons. The report acknowledged the lack of unanimity among Manhattan Project scientists, but concluded: 'We can propose no technical demonstration likely to bring an end to the war; we see no acceptable alternative to direct military use.' The report closed with some hand-washing worthy of Pilate:

With regard to these general aspects of the use of atomic energy, it is clear that we, as scientific men, have no proprietary rights. It is true that we are among the few citizens who have had occasion to give thoughtful consideration to these problems during the past few years. We have, however, no claim to special competence in solving the political, social, and military problems which are presented by the advent of atomic power.

Harrison raised the Franck report at a further meeting of the Interim Committee, held on 21 June, to which the scientific panel was not invited. The report was discussed in the context of the scientific panel's 16 June recommendations, and 'The Committee reaffirmed the position taken at the 31 May and 1 June meetings that the weapon be used against Japan at the earliest opportunity, that it be used without warning, and that it be used on a dual target, namely, a military installation or war plant surrounded by or adjacent to homes or other buildings most susceptible to damage'. A request for Harold Urey to be made a member of the scientific panel to represent the views of the Met Lab scientists was refused.

Szilard was beaten, but he was not yet ready to quit. In early July he organised a petition to the President, urging that he exercise his power as Commander-in-Chief to rule that America should not resort to the use of atomic weapons. It was unlikely to change any decision that had by now been made, but that was not its primary purpose. Szilard wanted to ensure that those scientists who opposed the use of the bomb on moral grounds went 'clearly and unmistakably' on record. The petition gained 59 signatures from Met Lab scientists, and Szilard distributed copies to Oak Ridge and Los Alamos.

Teller took a copy to Oppenheimer and, over subsequent years, recorded various versions of Oppenheimer's reaction. In an early version, Oppenheimer expressed the opinion that it was improper for a scientist to use his prestige as a platform for political pronouncements. Teller did not circulate the petition further. 'I should like to have the advice of all of you whether you think it is a crime to continue to work', he wrote to Szilard. 'But I feel I should do the wrong thing if I tried to say how to tie the little toe of the ghost to the bottle from which we just helped it to escape.' Teller's mind was in any case preoccupied with springing a rather larger ghost from the bottle.

MLAD and CHARLES

Byrnes had bolted the front door, but the back door was open and leaking atomic secrets to the Soviet Union at a rapid rate.

Sax had decided to return to Harvard in early 1945, having failed his liberal arts courses the previous year. He had signed on for courses in physics, chemistry, astronomy and engineering. This left the NKGB without a courier for Hall. Gold was briefly discussed as a possible candidate, but he was already working with Fuchs (and, shortly, Greenglass), and it was considered inadvisable to give him this further responsibility. Kvasnikov had recently reactivated Lona Cohen, an attractive, Massachusetts-born 32-year-old Communist Party member who had been inducted into Soviet espionage by her husband, Morris. Yatskov now assigned her the role of courier for Hall.

Cohen made her way to New Mexico sometime in late April or early May 1945. She did not meet with Hall, but nevertheless returned with a package which she handed to Yatskov in a Manhattan coffee shop. Feklisov waited outside.[4] The contents of the package have not been identified, but there is circumstantial evidence to suggest that it contained a detailed description of the Fat Man bomb design.

Kvasnikov now faced something of a dilemma. If they were genuine, these espionage materials were of incredible importance. But he was not about to rush the information to Moscow until he could find a way of corroborating it. He sent a coded and enciphered cable to Moscow on 26 May but this did not contain much new information. He explained that: 'The material has not been fully worked over. We shall let you know the contents of the rest later.'

That same day Yatskov met with Gold to brief him about his impending pre-arranged meeting with Fuchs in Santa Fe on 2 June. He also asked Gold to meet with Greenglass, whose wife Ruth had recently rented an

---

[4] Yatskov suffered from a partial colour-blindness, unable to tell the difference between red and green. Consequently, when driving to an important meeting he could either concentrate on traffic signals or on his FBI tail, but not both. Feklisov would sometimes accompany him to make sure he wasn't being followed.

apartment in Albuquerque, and gave him part of the cardboard top of a pack of processed dessert as a recognition signal. Gold baulked. This was not in the Soviet spy manual, and counter to his espionage training. But the mission was now of such importance that a deviation from strict protocol was deemed an acceptable risk. Yatskov lost his temper. 'I have been guiding you idiots every step. You don't realise how important this mission to Albuquerque is', he exclaimed.

Fuchs, in the meantime, was also preparing for his rendezvous, sitting quietly in his room at Los Alamos, writing out detailed descriptions of both the uranium and plutonium bombs. The Little Boy design would have been quickly dealt with. The Fat Man design required a little more attention. Fuchs described the solid plutonium core, the polonium–beryllium initiator, the tamper, the bomb casing and the materials used to form the explosive lenses. He described calculations of the efficiency of the bomb, and wrote that it was expected to yield about 5,000 tons of TNT equivalent. He drew a schematic giving all the most important dimensions. He mentioned the intention to use the bomb against Japan.

Gold was at the agreed rendezvous point, the Castillo Street Bridge on Alameda Street, when Fuchs pulled up in his grey, second-hand Buick.[5] Fuchs drove them to a quiet cul-de-sac. They talked briefly. Fuchs had a few things to add to what was in the document he had written. Everyone at Los Alamos was working flat-out on preparations for the Trinity test, scheduled for 10 July. The efficiency of the bomb had been revised, to 10,000 tons of TNT.

Fuchs suggested another meeting in August, but Gold had already experienced difficulties getting time off work to make the current rendezvous, and thought an August meeting would be impossible. They fixed on 19 September, 6:00pm, at a different location. Fuchs passed over the envelope containing his handwritten report, Gold stepped out of the car and Fuchs drove away.

---

[5] Feynman borrowed Fuchs' car to drive to Albuquerque two weeks later, on 16 June. He had received word that his wife, Arline, was very ill. After suffering delays resulting from three flat tyres, he arrived at the hospital just a few hours before she died.

Gold spent the night in Albuquerque, and the next morning headed for the address that Yatskov had given him. At the door of an apartment at 209 North High Street, he told Greenglass: 'Julius sent me', and gave him the cardboard top. Greenglass fetched a matching part from the kitchen. The correct recognition signals had been given. They met again later that afternoon, and Greenglass gave Gold an envelope containing a list of potential espionage recruits at Los Alamos and a sketch of a high-explosive lens mould of the type used in the Fat Man design.

Gold passed the envelopes he had received from Fuchs and Greenglass to Yatskov in New York on 4 June. Kvasnikov had his corroboration. Fuchs' report was even more detailed than the information he had received from Hall. He cabled a full report to Moscow on 13 June. Kurchatov was briefed on its contents a few weeks later, on 2 July.

Kvasnikov sent a further cable on 4 July, adding details that had been gleaned from a proper debriefing of Gold. The materials were now sufficiently important to warrant a summary for Beria himself. A letter to Beria, dated 10 July, cited two 'reliable agent sources', given the codenames MLAD and CHARLES, and summarised the information in Kvasnikov's cables of 13 June and 4 July. 'NKGB USSR received data that in the USA in July this year the first experimental explosion of the atomic bomb is scheduled. It is expected that the explosion will take place on the tenth of July.' MLAD and CHARLES were the codenames for Hall and Fuchs.

Despite this evidence from spies at the heart of the Manhattan Project, there was no immediate effort to expand the scale of the Soviet bomb programme. Beria, it seems, remained highly sceptical of the intelligence materials, suspecting they were disinformation. Beria didn't even trust his own scientists.

A small sun, shining in the desert

Bainbridge had constructed a small town out in the Alamogordo desert. His team had expanded from 25 to more than 250. At Point Zero a 110-foot tower had been constructed, at the top of which the bomb would be detonated. A concrete-reinforced command centre was situated 10,000 yards south of Point Zero. A series of observation bunkers, a field laboratory and a Base Camp had also been constructed. An array of instruments had been assembled to measure the blast, ground shock, intensity of neutrons and gamma rays and the characteristics of the bomb radiation. Bainbridge had to beg Groves for $125,000 just to build roads. During construction they had to contend with scorpions, tarantulas and poisonous snakes, murderous heat and dust and occasional attacks by American aeroplanes which mistook Base Camp for one of their practice targets.

Oppenheimer had recruited his brother Frank to provide administrative support to Bainbridge and to help troubleshoot the test. Frank Oppenheimer had been working with Lawrence at the Rad Lab. He now left his wife and children in Berkeley and made his way to New Mexico. He arrived at the Trinity test site in late May, and observed the feverish activity taking place in the desert. One of his administrative tasks was to work out escape routes in case disaster struck, 'making little maps so everybody could be evacuated'.

The plutonium for the test bomb had arrived at Los Alamos, also at the end of May. The efficacy of the solid core design was confirmed by Frisch's group on 24 June. The core was somewhat less than a critical mass but would be squeezed to a density equivalent to twice the critical mass by the implosion. Just five kilos was needed, less in physical size than a small orange. It was warm to the touch.

The Trinity test had originally been scheduled for 4 July, but at the end of June the Cowpuncher committee felt it had no choice but to push the timing back to 4:00am on 16 July, at the earliest. Truman, hoping to have news of a successful test in his back pocket at the next meeting of Allied heads of state in Potsdam, had negotiated a delay to the start of the conference, to 15 July. To ensure that Truman had what he needed, Groves insisted that the scientists adhere to the target date of 16 July.

The problem lay in the high-explosive lenses. The moulds for the lenses that had been delivered to Kistiakowsky's laboratory were found to be cracked and pitted. X-ray inspection of the explosives cast from the moulds revealed air cavities that would impair the lens performance and risk an unsymmetrical implosion. There were more faulty castings produced than acceptable ones, and by 9 July it looked as though there wouldn't be enough lenses. To make matters worse, Oppenheimer had insisted on a 'dry run' implosion test, which required a duplicate Fat Man structure without the plutonium core. Struggling to produce enough lenses even for one test, Kistiakowsky was now obliged to find enough for two.

Kistiakowsky laboured heroically through the night to fix some of the faulty castings using a dental drill and molten explosive. 'You don't worry about it', he said. 'I mean, if fifty pounds of explosives goes in your lap, you won't know it.'

The dry run was carried out five days later, on 14 July, at an isolated canyon near Los Alamos. It was a failure. Oppenheimer called an emergency meeting. The last few weeks had visibly taken their toll on the Los Alamos scientific director; his nerves were frayed and he was close to despair. He now ripped into Kistiakowsky. The failure of the dry run surely meant the impending failure of the Trinity test and, Oppenheimer now argued, with emotions running high, Kistiakowsky was therefore personally responsible for the failure of the entire project. When Groves and Conant arrived in Albuquerque shortly afterwards, Kistiakowsky was carpeted for what seemed like an age.

The former Cossack couldn't understand it. He questioned the results of the magnetic measurements used to assess the symmetry of the implosion, and was further accused of questioning the validity of Maxwell's equations, the basis of the theory of electromagnetism which had been established since 1860. But he stood firm: he bet Oppenheimer a month's salary to ten dollars that his part of the bomb would work. Oppenheimer accepted the bet. The mood over dinner that evening was gloomy, with Oppenheimer again seeking solace in snatches of the Bhagavad-Gita: 'in the depths of shame, The good deeds a man has done before defend him.'

There was some relief the next morning. Bethe had worked through the night analysing the theory behind the dry run experiment. There was no doubting the validity of Maxwell's equations but, he now reported, the results of the dry run magnetic measurements were meaningless. If the implosion had been perfectly symmetrical it would not have produced results different to those that had been observed. While he could not say categorically that the dry run implosion had worked, he could at least say that it had not failed.

The gloom lifted, but there was still much to worry about. Analysis of a test explosion of 100 tons of TNT and spent reactor slugs from Hanford, a 'calibration and rehearsal' shot on 7 May, suggested that high winds in the wrong direction might carry radioactive fallout from the Trinity test over populated areas of New Mexico. It was hoped that by detonating Fat Man 110 feet above the ground this risk would be reduced. Teller had worried about rattlesnakes, and asked Serber what he was going to do. He would take a bottle of whisky, he said. Teller was also worrying once more about the risk of setting fire to the atmosphere: what did Serber think of that? 'I'll take another bottle of whisky', Serber said. When Fermi offered to take bets on whether or not the atmosphere would be ignited, Groves was not impressed.

Even if the bomb worked, there was still considerable uncertainty about its potential yield. The scientists were running a pool. Teller had bet on an optimistic 45,000 tons of TNT equivalent. Serber 12,000 tons. Bethe 8,000 tons. Kistiakowsky 1,400 tons. Oppenheimer had bet on a pessimistic 300 tons. When Rabi arrived at Los Alamos he found that the only bet left open was for 18,000 tons, so he bought it.

As the morning of the test approached, the bad weather became the overriding worry. Storms had flared up on 14 July and were forecast to last for at least two days. At 2:00am on 16 July, thunderstorms were still raging, heavy rain and high winds lashing the Base Camp and command centre. The wooden structure at the top of the tower at Point Zero looked fragile in the midst of this elemental display. Some physicists worried that Fat Man might be detonated prematurely.

VIPs and scientists not directly involved in the final preparations or monitoring arrived to view the test on a cold and damp Compañia Hill, twenty miles north-west, at around 2:00am. Among them were Bethe, Chadwick, Fermi, Feynman, Frisch, Fuchs, Lawrence, Serber and Teller. They each held a plate of welder's glass to protect their eyes. Teller had brought sunscreen. There were about twenty in the crowded command centre, preparing for the test with barely suppressed excitement. Groves tried to keep Oppenheimer calm, taking him for quiet walks in the rain when he seemed about to explode.

The 4:00am deadline passed, but the weather gradually cleared and the time of the test was fixed for 5:30am. Bainbridge armed the bomb and returned to the command centre with Kistiakowsky and others who had spent the night playing nursemaid to Fat Man. At 5:10am on Monday, 16 July, the countdown began. Groves left the control bunker for Base Camp. Warning rockets were fired with five, two and then one minute to go.

The countdown reached zero. The firing circuit closed. Electronic detonators positioned symmetrically over Fat Man's surface simultaneously triggered 32 separate explosions. The explosions burned through the bomb towards the centre, the expanding shockwaves forced to turn, focusing their energy towards the core. The uranium tamper was collapsed in on itself, followed by the solid plutonium core. As the polonium and beryllium components of the initiator were crushed together, alpha particles released by the polonium chipped neutrons from the beryllium nuclei. The neutrons spilled into the small volume of high-density, super-critical plutonium. Fission upon fission of Pu-239 nuclei produced wave upon wave of neutrons, as matter converted into primordial energy.

Frisch described what happened next:

And then without a sound, the sun was shining; or so it looked. The sand hills at the edge of the desert were shimmering in a very bright light, almost colourless and shapeless. I turned round, but that object on the horizon which looked like a small sun was still too bright to look at. I kept blinking and trying to take looks, and after another ten seconds or so it had grown and dimmed into something more like a huge oil fire … It was an awesome spectacle; anybody who has ever seen an atomic explosion will never forget it. And all in complete silence; the bang came minutes later, quite loud though I had plugged my ears, and followed by a long rumble like heavy traffic very far away. I can still hear it.

The abstract scientific problem that Frisch and Meitner had wrestled with, that Christmas Eve in Kungalv, had become a horrifying reality.

Some laughed. Some cried. Most stayed silent. Fermi used a simple experiment, measuring how far some small pieces of paper were carried by the shockwave, to estimate that the bomb had produced a blast equivalent to 10,000 tons of TNT.[6]

Kistiakowsky, blown over by the shockwave from the blast, got back to his feet and demanded his ten dollars from Oppenheimer. But Oppenheimer's wallet was empty. 'There floated through my mind a line from the Bhagavad-Gita,' Oppenheimer later recalled, 'in which Krishna is trying to persuade the Prince that he should do his duty: "I am become death, the shatterer of worlds."' Bainbridge was more forthright: 'Oppie,' he said, 'now we're all sons of bitches.' At Base Camp, Groves, Conant and Bush acknowledged success with silent hand-shakes.

The atmosphere did not catch fire.

Raising the price

News of the successful test was passed to Truman and Byrnes in Potsdam. Secure in the knowledge that the Soviet Union was no longer needed to help end the war against Japan, Truman decided to let the Soviets in on the secret. A few days later he mentioned to Stalin that America had developed 'a new weapon of unusual destructive force'.

Stalin didn't seem very surprised. He told Molotov about the conversation. 'They're raising the price', said Molotov. Stalin laughed. 'Let them', he said. 'We'll have to have a talk with Kurchatov today about speeding up their work.'

---

[6] The actual yield was 18,600 tons. Rabi won the bet.

Chapter 16

HYPOCENTRE

July–August 1945

As Frisch watched the false dawn of a small, man-made sun over the New Mexico desert, preparations were under way at Hunters Point naval shipyard in San Francisco to send the physicists' first weapon to war. Furman, his artillery insignia incongruously stitched upside-down to his uniform, watched as the two components of the Little Boy uranium bomb were loaded on the USS *Indianapolis*. Inevitably, it was the large wooden crate that attracted all the attention. Some speculated that it contained Rita Hayworth's underwear.

The small but rather heavy bucket carried on board by two sailors was much less conspicuous. It was a lead bucket and it contained the U-235 shy that would be fired into the hollow cylinder formed from the uranium target rings at the other end of the bomb. The bucket was bolted to the floor in the flag lieutenant's cabin and placed under 24-hour armed guard.

The captain of the *Indianapolis* had not been advised of his mission, or his destination. Deke Parsons now told him that they were to head, unescorted, for Tinian Island in the Pacific. The captain eyed the lead bucket with some suspicion. 'I didn't think we were going to use bacteriological weapons in this war', he said. He did not get an answer. Just four hours after the successful Trinity test, the Little Boy uranium bomb left San Francisco bound for the Marianas.

Four days later, on Friday 20 July, a small team of physicists left Los Alamos for the same destination. Headed by Norman Ramsey, the team included Serber, Alvarez and Penney. Their job was to assemble the Little Boy and Fat Man bombs on Tinian Island and prepare them for delivery to their targets in Japan. From Albuquerque airport they flew first to Wendover airbase in Utah, headquarters of the 509th Composite Group on the edge of the Salt Lake desert. At Wendover the physicists were inducted into the US Army. Serber's assimilated rank was that of Colonel. Alvarez was rather disappointed to find he was to be a Lieutenant Colonel. They were issued with passports, uniforms, dog tags, and service kit. Serber looked the most at home in uniform, attracting occasional salutes from passing GIs.

The team crossed the Pacific on a C-54 transport plane, hopped islands and arrived at North Field on Tinian Island on 27 July. The *Indianapolis* had delivered its cargo the previous day.[1]

Tinian is a small island, measuring twelve miles by five miles at its widest. Yet, by this stage in the war, North Field had become the world's largest airport, consisting of six 8,500-foot runways that ended at the top of a sheer cliff perched 50 feet above the sea. The physicists would watch as, day after day, hundreds of B-29s took to the skies, following the 'Hirohito Highway' to rain death and destruction on Japanese cities in what Major General Curtis LeMay, commander of strategic air operations against Japan, called 'fire jobs'. The heavily loaded bombers would dip alarmingly after clearing the runways, before climbing slowly to their cruising altitude. Sometimes the planes didn't make it. They would 'go skimming horribly into the sea, or into the beach to burn like a huge torch'. They were so loaded with bombs that fully 1 per cent of the planes risked this fate.

---

[1] The *Indianapolis* was torpedoed on 30 July during its return to the Philippines. Only 317 of the 1,196 men on board were rescued. Whitetip sharks took many of the immediate survivors in wave after wave of attacks as they waited four days in the sea, an event memorably described by the character Quint in Steven Spielberg's film of Peter Benchley's *Jaws*.

LeMay had figured this was an acceptable loss compared to flying more sorties with lighter bomb loads.

On Tinian the physicists settled into a routine, as Serber explained to his wife Charlotte, waiting back in Los Alamos:

> The form of life out here is quickly taking shape. We get up about 6:00am, and have breakfast at 7:00. Then everybody goes to work until 11:00. Lunch then, and lie around in the sun (if it's out) till 1:00. Work till about 4:30, dinner about 5:00, kill time until 7:15 when there's a movie or a show. The movies are preceded by 15 minutes of news and combat reports. After the movies to the officer's club, for a drink or a beer or a coke, and to bed by about 10:00.

The physicists would go swimming from Yellow Beach in the afternoons. Serber was struck by the number of .50-calibre machine gun shells that littered the sea floor, testimony to the firepower of the American 2nd and 4th Marine Divisions that had invaded and captured Tinian in July–August 1944. Napalm had been used for the first time to firebomb Japanese strongholds hidden beneath the island's foliage.

Tibbets and the crews of his fifteen specially modified B-29s had spent their time since arriving on Tinian practising flying to Iwo Jima and back. They practised dropping standard bombs and 'pumpkins', crude imitation bombs the size of Fat Man, made of concrete, each filled with 6,300 pounds of high explosive and painted bright orange. Only Tibbets knew what they were training for, and because of this he was himself forbidden from flying practice missions over enemy territory.

As nobody had ever dropped an atomic bomb before, Tibbets was understandably concerned for the safety of his crew, and himself. He had worked out an elaborate manoeuvre to avoid the likely effects of the blast, turning at 30,000 feet through about 150 degrees in 30 seconds, at 200 to 250 miles per hour, ending 1,500 feet lower than at the start of the turn. He asked Serber what he thought. Serber did some quick calculations and assured him that he would be perfectly safe.

No surrender, no concessions

As Little Boy and the team of physicists made their various ways to Tinian, political and military debates were raging in Potsdam and Tokyo. Decisions were being taken which would seal Japan's fate and determine the role that atomic bombs would play in this fate.

Eisenhower learned about the success of the Trinity test from Stimson over dinner at Allied headquarters in Germany. He expressed the opinion that the bomb should not be used. The Japanese were ready to surrender, he claimed. 'I was one of those who felt that there were a number of cogent reasons to question the wisdom of such an act.' He did not want his country to be the first to use such a weapon. Stimson was furious. He was the one who had agreed to a $2 billion expenditure to build the atomic bomb, and he now felt burdened with the moral accountability for its use. Other American military leaders echoed Eisenhower's sentiments, though some, like LeMay, more from practicality than from moral concerns. LeMay was systematically wiping out Japanese cities with his incendiary bombs. The war could be ended without atomic weapons. It was surely only a matter of time.

But how much time, and at what cost in human lives? Japan's willingness to surrender was actually highly debatable. The US Army Signals Intelligence Service had towards the end of 1940 managed to build a replica of the PURPLE cipher machine used by the Japanese to encipher their diplomatic messages. Truman was receiving daily summaries of the contents of radio communications between Tokyo and its embassies. This deciphered message traffic, codenamed 'Magic', gave the Allies a window on the drama unfolding in Tokyo.

The Magic intelligence revealed that attempts to seek a negotiated peace by Japanese diplomats in Europe were being conducted without authority from Tokyo. The only authorised diplomatic effort was that of ambassador Naotake Sato in Moscow, who had been charged initially with the task of improving Soviet–Japanese relations. A meeting with Molotov on 11 July had produced no result. According to Sato, Molotov was non-committal on his proposals.

Foreign Minister Shigenori Togo had not yet received Sato's report of his meeting with Molotov when on 12 July he sent a message recommending that Sato take a further step, informing the Soviets of the 'Imperial Will' to see an end to the war. In essence, Togo now sought help from the Soviet Union to broker an end to the war on terms that would be acceptable to the Japanese. But his message betrayed little room for compromise:

> His Majesty the Emperor, mindful of the fact that the present war daily brings greater evil and sacrifice upon the peoples of all belligerent powers, desires from his heart that it may be quickly terminated. But so long as England and the United States insist upon unconditional surrender the Japanese Empire has no alternative but to fight on with all its strength for the honour and the existence of the motherland.

On 18 July Sato argued in his reply to Togo that an unconditional surrender, or something very close to this, was all Japan could hope for. In a further message dated 20 July he suggested that an unconditional surrender in which Japan's imperial institutions were allowed to continue might be acceptable to the Allies. He argued that while loyalty to the obligations of honour was good, it was surely meaningless to prove such devotion by wrecking the state.

In his reply on 21 July, Togo was emphatic:

> With regard to unconditional surrender (I have been informed of your 18 July message) we are unable to consent to it under any circumstances whatever. Even if the war drags on and it becomes clear that it will take much more bloodshed, the whole country as one man will pit itself against the enemy in accordance with the Imperial Will so long as the enemy demands unconditional surrender. It is in order to avoid such a state of affairs that we are seeking a peace which is not so-called unconditional surrender through the good offices of Russia.

Togo's vision of the whole country rising as one against the enemy was a central plank of *Ketsu Go*, a military strategy designed to weaken American resolve by inflicting enormous losses during the initial stages of the threatened invasion of Japan's home islands. In preparation, Japan had built up its homeland defences, planned for 'special attack' or suicide missions against invading forces, and formed a National Resistance Programme, consisting of all able-bodied citizens, male and female, armed with bamboo spears.

Operation Downfall, the planned invasion of Japan, was scheduled to begin with Operation Olympic, the invasion of Kyushu, southernmost of Japan's home islands, on 1 November 1945. After its capture, Kyushu would provide the air and naval bases for the launch of Operation Coronet, the capture of Tokyo itself, tentatively scheduled for 1 March 1946. In the planning of Operation Olympic, it had been assumed that the Japanese would be able to muster no more than six divisions on Kyushu, with just half of these confronting the American invasion forces, and between 2,500 and 3,000 aircraft. Intercepted military (as opposed to diplomatic) radio messages now revealed the massing of considerably greater Japanese forces on Kyushu, completely undermining the assumptions on which Olympic was based.

Estimates of American casualties varied greatly, but the battle for Okinawa had left 12,500 Americans killed or missing and an estimated 100,000 Japanese dead. Nobody was under any illusions that a high price would have to be paid for the conquest of Japan.

Stimson nevertheless believed that the Japanese were susceptible to reason. Roosevelt had set the unconditional surrender of Japanese forces as a national war aim in 1943. Stimson had earlier argued for a softening of the American position through the offer of concessions regarding the Emperor and his imperial institutions which he characterised as 'the equivalent of an unconditional surrender'. This, he believed, could make all the difference between Japan's acceptance or rejection of the ultimatum now being drafted in Potsdam. At issue, however, was the very use of the word 'unconditional'. The 22 July Diplomatic Summary of the Magic intelligence had revealed that if this word was to be used in any final ultimatum, then Togo had already rejected it.

What further complicated matters was the Soviet Union. Despite the representations now being received from Sato, Stalin had agreed to Truman's request to declare war on Japan by 15 August. This was a move that in itself would certainly end the war, most probably with the unconditional surrender of Japan, but at the same time would give the Soviets licence to make more territorial gains in the Pacific region.

There was another way, Byrnes argued. The atomic bomb offered a way to end the war against Japan before the Soviets could get involved. American lives could be saved, a war that had dragged on far too long could finally be ended, Soviet aspirations could be thwarted and American superiority in military technology could be unequivocally demonstrated, thereby establishing a strong position in the post-war world. There was a further consideration. To spend $2 billion developing a weapon that would never be used was simply unprecedented in the history of warfare.

For Truman and Byrnes, this was an easy decision. On 25 July, Truman wrote in his diary:

This weapon is to be used against Japan between now and August 10th. I have told the Sec. of War, Mr Stimson to use it so that military objectives and soldiers and sailors are the target and not women and children. Even if the Japs are savages, ruthless, merciless and fanatic, we as the leader of the world for the common welfare cannot drop this terrible bomb on the old Capitol or the new.

He and I are in accord. The target will be a purely military one and we will issue a warning statement asking the Japs to surrender and save lives. I'm sure they will not do that, but we will have given them the chance. It is certainly a good thing for the world that Hitler's crowd or Stalin's did not discover this atomic bomb. It seems to be the most terrible thing ever discovered, but it can be made the most useful.

Stimson was a tired insomniac. Truman and Byrnes, confident that they now had a weapon that would be decisive, were not interested in concessions. On 26 July, as Little Boy was being unloaded on Tinian, Truman,

Churchill[2] and the Chinese leader Chiang Kai-Shek released the Potsdam Declaration to the waiting press. The Allied position remained firm:

> We call upon the government of Japan to proclaim now the unconditional surrender of all Japanese armed forces and to provide proper and adequate assurances of their good faith in such action. The alternative for Japan is prompt and utter destruction.

There were no concessions.

Japan was not at war with the Soviet Union and Stalin was not a signatory to the declaration. Togo reasoned that it might yet be possible to get better terms through his diplomatic channels to Moscow. He appears to have been unaware that Soviet forces were already massing on the border of Japanese-occupied Manchuria, preparing for invasion.

The Supreme Council for the Direction of the War met the next day to discuss the declaration. These were the so-called 'Big Six' – Prime Minister Kantaro Suzuki, Army Minister Korechika Anami, Navy Minister Mitsumasa Yonai, General Yoshijiro Umezu and Admiral Soemu Toyoda, the Chiefs of Staff of the Imperial Army and Navy, and Foreign Minister Togo. Unsurprisingly, the militarists Anami and Umezu advocated outright rejection. Togo argued that they should instead seek to buy some time, wait until Stalin had returned to Moscow from Potsdam, and press for a mediated peace on better terms.

The outcome of the ensuing debate was, perhaps, predictable. Suzuki delivered the Japanese response the next day, 28 July, during a press conference at his official residence in Tokyo. When asked by journalists for his view regarding the Potsdam Declaration, he claimed that his government

---

[2] Churchill represented British interests at the Postdam conference, although he was about to suffer a shock defeat in the snap General Election he had called on 5 July. No doubt anticipating that a British victory in the war in Europe and Churchill's own status as nothing short of a hero would make election of Churchill's Conservative Party certain, the Conservatives had run a poor election campaign. The result, declared on 26 July – the day of the Potsdam Declaration – was a landslide victory for Clement Attlee's Labour Party. The British electorate, heartily sick of war, favoured a political party they perceived to be more committed to managing the peace.

found no value in it. He said that there was no other recourse but to *moku-satsu* the declaration, variously interpreted as 'ignore', 'withhold comment', or 'treat with silent contempt'. The Japanese would 'resolutely fight for the successful conclusion of this war'.

For the Allies, there was no ambiguity in Suzuki's reply. The die was now cast.

Petitions to the President

Szilard had not been invited to the Trinity test, and was not even aware that it was taking place. On 16 July has was busy redrafting his petition to the President. As before, the purpose of the new petition was not necessarily to try to change any decisions that had already been made. Its purpose was rather to record the dissent among the community of physicists that had helped to build the bomb. The petition acknowledged the realities of the situation but urged that Japan be given the opportunity to surrender:

> The war has to be brought speedily to a successful conclusion and attacks by atomic bombs may very well be an effective method of warfare. We feel, however, that such attacks on Japan could not be justified, at least not until the terms which will be imposed after the war on Japan were made public in detail and Japan were given an opportunity to surrender.
>
> If such public announcement gave assurance to the Japanese that they could look forward to a life devoted to peaceful pursuit in their homeland and if Japan still refused to surrender, our action might then, in certain circumstances, find itself forced to resort to the use of atomic bombs. Such a step, however, ought not to be made at any time without seriously considering the moral responsibilities which are involved.

Neither the 3 July nor 17 July petitions referred to a demonstration of the bomb as part of an ultimatum to the Japanese. Rather, the petitions sought to appeal to Truman's sense of moral responsibility. The decision to use atomic bombs against Japan would usher in a new era of warfare based on weapons of mass destruction. It was within the gift of the President of the United States – singled out by virtue of America's lead in atomic power – to prevent this from happening.

The 17 July petition attracted 68 signatures and was passed to Compton on 19 July. Compton checked first with Groves before passing it on to Nichols on 24 July. It was couriered to Groves the next day. Groves then sat on it until 1 August, when he forwarded it to Stimson's office. But Stimson was still in Potsdam, and he would not see the petition until he returned.

The wheels were already spinning. In the Potsdam Declaration, the demand for unconditional surrender was clear enough, though there was no clarity on the terms to be imposed on Japan after the surrender. The Allies threatened 'prompt and utter destruction', but there was no clarity on how this might be achieved.

There is no telling what the Big Six would have decided if, as Stimson had suggested, the 'equivalent' of unconditional surrender had been demanded without use of the actual words and if, as the physicists had urged, the terms to be imposed on Japan and the means of destruction had been made clear. However, the interpretation of subsequent events suggests that even these efforts would not have been enough to avert the catastrophe that now loomed.

The moment has arrived

The team of physicists on Tinian had all but prepared Little Boy for battle by 31 July. The final arming of the bomb would be completed by Parsons on board the B-29 that would deliver the weapon to its target. On the same day, three B-29s of the 509th Composite Group completed a final training mission to Iwo Jima, and practised the manoeuvre that Tibbets had worked out. The mission was successful. Everything was ready.

Only the weather forecast prevented the first atomic bomb from being dropped on 1 August. On 2 August, three B-29s delivered the components of a second, Fat Man, bomb to Tinian. Tibbets and his crews monitored the weather forecasts as the tension grew.

At 3:00pm on 4 August, Tibbets called a briefing for the crews of the seven B-29s that would be involved in the first atomic bombing mission. Three planes would fly about an hour ahead to determine the weather conditions and level of cloud cover over the various target choices. The B-29 carrying Little Boy would be flown by Tibbets himself, accompanied by two planes to observe and photograph the mission. The seventh B-29 would remain on the ground on Tinian as a contingency in the event that Tibbets' plane ran into problems.

The crews assembling for the briefing were surprised to find the combat room guarded by military police. Tibbets stood on the platform in front of two blackboards – both covered – and a projector screen. 'The moment has arrived', he said. 'This is what we have all been working towards. Very recently the weapon we are about to deliver was successfully tested in the States. We have received orders to drop it on the enemy.' The blackboards were uncovered. They bore maps of three Japanese cities, Hiroshima, Kokura and Nagasaki, in order of priority.[3] Forecasts predicted a break in the weather over southern Japan within the next few days, and the attack had been set provisionally for the morning of 6 August.

Tibbets then introduced Parsons, who proceeded to explain that the new weapon was unprecedented in the history of warfare. 'It is the most destructive weapon ever produced. We think it will knock out everything within a three-mile area.' It would deliver the kind of devastation that would typically require *2,000* fully loaded B-29s. Parsons now had every-body's attention. He described the Trinity test, which he had witnessed from a B-29, flying overhead, but was unable to show film of the test because the projector malfunctioned. His descriptions were enough, however. He handed out welder's goggles, and explained that the blast was expected to be brighter than the sun and might blind anyone looking directly at it.

Abe Spitzer, radio operator aboard the B-29 *The Great Artiste*, kept an illicit diary during his time on Tinian. 'We snickered here and back in the States,' he wrote, 'when informed that if this development was successful,

---

[3] It appears that Niigata had been ruled out because of the weather.

the war would be shortened by at least six months. We did no snickering now.'

Tibbets made some emotional closing remarks, and the briefing broke up in stunned silence.

Unpleasant duty

Lieutenant Shuntaro Hida was 28 years old. He had taken up his new post as a medical officer at the Hiroshima Military Hospital in 1944. Back then the progress of the war had appeared favourable. However, since the beginning of 1945 he and his colleagues had felt a growing unease about the prospects for the future of their country, despite continuous government reports of victories. His patients were witness to the battles that Japan had fought, and lost. He knew that many large Japanese cities had been heavily bombed by American forces, although, despite the frequent sighting of B-29s in the skies above Hiroshima, his city had so far been spared.

He had spent the last three months stationed at Hesaka village, near the mountain about four miles north of Hiroshima, building an underground shelter for the central hospital. On completion of this task, he and about 300 soldiers were recalled to Hiroshima on 5 August. When he arrived back at the hospital around 8:00pm there was nobody available to give him new orders, but he was asked by the duty officer to wait upon some senior medical officers who planned to stay the night, using the hospital as though it were a hotel. Hida served dinner in the X-ray room.

It was an unpleasant duty. When he was sure all his guests were drunk with *sake*, he got drunk himself.

He was woken in the middle of the night by an old farmer from Hesaka village. Hida had previously treated the farmer's granddaughter, who was suffering from heart disease. The granddaughter had fallen ill once more, and the farmer was desperate for Hida to go back with him and treat her again.

Hida remembered little of the journey, on the back of the farmer's bicycle, still very drunk, strapped to the farmer around the waist to prevent him falling off.

Enola Gay

Tibbets chose the plane that he would fly. This was Victor 82, normally piloted by Robert Lewis. For luck, and much to Lewis' annoyance, Tibbets had named the B-29 for his mother, *Enola Gay*. A final briefing was held at midnight on 5–6 August. A Protestant chaplain spoke a prayer.

The *Enola Gay* began take-off at 2:45am on 6 August. Little Boy was squeezed tightly in the bomb-bay. The plane was almost seven tons over-weight, and carrying a completely untried, though unarmed, four-ton weapon that looked like 'an elongated trash can with fins'. Tibbets, his co-pilot Lewis casting him nervous glances, chewed up all the available runway and lifted the plane gracefully into the air.

The flight to Japan's home islands passed without incident. Parsons armed the bomb at 7:30am, and advised Tibbets that it was now 'final'. At 8:15am the weather plane over Hiroshima reported light cloud. Tibbets announced that they were heading for the primary target, and climbed to 31,000 feet. Flying over Kyushu, they encountered no Japanese fighters or anti-aircraft fire.

Tibbets' bombardier, Major Thomas Ferebee, had chosen his aiming point. This was the Aioi Bridge, a T-shaped bridge at a fork in the Ohta River in central Hiroshima. The bomb bay doors opened. On Ferebee's signal, the radio operator gave a warning to the other B-29s, a continuous, low-pitched tone indicating fifteen seconds to release. Lewis scribbled in his log: 'There will be a short intermission while we bomb our target.'

The radio tone ended. The bomb was gone. The *Enola Gay* jumped. Tibbets took manual control and made his turn. He pulled on the welder's goggles only to find they were useless – he could see nothing through them at all.

He cast the goggles aside, as a bright light filled the plane.

Numberless ranks of living dead

Hida woke on the morning of 6 August. He was attending to the young girl when he noticed another B-29 high up in the sky. He thought little more about it, and turned his attention back to his patient.

At that moment, a dazzling flash struck my face and penetrated my eye. Violent heat blew against my face and both arms ... In an instant, I crept on the mat, covering my face with both my hands by instinct, and tried to flee outside by creeping. 'Fire!', I expected, but saw only the blue sky between my fingers. The tips of the leaves on the porch did not move one inch. It was entirely quiet.

Just then I saw a great fire ring floating in the sky of Hiroshima, as if a giant ring lay over the city. In a moment, a mass of deep white cloud grew out in the centre of the ring. It grew quickly, extending itself more and more in the centre of the red ring. At the same time, a long black cloud appeared spreading over the entire width of the city, spread along the side of the hill and began to surge over the valley of Ohta toward the Hesaka village, enveloping all woods, groves, rice fields, farms and houses. It was an enormous blast storm rolling up the mud and sand of the city. The delay of only several seconds after the monumental flash and heat-rays permitted me to observe the whole aspect of the black tidal wave.

I saw the roof of the primary school below the farmer's house stripped easily by the cloud of dust. My whole body suddenly flew up in the air before I could guard myself. The shutters and screens flew up around me as if they were scrap paper. The heavy straw-thatched roof of the farmer's house was blown through and was lifted up with the ceiling, and the next moment the blue sky was seen through the newly formed hole. I flew ten metres through two rooms, shutting my eyes and bending my back, and was thrown against the big Buddhist altar at the inner part of the room. The huge roof and large quantity of mud tumbled down with a terrible sound upon my body. I felt some pain here and there, but there was not time to take heed. I crept outside, groping to find my way. My eyes, ears, nose and even my mouth were filled with mud.

Checking first to make sure the young girl was still alive, Hida then borrowed the farmer's bicycle and headed back towards the city along the banks of the Ohta River. On the road he was confronted with the first of a long procession of nightmarish visions.

It was anything but 'a man'. The strange figure came up to me little by little, unsteady on its feet. It surely seemed like a man form but was wholly naked, bloody and covered with mud. The body was completely swollen. Many pieces of ragged cloth hung down from its bare breast and waist. The hands were held before the breasts with palms turned down. Water drops dripped from all the tips of the rags. Indeed, it was human skin which I thought was ragged cloth, and the water drops were human blood. I couldn't distinguish between male and female or soldier or citizen. It had a curious large head, swollen eyelids and big projected lips grew as if they formed half of its face. There was no hair on its burned head. I stepped backwards in spite of myself. Surely this strange thing was a 'man'. But it was a mass of burnt flesh hanging like rawhide, and it was covered with blood and mud.

The figure fell to the ground and convulsed. Hida tried to find a pulse but the vital signs were gone. He turned and was now confronted by more burnt and bloody survivors. Numberless ranks of living dead, staggering, on their knees, crawling on all fours.

Legacy of the bomb

The *Enola Gay* dropped the Little Boy U-235 bomb on Hiroshima at 9:15am (8:15am local time) on 6 August 1945. It exploded 43 seconds later, 1,900 feet above the city, with a yield of 12,500 tons of TNT equivalent. Ferebee missed his aiming point by 550 feet. The temperature at the burst point reached 60 million degrees, about four times the temperature at the sun's core. Those caught within a half-mile of the blast zone were carbonised in a fraction of a second. Some were simply evaporated, the only evidence of their previous existence being the shadow they had cast as they were overwhelmed by the thermal flash.

After the flash, there came the shockwave. It thrust outwards from the hypocentre – the point on the ground immediately below the burst point – at 10,000 feet per second and with a force of seven tons per square metre. In an instant it destroyed 60,000 buildings. Skin blistered by the flash was torn loose by the shockwave, left to dangle like ragged clothing.

The flash and the shockwave killed between 70,000 and 80,000 people.

Many did not die straight away. At the hastily organised field hospital in Hesaka, Hida watched as every day the most badly burned and mutilated survivors died. After a week, those who were not expected to live had died, and the remaining survivors began to show signs of recovery. Until one of the nurses noticed that her patients had suddenly developed a high fever.

> We ran up in a hurry. The patients sweated like a waterfall, and their tonsils became necrotic. While we were confused by the severity and violence of their symptoms, they began to bleed from their mucous membranes and soon spat a quantity of blood.

The doctors suspected typhoid or dysentery. But it was neither. These were the first symptoms of the radiation poisoning that was to kill at least another 60,000 people before the end of the year.

All those kids

Confusion reigned. Little Boy had been too successful. A single bomb had in one instant reduced an entire city to ruins, its infrastructure so badly damaged that it took over a day for word to get back to Tokyo.

Truman issued a press release in which he declared that the Americans had used an atomic bomb, a bomb which harnessed the basic power of the universe. But even now the militarists were in denial. Japan had had a small atomic research effort of its own, under the leadership of physicist

Yoshio Nishina, but attempts to separate U-235 using gaseous diffusion methods had been unsuccessful. Japan's military leaders understood that the production of fissionable materials for an atomic bomb was extremely difficult, if not impossible. Some argued that, whatever had been dropped on Hiroshima, it had not been an atomic bomb. Toyoda argued that even if the Americans had succeeded in making atomic bombs, they surely couldn't have too many of these weapons in their arsenal. And public opinion would prevent further use of any bombs they might possess.

On 8 August Sato renewed his efforts to persuade the Soviet Union to act as mediator, only to be told by Molotov that the Soviets had declared war on Japan and hostilities would commence the next day. One hour after midnight, Soviet troops poured across the Manchurian border and attacked Japanese positions.

The Big Six met at 10:30am local time on 9 August. They remained deadlocked. Both the Supreme Council and the cabinet were evenly divided. The militarists argued that the entry into the conflict of the Soviet Union did not invalidate *Ketsu Go*. Anami, Umeda and Toyoda pushed for concessions: there should be no occupation of the home islands, Japan should manage its own disarmament and deal with its own war criminals. Togo now argued that they should accept the Potsdam Declaration, provided they could secure assurances concerning the fate of the Emperor. Suzuki and Yonai agreed.

As the debate raged in Tokyo, Major Charles Sweeney piloted the B-29 *Bock's Car* carrying the Fat Man plutonium bomb over his primary target, Kokura Arsenal. But, although the weather plane had reported favourable conditions over the target, it was found to be obscured by haze and smoke from an earlier attack on a nearby city. Now attracting the attentions of fighters and bursts of anti-aircraft fire, and running low on fuel, Sweeney turned to his secondary target, Nagasaki, rather than bring the bomb back to Tinian or ditch it in the ocean. A brief hole in the cloud cover over Nagasaki gave the bombardier just enough time to do his job.

The Fat Man bomb was dropped on Nagasaki at 12:02pm, 11:02am local time. It exploded 1,650 feet above the city with a yield of about 22,000 tons of TNT equivalent. The steep hills surrounding the city helped to confine the explosion and reduce its impact, but 70,000 died in the blast.

The Big Six were still vacillating when news of the attack on Nagasaki reached them. Later that day Emperor Hirohito finally intervened to end the stalemate and force a decision on the terms of Japan's surrender. A formal offer of surrender was submitted to Washington via neutral Switzerland and Sweden on 10 August. In essence, it reflected the terms of the Potsdam Declaration except for one important caveat, that the declaration 'does not comprise any demand which prejudices the prerogatives of His Majesty as a sovereign ruler'.

On the surface, this appeared to be the concession that Stimson had earlier recommended, and Truman now seemed ready to accept it. But Byrnes was not. The vague wording reflected the views of the nationalist president of the Privy Council Kiichiro Hiranuma, in effect reaffirming the theocratic powers of the Emperor as a 'living god', above the rule of man-made law. So what, precisely, were the 'prerogatives' of the Emperor? Did they allow an imperial veto over any Allied decisions regarding occupation and reform in post-war Japan? Why agree to such a concession now, Byrnes argued, after the atomic bombs and the Soviet Union's entry into the war had effectively beaten the Japanese into submission?

At Truman's suggestion, Byrnes drafted a response. He was forthright, but left interpretation sufficiently open to suggest that the Japanese could determine the future role of the Emperor for themselves. 'From the moment of surrender,' he wrote, 'the authority of the Emperor and the Japanese Government to rule the state shall be subject to the supreme commander of the Allied powers, who will take such steps as he deems proper to effectuate the surrender terms.'

Although Groves was preparing a third atomic bomb for delivery to Tinian, in readiness for use after 17 August, Truman had lost his appetite for atomic carnage. 'Truman said he had given orders to stop atomic bombing. He said the thought of wiping out another 100,000 people was too horrible. He didn't like the idea of killing, as he said, "all those kids".'

In Tokyo, the Byrnes note had not helped to break the deadlock. Anami argued that they still had some power left to fight. Intercepted radio messages again betrayed a determination to fight to the bitter end: 'However, the Imperial Army and Navy are resolutely determined to continue their efforts to preserve the national structure – [one or two words missing] even if it means the destruction of the Army and Navy.' On 13 August, Truman ordered the Army Air Force to resume conventional incendiary attacks on Japanese targets, as debate about the use of a third atomic bomb on Tokyo intensified.

Finally, on 14 August, the Emperor intervened once more. Hirohito spoke of 'tolerating the intolerable', and directed his ministers to draft an Imperial Rescript (a formal edict) accepting the terms of the Potsdam Declaration. The rescript acknowledged that: 'The enemy now possesses a new and terrible weapon with the power to destroy many innocent lives and do incalculable damage.' The Emperor made a recording of his speech so that it could be broadcast to the nation the next day. Army officers launched a coup attempt that evening and tried to seize the recording to prevent its broadcast. The coup failed, and Anami committed suicide. The message was broadcast on the morning of 15 August 1945.

The war was ended.

Nausea

On Tinian, the news of Japan's surrender was received quite soberly. There was little celebration. But the physicists who had assembled and prepared the bombs for delivery were greeted as heroes. 'There were an awful lot of guys who weren't looking forward to landing on the Japanese beaches in October', Serber wrote. 'And there were about three million men whose main desire in life was to get back home. They thought we were great.'

News of the atomic bombings and their aftermath affected many Manhattan Project physicists. The bombs had helped to end the war, arguably saving the lives of many on both sides that would otherwise have been lost if Japan had been invaded. But for some physicists there was no denying the stomach-churning awfulness of the weapon they had helped create. Walking back from the celebrations at Los Alamos, Oppenheimer found one young physicist, perfectly sober, retching in the bushes.

'I still remember the feeling of unease, indeed nausea,' Frisch wrote, 'when I saw how many of my friends were rushing to the telephone to book tables at the La Fonda hotel in Santa Fe, in order to celebrate. Of course they were exalted by the success of their work, but it seemed rather ghoulish to celebrate the death of one hundred thousand people, even if they were "enemies".'

Frisch's aunt, Lise Meitner, had been staying in a small hotel in Leksand, Dalecarlia, in central Sweden, when she heard news of the bombing of Hiroshima on 7 August. A journalist from the Stockholm newspaper *Expressen* had telephoned her for a reaction. She left the hotel and walked alone in the countryside for the next five hours.

Although some effects from exposure to radiation were anticipated by Manhattan Project scientists, Groves had dismissed as propaganda the initial reports of deaths from radiation poisoning following the bombings of Hiroshima and Nagasaki. But on 21 August the full horror of death by radiation poisoning was bought home directly to Los Alamos. Twenty-four-year-old physicist Harry Daghlian was working alone late at night on a variation of the Dragon experiment involving the use of blocks of tungsten carbide reflector surrounding a six-kilo plutonium bomb core. As he put the last brick in place, it slipped and fell into the centre. The core, now exposed to extra neutrons reflected by the brick, immediately went critical. The laboratory was bathed in the blue glow of ionised air as the apparatus sprayed a lethal dose of radiation.

Daghlian developed second-degree burns on his hands and chest. As the burns blistered and his hair fell out he developed a fever. Within 26 days of the accident he was dead.

The mood at Los Alamos changed. Physicists began to leave the project and return to academia. Some left physics for good. Oppenheimer himself returned to academic life shortly after a ceremony on 16 November 1945, at which Los Alamos was presented with a Certificate of Appreciation by the US Army.

In his acceptance speech, Oppenheimer had this warning: 'If atomic bombs are to be added as new weapons to the arsenals of a warring world, or to the arsenals of the nations preparing for war, then the time will come when mankind will curse the name of Los Alamos and Hiroshima.'

Chapter 17

OPERATION EPSILON

April 1945–January 1946

'I wonder whether there are microphones installed here?' Diebner asked.

Heisenberg laughed.

'Microphones installed?' he said. 'Oh no, they're not as cute as all that. I don't think they know the real Gestapo methods; they're a bit old-fashioned in that respect.'

The War Between Men and Women

In April 1945 it had fallen to Goudsmit to decide the fates of the captured German physicists. On his recommendation, a selection were to be interned. The reasons for their detention were never clearly stated. The Uranverein physicists had been targets for so long it no doubt seemed senseless to let them go free after all the effort that had been made to capture them. Goudsmit wondered if the Allies had spent more money searching for the German physicists than the Germans had spent on the whole of their nuclear programme.

Alsos and AZUSA had followed hard on the heels of the Allied armies into Italy, France and Germany. Heisenberg had been pursued across Europe, tracked down and confronted by an armed OSS operative in

Zurich with orders to assassinate him if the wrong signals were given. Many lives had been lost to deny the German physicists access to the heavy water they had needed for their research. The Allies had indeed paid a high price and there might yet be things that could be learned from the captured physicists, even, as Goudsmit suggested, 'if it were only to convince the colleagues back home that our deductions were correct'.

And, of course, the physicists had to be kept out of Soviet hands.

It was not possible to intern them all, and Goudsmit wrestled with his selection. There was clearly nothing further to be learned from Bothe. Hahn had to be included because to omit the man who had discovered nuclear fission in 1939 would no doubt invite some severe criticism. From the documents Goudsmit had been able to review, it was clear that Weizsäcker and Wirtz had played key roles in the programme. Bagge and Korsching were junior physicists and their inclusion puzzled Weizsäcker at first, but Goudsmit reasoned that they had been involved in some novel research on isotope separation.

The most difficult choice was Laue. Early rumours of Laue's involvement in the Uranverein had turned out to be unfounded – the Nobel laureate had never worked on the project. Goudsmit was aware of Laue's brave, long-standing opposition to Nazi authority. 'Here was a man who had virtually been on our side throughout the war,' Goudsmit wrote, 'who demanded the respect of his colleagues all over the world for his science as well as his personality. Such a man was indeed rare in Germany.' Goudsmit elected to include Laue in the hope that he would be able to discuss the future of German physics with Allied scientists. Rather to Goudsmit's dismay, Laue was treated no differently from his fellow captives.

Goudsmit watched as his six 'prisoners' were bundled off to Heidelberg. Their appearance reminded him of a cartoon from James Thurber's *The War Between Men and Women*, in which a fierce female with a gun captures three sad, sorry-looking physicists.

From Heidelberg the six were transported to Rheims, France, escorted by Furman and guarded by soldiers armed with machine-guns. They were joined on 2 May by Major T.H. Rittner, assigned the task of babysitting the physicists by British intelligence with Groves' agreement. Rittner had been ordered to treat the physicists as 'guests' and ensure they were in contact with no one unless authorised by Perrin or Welsh.

From Rheims the group relocated to Versailles on 7 May. Rittner was disappointed to discover that his captives were to be held at a concentration camp at the Château du Chesnay, known colloquially as the 'Dustbin'. As the end of the war in Europe was celebrated, he protested against the conditions and urged that alternative accommodation be found. Diebner and Heisenberg joined the group on 9 May. Two days later they relocated again, to considerably better accommodation at the Villa Argentina in Le Vésinet. Furman delivered Harteck to the group later that same evening.

Although they grumbled about their circumstances, and particularly about the limited contact they had with their families in Germany, the captive physicists settled into a routine. They worked in their rooms, sat sunbathing in the garden, developed a passion for physical exercise and organised colloquia among themselves on current topics in physics. Rittner arranged for books, technical journals and games to keep them amused.

When on 3 June they were advised that they were to relocate again, Laue protested. 'That's impossible!' he declared, 'Because I have my colloquium then.' When asked if the colloquium could not be arranged for another time, Laue responded: 'But could you not have the aeroplane come some other time?'

They moved to the Château de Facqueval, in Huy, Belgium, on 4 June. There was more grumbling. The physicists were growing increasingly concerned for the welfare of their families. News reports of the expansion of the Soviet occupation zone were particularly worrying for Diebner, whose wife and son in Stadtilm were now threatened. Diebner first declared his intention to escape, then threatened suicide. Heisenberg intervened. He argued that Diebner's wife had worked with her husband on the uranium project and therefore knew too much for the Allies to risk her capture by the Soviets. Rittner was able to arrange for Diebner's family to be moved to a safer location. Although Diebner was not a religious man, with the situation happily resolved he felt the need to express his relief by visiting church. Rittner took him to a mass the following Sunday. Diebner caused something of a stir among the local churchgoers by turning up formally dressed, as though for a church parade.

Gerlach joined the group on 14 June, bringing the number of captive German physicists to ten. On 3 July they moved once more, this time to their final destination in England. As they made their approach by road from RAF Tempsford in Bedfordshire, Harteck recognised the cathedral in nearby Ely. He had once worked with Rutherford in Cambridge, and was familiar with the local countryside. In fact, they were headed for STS-61, the SOE's country house known as Farm Hall, set in spacious grounds in the Cambridgeshire village of Godmanchester.

Jens Poulsson and his team of Norwegian commandos had stayed here in 1942, preparing for their operation against the heavy water plant at Vemork. The house was riddled with hidden microphones, and a team of translators stood ready to transcribe every overheard word into English as part of an intelligence-gathering operation codenamed Epsilon.

On 6 July, a few days after their arrival, Diebner wondered if there were any hidden microphones installed there. Heisenberg laughed.

Guests at Farm Hall

The principal purpose of Operation Epsilon was to discover the full extent of the German atomic programme. It was anticipated that in informal conversation between themselves the Uranverein physicists might reveal details of the programme that they had perhaps chosen to withhold during interrogation and which were not revealed in the many documents seized by the Alsos mission.

Rittner prepared detailed, top secret reports to accompany the transcripts of the physicists' conversations. The reports commented on the morale of the physicists, their personalities, attitudes to the Allies, allegiance to the Nazi Party, their hopes and fears for the future, speculation on the reasons for their detention, their views on current events and, of course, their discussions on atomic physics.

The physicists were still 'guests'. They could not be considered as prisoners of war, as none of the ten scientists had been in the German military. They were neither suspected nor accused of any crime. The British authorities had instead cited a particularly flexible wartime law which allowed detention of specified individuals for up to six months at 'His Majesty's pleasure'.

The physicists had committed in writing to remain at Farm Hall and understood that if any one of them attempted to escape the result would be a considerable restriction of their liberty. But life at Farm Hall was comfortable. Each internee had a prisoner-of-war batman. The food was good. There was a small library and a tennis court in the grounds. There was a piano in the common room that Heisenberg would play. The physicists once again settled into a routine.

It was still not clear to them why they had been detained, and speculation inevitably inflated their sense of their own importance. They knew nothing of the Manhattan Project and had no idea of the frantic energy being expended in New Mexico as their erstwhile scientific colleagues prepared for the Trinity test. They imagined their work on the 'uranium problem' to be so far advanced and therefore so important that their fates were surely to be discussed and decided at the forthcoming meeting of the Allied heads of state at Potsdam.

'It's quite possible that they just don't want to say anything', Diebner said on 6 July.

'Then of course they will have to wait until everything has been settled by the "Big Three",' said Korsching, a reference to the impending meeting of Truman, Churchill and Stalin.

'I think the right thing in that case,' continued Diebner, 'would be for the English to give us a hint in some way. They may not be able to say it openly because of Comrade Stalin.' Diebner's biggest fear was that they would be sent back to Germany, where they might be forced to work on atomic physics for the Soviet Union.[1]

Heisenberg echoed Diebner's fears. 'It is possible that the "Big Three" will decide it at Potsdam,' he said, 'and that Churchill will come back and

---

[1] Diebner's fears were of course well-founded. Ardenne, Riehl and many other German scientists were by this time already in Moscow.

say: "Off you go, the whole group is to return to Berlin" and then we'll be in the soup.' Berlin was in the Soviet occupation zone.

That they led the world in atomic physics was an impression further emphasised by the young physicists Bagge and Korsching on 21 July, five days after the successful Trinity test.

'I am convinced [the Anglo-Americans] have used these last three months mainly to imitate our experiments', said Bagge.

'Not even that', said Korsching. 'They used them to discuss with their experts their possibilities and to study the secret documents. They probably examined a few specimens of our uranium blocks. From these specimens they can see for instance whether the [reactor] has been running already. It could have been run; the blocks must have undergone some internal chemical change.'

Korsching's logic was that they were being detained because their knowledge presented the Allies with a threat. 'But there are many military men in England,' he continued, after some further exchanges with Bagge, 'who say "Once we let those swine go back then they'll construct the uranium [reactor] and in the end they'll blow it up." They might also say "These people are so clever that our guard troops will be blown up with it, but not they themselves."'

Their illusions were about to be completely shattered.

The announcement

Just before dinner on 6 August, Rittner took Hahn to one side and advised him that the Allies had used an atomic bomb against Japan earlier that day. Hahn was distraught. As one of the principal discoverers of nuclear fission he felt personally responsible for the deaths of hundreds of thousands of people. He contemplated suicide, before calming down with a few drinks. 'If the Americans have a uranium bomb then you're all second raters', he told the rest of the group over dinner. 'Poor old Heisenberg.'

Their reactions went through the classic cycle of shock, denial, dawning realisation, incredulity and, eventually, comprehension. At 9:00pm, they listened to the radio:

Here is the news ... The greatest destructive power devised by man went into action this morning – the atomic bomb. British, American and Canadian scientists have succeeded, where Germans failed, in harnessing the basic power of the universe ... The bomb, dropped on the Japanese war base of Hiroshima, was designed for a detonation equal to twenty thousand tons of high explosive[2] ... The Allies have spent five hundred million pounds[3] on what President Truman calls the greatest scientific gamble in history – and they've won ... Up to a hundred and twenty-five thousand people helped to build the factories ... Mr Stimson, American Secretary for War, announces that *uranium* is used in making the bomb ...

The speculation now raged as to how the Allies had done it. Was it really true? Was it actually a uranium bomb? Or had they separated enough plutonium to make a plutonium bomb? If so, then surely they must have succeeded in getting a uranium reactor to work some time ago?

'I think it is dreadful of the Americans to have done it', said Weizsäcker. 'I think it is madness on their part.'

'One can't say that', Heisenberg replied. 'One could equally say "That's the quickest way of ending the war."'

'That's what consoles me', said Hahn.

Their discussion turned to practicalities. Heisenberg remained doubtful. 'I still don't believe a word about the bomb but I may be wrong. I consider it perfectly possible that they have about ten tons of enriched uranium, but not that they can have ten tons of pure U-235.'

Heisenberg's comment may have belied a degree of forgetfulness on his part. In their February 1942 report to German Army Ordnance, the Uranverein physicists had correctly estimated that a fission weapon could be constructed from between ten and 100 kilos of fissile material. Heisenberg himself was reported to have told the audience of high-ranking military figures at Harnack House in June 1942 that the mass of active

---

[2] This was an overstatement. The yield of the Hiroshima bomb was only 12,500 tons of TNT equivalent.
[3] Equivalent to $2 billion based on the prevailing exchange rate.

material for an atomic bomb would need to be about the size of a pineapple. And yet now he seemed to have drifted back to a much earlier, even pre-war conclusion, that a bomb would require tons of U-235. He doubted that the Americans had managed to separate anything like this quantity.

Hahn, for one, was puzzled. 'I thought that one needed only very little "235"', he said.

'If they enrich it slightly,' Heisenberg replied, 'they can build [a reactor] which will go but with that they can't make an explosive which will—'

'But if they have, let us say, 30 kilograms of pure "235"', Hahn interjected, 'couldn't they make a bomb with it?'

Hahn was right, of course. Frisch and Peierls had discovered this much in March 1940. And, just as clearly, someone in the Uranverein had discovered much the same by early 1942. But Heisenberg was still uncertain.

'But it wouldn't go off,' he declared, 'as the mean free path is still too big.' This was a reference to an altogether different – and erroneous – approach to calculating the critical mass, based on so-called random-walk diffusion theory. Heisenberg elaborated this approach in the subsequent discussion. If followed to its logical conclusion, the method suggests a totally impractical critical mass of about thirteen tons of U-235. Heisenberg was clearly confused, and it seems that he had at some stage worked on a different line of reasoning, as Hahn now reminded him.

'But tell me why you used to tell me that one needed 50 kilograms of "235" in order to do anything', Hahn asked.[4] 'Now you say one needs two tons.'

Heisenberg needed more time to think. 'I wouldn't like to commit myself for the moment,' he said, 'but it is certainly a fact that the mean free paths are pretty big.' His confusion and his doubts probably reflected the fact that he had paid relatively little attention to the nuclear programme in the latter stages of the war. He may have given no further thought to atomic weapons after mid-1942.

---

[4] The Little Boy bomb dropped on Hiroshima actually contained about 56 kilos of uranium, enriched to about 90 per cent U-235.

With the possible exception of Diebner (and Gerlach), the German physicists had after 1942 concluded that an atomic bomb was out of reach in any timeframe likely to have an impact on the war. By the modest objectives they had set for themselves, the German nuclear research programme could be considered relatively successful. But when compared to the achievements of the Manhattan Project, it was difficult to see the German programme as anything more than a failure. The German physicists now started to grasp for reasons for this failure.

'We wouldn't have had the moral courage to recommend to the government in the spring of 1942 that they should employ 120,000 men just for building the thing up', Heisenberg remarked, a reference to the BBC news broadcast which had mentioned that up to 125,000 people had helped to build the atomic industry in America.

Weizsäcker followed with a comment that was eventually to shape the *Lesart*, literally the 'version' or 'party line', on the reason for their lack of success, a comment that was to provoke highly-charged debate for the next 60 years.

'I believe the reason we didn't do it,' he said, 'was because all the physicists didn't want to do it, on principle. If we had all wanted Germany to win the war we would have succeeded.'

Hahn didn't accept Weizsäcker's argument. 'I don't believe that but I'm thankful we didn't succeed', he said.

As they cast about in the search for someone other than themselves to blame, Heisenberg pointed the finger at their political and military masters. 'The point is that the whole structure of the relationship between the scientist and the state in Germany,' he said, 'was such that although we were not one hundred per cent anxious to do it, on the other hand we were so little trusted by the state that even if we had wanted to do it, it would not have been easy to get it through.'

Diebner agreed. 'Because the official people were only interested in immediate results. They didn't want to work on a long-term policy as America did.'

'Even if we had gotten everything we wanted,' Weizsäcker said, 'it is by no means certain whether we would have gotten as far as the Americans and English have now. It is not a question that we were very nearly as far as they were but it is a fact that we were all convinced that the thing could not be completed during the war.'

Heisenberg disagreed. 'Well that's not quite right', he said. 'I would say that I was absolutely convinced of the possibility of our making a uranium [reactor] but I never thought that we could make a bomb and at the bottom of my heart I was really glad that it was to be [a reactor] and not a bomb. I must admit that.'

'If you had wanted to make a bomb we would probably have concentrated more on the separation of isotopes and less on heavy water', said Weizsäcker.

At this point Hahn left the common room. Weizsäcker continued: 'If we had started this business soon enough we could have got somewhere. It they were able to complete it in the summer of 1945, we might have had the luck to complete it in the winter of 1944–45.'

'The result would have been that we would have obliterated London but still would not have conquered the world,' remarked Wirtz, 'and then they would have dropped them on us.'

Weizsäcker responded, picking up his earlier theme. 'I don't think we ought to make excuses now because we did not succeed, but we must admit that we didn't want to succeed. If we had put the same energy into it as the Americans and had wanted it as they did, it is quite certain that we would have not succeeded as they would have smashed up the factories.' This was probably a reference to the Allied sabotage operations against the Vemork plant, and subsequent bombing of German factories towards the end of the war.

Later in the same conversation, Wirtz further elaborated Weizsäcker's *Lesart*. 'I think it is characteristic that the Germans made the discovery and didn't use it,' Wirtz said, 'whereas the Americans have used it. I must say I didn't think the Americans would dare to use it.'

Upset by Korsching's criticism of the project's leadership, Gerlach had left the common room and headed for his bedroom, where he was later heard sobbing. Hahn, Laue and Harteck joined him to offer their consolations. In his report, Rittner observed that Gerlach, who had once hoped that their work on uranium would help to 'win the peace', now acted like a defeated general, whose only remaining option was to shoot himself.

As Harteck entered Gerlach's room, Gerlach asked: 'Tell me, Harteck, isn't it a pity that the others have done it?'

'I am delighted', he replied.

'Yes,' said Gerlach, 'but what were we working for?'

'To build [a reactor],' said Hahn, 'to produce elements, to calculate the weight of atoms, to have a mass spectrograph and radioactive elements to take the place of radium.'

'We could not have produced the bomb but we could have produced [a reactor],' said Harteck, 'and I am sorry about that. If you had come a year earlier, Gerlach, we might have done it, if not with heavy water, then with low temperatures. But when you came it was already too late. The enemy's air superiority was too great and we could do nothing.'

Lesart

Few of them appeared to get any sleep on the night of the announcement. Laue in particular was concerned about Hahn's state of mind. Others were concerned for Gerlach. They stayed awake until the early hours of the next morning, so that they could be sure that Hahn would not take his own life. In his diary, Bagge wrote of a conversation with Laue at one o'clock in the morning: 'When I was young,' Laue had told him, 'I wanted to do physics and experience world history. I have done physics, and I have witnessed world history. I can really now say that, in my old age.'

Heisenberg's argument, that they hadn't tried to build a bomb because they didn't think it could be done within the likely timeframe of the war, was too morally ambiguous for Weizsäcker. The next day, he further elaborated the *Lesart*. 'History will record that the Americans and the English made a bomb,' he said, 'and that at the same time the Germans, under the Hitler regime, produced a workable [reactor].[5] In other words, the peaceful

---

[5] Not true, of course. At this stage the German physicists did not know that the Allies had built a working reactor as early as December 1942.

development of the uranium [reactor] was made in Germany under the Hitler regime, whereas the Americans and the English developed this ghastly weapon of war.'

So, Weizsäcker's argument went, the Allies had succeeded in making and using an 'immoral' weapon. The German physicists had not wanted to do this on moral grounds, but they could have done it if they had really wanted to.

They now became concerned that the reports of their work appearing in the press were inaccurate and, at Rittner's suggestion, agreed to draft a memorandum to set the record straight. This memorandum, dated 8 August, explained that the Uranverein had never seriously pursued the possibility of a bomb:

> At the beginning of the war a group of research workers was formed with instructions to investigate the practical application of [nuclear] energies. Towards the end of 1941 the preliminary scientific work had shown that it would be possible to use the nuclear energies for the production of heat and thereby to drive machinery. On the other hand, it did not appear feasible at the time to produce a bomb with the technical possibilities available in Germany. Therefore the subsequent work was concentrated on the problem of the [reactor] for which, apart from uranium, heavy water is necessary.

The memorandum also sought to establish Hahn's priority as the discoverer of nuclear fission, playing down Lise Meitner's role:

> The Hahn discovery was checked in many laboratories, particularly in the United States, shortly after publication. Various research workers – Meitner and Frisch were probably the first – pointed out the enormous energies which were released by the fission of uranium. On the other hand, Meitner had left Berlin six months before the discovery and was not concerned herself in the discovery.

No reason was given for Meitner's departure from Berlin. And Hahn had conveniently forgotten the letters he had exchanged with his much-missed associate.

The memorandum was signed by all ten physicists, though Heisenberg had had to lean on Bagge, Diebner, Korsching, Weizsäcker and Wirtz to get them to sign. Laue signed to endorse the accuracy of the statement, but emphasised that he had played no part in the work described.

A confused explanation

In the days that followed, Heisenberg set about the task of working out how the Allies had done it. He gave a seminar to the group on 14 August. This seminar reveals quite starkly the level of ignorance among the German physicists concerning even some of the most basic principles of atom bomb physics.

By this time Heisenberg had stopped using the approach which had earlier led him to conclude that tons of U-235 would be required, but he was still far from working out a critical mass based on a fast-neutron chain reaction in the way that Frisch and Peierls had done. Whatever methodology had been used to conclude that between ten and 100 kilos of fissile material would be required appeared to have been forgotten. But, although Heisenberg was now at least heading in the right direction, in his seminar he was still unable to distinguish clearly between the physics of a bomb and the physics of a reactor.

In fact, this lack of distinction dogged the seminar right from the beginning, which Heisenberg opened by saying: 'I should like to consider the U-235 bomb following the methods we have always used for our uranium machine.' He goes on to say: 'It then turns out in fact that we can understand all the details of this bomb very well.' In very general terms, Heisenberg was able to fathom some of the principles of the Little Boy uranium bomb, but by the end of the seminar there appears in fact to have been very little real understanding. The remarks made by Heisenberg's colleagues only added to the confusion.

As he had sat facing Heisenberg across the table in Heidelberg, Goudsmit had found it all rather sad and ironic. In the early stages of the war the Allied physicists had held their German colleagues in high esteem and feared for what such combined talent could deliver to Hitler's arsenal. Bohr had come away from his meeting with Heisenberg in Copenhagen in September 1941 with the clear impression that everything was being done in Germany to develop atomic weapons. That fear had led, eventually, to Tube Alloys and the Manhattan Project. And, ultimately, to Hiroshima and Nagasaki.

The stark reality was that the German physicists had made relatively little progress. The programme had never been escalated beyond a somewhat loose association of individual academic research projects, within which the physicists would occasionally squabble over scarce resources. Although they had carried out some limited research and indulged in some speculation, from mid-1942 the physicists at least in Heisenberg's circle had never really attempted to build a bomb.

However, they *had* attempted to build a working reactor. They had failed, in part because the Allies were relentless in denying them access to sufficient quantities of heavy water for their experiments. But they had also failed in part because they had not been able to work together effectively. While detained in Huy, Heisenberg had argued that the physicists had now been afforded an opportunity to collaborate and pool their information, advancing their work beyond what the Alsos mission had uncovered from the documents that had been confiscated. It was a remarkable admission. Useful collaboration between the disparate groups within the Uranverein had happened only because they had been captured and interned by the Allies.

Then there was the role that Heisenberg himself had played. Bagge and Diebner acknowledged Heisenberg's influence as they speculated on what would now happen to them after the announcement on 6 August.

'They won't let us go back to Germany', Diebner said. 'Otherwise the Russians will take us. It is quite obvious what [the Americans] have done, they have just got some system other than ours. If a man like Gerlach had been there earlier, things would have been different.'

Bagge was not so sure. 'Gerlach is not responsible,' he said, 'he took the thing over too late. On the other hand it is quite obvious that Heisenberg was not the right man for it … Heisenberg could not convince anyone that the whole thing depended on the separation of isotopes. The whole separation of isotopes was looked on as a secondary thing. When I think of my own apparatus – it was done against Heisenberg's wishes.'

Although he had never led the programme, deference to Heisenberg's authority as Germany's leading theoretical physicist meant that he remained highly influential throughout the programme and was rarely challenged on his views. He had stubbornly persisted with inferior reactor configurations 'for the sake of method'. This, combined with his tendency to favour a rather loose approach to problem-solving, his unwillingness to let go of what he perceived to be an elegant theoretical solution, animosity towards Diebner and his experimental approach, and a general lack of experimental or engineering experience, were all ultimately telling.[6]

It seems that, at least in his own mind, Heisenberg had throughout the war maintained his commitment to the ideal of the 'apolitical' scientist, rising above the day-to-day political concerns of ordinary German citizens. In this way he believed he had remained uncontaminated by Nazi ideology. Whether by design or default, this attitude would allow him subsequently to distance himself from the brutal acts that had been perpetrated in the name of that ideology, acts that were now being exposed to a disbelieving world.

This aloofness was transmitted through to his work for the Uranverein. He had remained above it all, largely ambivalent to the work on nuclear problems[7] and, in the latter stages of the war, happy to exploit the advantages of the nuclear programme as a means for self-preservation, for himself and for his colleagues. In truth, in the final years of the war Heisenberg

---

[6] Heisenberg had almost failed to secure his doctorate at the University of Munich because he had been unable to derive the simple mathematical expressions for the resolving power of a microscope.

[7] This characterisation is derived from recent conclusions by historian Mark Walker, who takes some pains to remind us of the formal definition of ambivalence: contradictory emotional or psychological attitudes toward a particular person or object, often with one attitude inhibiting the expression of another. See Walker, 'Nuclear Weapons and Reactor Research at the Kaiser Wilhelm Institute for Physics'.

had focused more of his attention on his academic research on cosmic rays and on his foreign lecture tours, as an ambassador for German culture.

It was this latter activity that had betrayed him. To his former colleagues in Nazi-occupied Europe, Heisenberg was perceived to be a very willing representative of an oppressive, hateful, evil regime. He was seen as fully engaged in that regime's strategy of cultural imperialism. His own brand of nationalism, which was no doubt in his own mind very distinct from Nazi ideology, was not so readily distinguished by those he encountered living under the Nazi yoke. He might have thought he had maintained an apolitical stance, but his colleagues saw otherwise.

And this was what Heisenberg had ultimately failed to appreciate. In the end it did not matter what Heisenberg *thought*. What really mattered was what he *did*.

The German programme was never escalated to an industrial scale because the physicists believed the bomb to be out of reach. They had not produced a bomb because they believed it to be technically unfeasible, *not* because it was an immoral weapon that should not be made. It was, as Heisenberg himself acknowledged, a *lack* of courage that had prevented the physicists from sticking their necks out and asking for funding for an industrial-scale effort when they were presented with an opportunity to do just this on 4 June 1942.

Nobel prize song

Groves received copies of Rittner's reports from Farm Hall and read them with great interest, often making notes in the margins. In truth, there was little to be learned except the reasons for the German physicists' failure. The recorded conversations yielded some insights into the physicists' thinking, their attitudes and their aspirations for the future, but there were no further secrets to be revealed.

As their captivity at Farm Hall continued into the winter of 1945, the German physicists became increasingly restive. Delays in reaching a decision on their fate and the perceived intransigence of the British authorities sometimes led to heated exchanges between the physicists and their captors. They threatened to break the terms of their parole, and drafted letters calling for an immediate release and an opportunity to return to Germany so that they could pursue their scientific work.

There was one piece of news that served to lighten the gloom, however. On 16 November the *Daily Telegraph* reported that the 1944 Nobel prize for chemistry had been awarded to Hahn, for his discovery of nuclear fission. The Swedish Academy had no idea where Hahn was.

Although the physicists were initially doubtful that the announcement was genuine, Rittner promised to try to verify the report via London. The physicists nevertheless celebrated in style. Laue made an emotional speech which he ended thus:

> But my speech would be grossly incomplete if I did not also come to mention yet another person: your wife. She must have received the news also; what conflicting feelings must be assailing her this evening! But I hope, indeed, that joy will finally predominate with her, the proud joy to be the wife of such a man. Gentlemen! We lift our glasses and drink to the health of Otto and Edith Hahn. Three cheers for them.

Both Laue and Hahn were in tears.

Aside from speeches, the celebrations also included the hastily-written 'Farmhaller Nobel-prize song', sung by Diebner and Wirtz, which opened as follows:[8]

> Detained since more than half a year
> Sind Hahn und wir in Farm Hall hier.
> Und fragt man wer is Shuld daran
> So ist die Antwort: Otto Hahn.

For her part, Lise Meitner had fallen victim to selective memory and the jealousy of a scientific rival. She undoubtedly deserved either a share of

---

[8] 'Detained since more than half a year/Are Hahn and we in Farm Hall here./If you ask who bears the blame/Otto Hahn's the culprit's name.'

the chemistry prize with Hahn or award of the physics prize. She received neither. In their 8 August memorandum the German physicists at Farm Hall had already forged another *Lesart*: that nuclear fission had been discovered by a German chemist without the help of physics or of Meitner. The Germans needed a new hero. It was to be Hahn, and Hahn alone.[9]

At the same time, Manne Siegbahn had blocked the award of the 1945 physics prize to Meitner. It had gone instead to Wolfgang Pauli. From being hideously lauded in the press as the 'Jewish mother of the bomb', Meitner was now relegated to a footnote in history as Hahn's *Mitarbeiter*, or subordinate.

On 3 January 1946 the German physicists were flown to Lübeck and then transported by bus to Alswede, in the British zone of occupation in northern Germany. There they were released, but forbidden from travelling outside the British zone. Exactly six months had elapsed since their arrival at Farm Hall.

Their months of comfortable detention made the shock of a ruined Germany all the more palpable. Laue wrote to his son in Princeton that: 'The complete suffering of war makes itself felt only now.' Heisenberg had learned of the death of his mother while at Farm Hall, and he now made an emotional journey to her graveside.

The physicists eventually went their separate ways. Diebner and Harteck went to Hamburg, Gerlach went first to Bonn, then to Munich. Hahn and Heisenberg moved to Göttingen, where Max Planck had sought refuge at the end of the war and which had been designated by the British occupiers as a centre for revival of the fortunes of German science. They were soon joined by Bagge, Korsching, Laue, Weizsäcker and Wirtz. The physicists slowly began to pick up the pieces of their lives, and their science, though they were forbidden from working on nuclear physics.

The first war of physics was finally over.

---

[9] Fritz Strassman's role had also been overlooked, although the Swedish Academy tends to recognise and reward only the leaders of notable scientific endeavours, not their research assistants, students or subordinates.

PART IV

PROLIFERATION

Chapter 18

ДОГНАТЬ И ПЕРЕГНАТЬ!

August 1945–February 1946

But another, very different kind of war was about to begin.

The Allies had argued that the atomic bombing of Hiroshima and Nagasaki had been necessary in order to end the war quickly and save potentially hundreds of thousands of lives. Stalin saw it somewhat differently. The Soviet Union had been poised to strengthen its position in the Far East, not only in Manchuria but also in Japan itself. Stalin had written to Truman on 16 August 1945 requesting that Soviet forces be allowed to occupy Hokkaido, the northernmost of Japan's home islands, and accept the surrender of Japanese forces on that island. Truman refused, insisting that Japanese forces on all the home islands surrender to the United States.

Despite Truman's objections, on 19 August the order was given for Soviet forces to occupy the northern half of Hokkaido. The order was rescinded three days later. Stalin had thought better of the move. He decided that it would risk a major political row with the United States, and possibly even precipitate direct conflict. 'To avoid the creation of conflicts and misunderstanding with respect to the Allies,' the order of 22 August stated, 'it is categorically forbidden to send any ships or planes at all in the direction of the island of Hokkaido.'

Stalin recognised that the use of atomic bombs against Japan had been as much about limiting Soviet ambitions in the region as ending the war. He did not fear that America would use the atomic bomb directly against his country, but he was nevertheless shaken by the weapon's destructiveness. He recognised that the balance of power had shifted. 'Hiroshima has shaken the whole world', he is reported to have said. 'The balance has been destroyed.' He anticipated that America would use the veiled threat of the bomb as a bargaining tool in post-war negotiations with the Soviet Union. This he could not accept.

The bombing of Hiroshima was given a muted fanfare in the Soviet press. The young physicist Andrei Sakharov was on his way to the bakery on the morning of the announcement. He stopped to glance at the newspaper headline. 'I was so stunned that my legs practically gave way', he later wrote. 'There could be no doubt that my fate and the fate of many others, perhaps of the entire world, had changed overnight. Something new and awesome had entered our lives, a product of the greatest of the sciences, of the discipline I revered.'

Even before the formal Japanese surrender, Stalin had decided that the Soviet Union must have atomic weapons in its arsenal. No amount of openness of the kind advocated by Bohr and Oppenheimer would have changed the simple fact that Stalin wanted a bomb of his own.

On 20 August the State Defence Committee issued an edict establishing a Special State Committee on Problem Number One charged with, among other tasks, 'the construction of atomic energy facilities, and the development and production of an atomic bomb'. The committee was to be chaired by Beria, Stalin's 'whip'. It reported initially to the State Defence Committee, and when this organisation was disbanded on 4 September it reported instead to the USSR Council of People's Commissars.

The appointment of Stalin's notoriously brutal henchman to chair the Special State Committee was probably not universally applauded by the Soviet physicists at the time. Beria was no scientist or engineer and he was immensely distrustful of intellectuals. Yet he also brought some strong, positive characteristics to the leadership of the Soviet atomic programme, as Pavel Sudoplatov, responsible for the administration of NKVD 'special tasks' (including sabotage and assassination),[1] noted:

> Beria was harsh and rude to his subordinates but at the same time attentive and supportive in every way to the people doing the real work. He protected them from the intrigues of the local NKVD and party bosses. Beria always warned every manager about his total responsibility for the fulfilment of his assignment. Beria had the singular ability to inspire both fear and enthusiasm.

Kurchatov had struggled under Molotov's leadership and had made no secret of his dissatisfaction. Despite their reservations, the physicists found that Beria was someone they could deal with, as Yuli Khariton and Yuli Smirnov acknowledged many years later:

> Beria understood the necessary scope and dynamics of the research. This man, who was the personification of evil in modern Russian history, also possessed great energy and capacity for work. The scientists who met him could not fail to recognise his intelligence, his will power, and his purposefulness. They found him a first-class administrator who could carry a job through to completion.

The committee included Georgei Malenkov, a member of the State Defence Committee and a rising star in the Politburo, Nikolai Voznesensky, head of the State Planning Committee (Gosplan), Boris Vannikov, People's Commissar of Munitions, Zavenyagin, who in May of that year had led the Soviet Union's equivalent of Alsos, the chemical industry commissar Mikhail Pervukhin, NKVD General Vitaly Makhnev, and the physicists Kapitza and Kurchatov. Perhaps surprisingly, there were to be no military members of the committee.

---

[1] While deputy director of the NKVD's Foreign Department, Sudoplatov had been responsible for overseeing the assassination of Leon Trotsky in August 1941.

Kurchatov was to continue as scientific director of the programme. The decree also established an eleven-man Technical Council, chaired by Vannikov and consisting of distinguished Soviet physicists such as Kapitza, Kurchatov, Abram Ioffe, Khariton, Alikhanov and Kikoin.

Stalin's immediate reaction to the news of the bombing of Hiroshima had been to blame his scientists for failing where the Americans had succeeded. He lost his temper, banging his fists on the table and stamping his feet. He accused Kurchatov of not being demanding enough. Kurchatov simply pointed out that their country had been devastated by war, a war that had killed between 25 and 26 million Soviet citizens and laid waste to much of the country's infrastructure. Irritated, Stalin muttered: 'If a child doesn't cry, the mother doesn't know what he needs. Ask for whatever you like. You won't be refused.'

Kurchatov was told to build a Soviet atomic bomb as quickly as possible and not to count the cost.

Department S

Pavel Sudoplatov was appointed as head of a new autonomous intelligence department – called Department S – dedicated to atomic espionage and combining the efforts of the GRU and NKVD. The Soviet foreign intelligence services were now told to redouble their efforts to acquire documentary materials on the atomic bomb. Towards the end of August, Colonel Nikolai Zabotin in Ottawa received the following urgent message from the GRU in Moscow: 'Take measures to organise acquisition of documentary materials on the atomic bomb! The technical process, drawings, calculations.' Zabotin had laboured for some time to establish contacts with scientists working on the Canadian atomic research project at Montreal, only to be told in early 1945 that a British GRU spy had been quietly working there for over two years.

On Zabotin's instructions, in May 1945 Alan Nunn May had been approached at his home on Swail Avenue in Montreal by GRU agent Pavel Angelov. May had been worried that he was under surveillance by the Royal Canadian Mounted Police (RCMP) and had initially expressed no wish to re-establish contact with Soviet intelligence. Angelov insisted and eventually, although reluctantly, May agreed to be reactivated. Through the late spring and early summer of 1945 May had provided the Soviets with reports on the Montreal project. He had made a total of four trips to Chicago in 1944 to liaise with the Met Lab physicists before Groves had become concerned about how much information the British physicist was gaining. Groves had no reason to be suspicious of May, but his obsession for compartmentalisation meant that a request for a further visit in the spring of 1945 was declined.

The spent fuel rods that had been sent to Montreal in July 1944 contained plutonium and traces of another radioactive isotope of uranium, U-233, produced by neutron bombardment of an isotope of thorium, Th-232. U-233 was being investigated as a potential alternative bomb material, and earlier in August 1945 May had provided small samples of both U-233 and enriched uranium which were eagerly despatched to Moscow. Unaware of the dangers from these radioactive materials, the courier who had delivered the samples to Moscow suffered painful lesions and was obliged to have regular blood transfusions for the rest of his life.

But Moscow had been largely unimpressed. May's espionage materials told them nothing particularly new. Now Zabotin faced even more urgent pressure to gather information on the bomb, and his principal source was about to leave Canada and return to Britain. May was to return in September to take up a lectureship in physics at King's College London. On 22 August Zabotin was given detailed instructions to pass on to May, specifying the time and place of his rendezvous with a London espionage contact, including the code phrase 'Best regards from Mikel'. May was scheduled to meet his new contact on 7 October, in front of the British Museum.

A complete picture

Although there would be no further breakthrough from Ottawa, Soviet spies much closer to the Manhattan Project at Los Alamos were about to deliver the goods. Just as the Special State Committee was being formed and Stalin was giving his instructions to Kurchatov, Lona Cohen was busy doing her bit for the cause.

In early August she had taken lodgings in the quiet, unassuming spa town of Las Vegas, New Mexico, in preparation for her first clandestine meeting with Hall. By this time the news about Hiroshima and Nagasaki had taken the lid off the secrecy surrounding Los Alamos. Now everyone knew what had been going on up on the Hill and security had been even further tightened. Hall and Cohen had arranged to meet on a Sunday in early or mid-August on the campus of the University of New Mexico in Albuquerque. Cohen had made the 120-mile trip from Las Vegas to Albuquerque on three consecutive Sundays, but Hall had not made an appearance. She decided to make the trip one last time.

She found a young man who didn't appear to be doing very much on a campus that was largely deserted. She figured this was her man. They spoke for about half an hour as they wandered aimlessly around the campus. Hall knew Cohen only as 'Helen', and was rather disconcerted by her frank physicality. As they walked past a pretty woman Cohen nodded in her direction and wondered aloud how much the young spy would enjoy spending time with her. Cohen explained that the spy network operated by the Soviets had ways to protect them if things got too 'hot'. She told him that they could look forward to a new life in Moscow should they be compromised. Hall didn't share Cohen's eagerness to embrace life in the Soviet Union. He told her he found the prospect grim.

At the end of their discussion Hall passed to Cohen about half a dozen sheets covered with writing and diagrams. He was somewhat concerned that there was little more information in them than he had already passed to the Soviets via his friend Saville Sax.

Cohen headed back to Las Vegas, and placed Hall's papers at the bottom of a Kleenex box, hidden beneath a pad of tissues. At the railway station, she was surprised and perturbed to discover that security was now considerably tighter. Each rail carriage was being checked by two plainclothes agents, presumably FBI, asking questions and conducting searches. Cohen had to think fast.

She decided to play the dumb blonde. She stood on the platform and rummaged through her belongings, as though looking for her ticket, a task hindered by the Kleenex box she was holding in her hand. She pulled at the zip on one of her bags. It stuck. Seeing her desperation, a conductor came over to help. She handed him the Kleenex box, freeing up both hands to search through her bags, and eventually found the ticket. She answered the agents' questions and the conductor directed her towards her rail carriage. She acted as though she had forgotten all about the box of Kleenex tissues. But the conductor had not forgotten. Without paying it the slightest attention, he handed the box back to her once she was on the train.

She told the story to Yatskov when she met him in New York to hand over Hall's materials. Yatskov told her that the material in the Kleenex box could have meant a trip to the electric chair if it had been discovered. She joked that it 'had been in the hands of the police'.[2] This episode, and Cohen's quick thinking, was to enter into Soviet intelligence legend, to be discussed and elaborated by Soviet spies for the next 60 years.

In September, Greenglass received an early furlough and headed for New York with his wife, Ruth. The day after their arrival they received a visit from Julius Rosenberg. No doubt under pressure to step up intelligence-gathering by Feklisov, his Soviet controller, he was keen to get whatever further information Greenglass could give him about the atomic bomb. He offered Greenglass $200. Now that she understood what this was all about, Ruth privately protested to her husband, but Greenglass was by now in too deep. 'I have gone this far and I will do the rest of it, too', he told her.

He wrote out a report containing as many details of the Fat Man bomb design as he had been able to glean from his work on the moulds for the explosive lenses, and by keeping his eyes and ears open. His account was wrong on several points, but it contained a description of the polonium–beryllium initiator and new information about improved implosion designs that would require less plutonium. These were based on ideas

---

[2] Yatskov later recalled that he had received a 'thick pile' of documents from Cohen, yet Hall recalled handing over only a few pages. It seems unlikely that Cohen could have hidden a thick pile of papers in a Kleenex box without this being noticed.

developed at Los Alamos towards the end of the war: hollow shell designs with an outer layer of plutonium or U-235 which was to be compressed by implosion onto a solid core of plutonium suspended or 'levitated' at the centre. If the Fat Man solid core implosion design was the equivalent of pushing a nail home, the levitated core design was the equivalent of hitting it home with a hammer.

Back at Los Alamos, the British delegation planned to hold a party to celebrate the end of the war, the birth of the atomic era and the physicists' impending return to Britain. Peierls had been awarded an OBE. Oppenheimer was to receive a Medal of Merit from Truman. There was much to celebrate. A performance of *Babes in the Wood* was rehearsed, with the physicists as the babes and a security officer as the wicked witch. Frisch was to play an Indian maiden. James Tuck was to play the devil, with a luminous red tail and impressive moustaches. Steak and kidney pie was to be served, followed by trifle – a dish unknown to the Americans. Engraved invitations were sent out.

Fuchs had not auditioned for a part, but he did offer to drive to Santa Fe to fetch beer a few days before the party. He set out on 19 September, the day he had arranged to meet Gold. Now wary of the tightened security at Los Alamos, he had not prepared notes for Gold beforehand. Instead he parked his dilapidated Buick at a quiet spot in the empty scrublands between Los Alamos and Santa Fe and wrote out his report in the car.

When he met Gold late in the afternoon on the outskirts of Santa Fe he told him that he had been appalled by the destruction and death that the atomic bombs had wrought on Hiroshima and Nagasaki. He also explained that there was no longer a free exchange of information between American and British scientists at Los Alamos. The British were being frozen out. He anticipated returning to Britain to continue working on atomic energy before the end of the year or in early 1946. Yatskov had himself already anticipated this, and had given Gold a protocol for Fuchs to use to establish a new Soviet espionage contact in London.

Aware that the Soviet Union was being increasingly considered as the enemy in place of Germany and Japan, Fuchs was more determined than ever to find ways to help the country to which he had given his allegiance.

He figured that the Soviets needed to know what they were up against, and had worked out how quickly America could build its arsenal of atomic weapons based on the prevailing rates of U-235 and plutonium production. He provided Gold with some further bomb design details, and information about the composite bomb design now being considered by the Los Alamos physicists.

Fuchs drove Gold back to Santa Fe. He handed over the report he had written and they parted company, never to meet again. Fuchs still didn't know Gold's real name.

Although the reports from Hall, Greenglass and Fuchs differed in their details, they provided essential corroboration. Had there been only one spy working at Los Alamos, there would always have been doubts about the veracity of the information he provided (and Beria was never anything less than deeply suspicious). Three spies independently reporting similar design details helped reassure the Soviets that they were not being fed disinformation.

When combined with the report on the development of the bomb by Princeton physicist Henry D. Smyth, entitled *Atomic Energy for Military Purposes*, which was openly published on 12 August, the espionage materials painted a believable picture.[3] The reports were consolidated and distilled into a summary that was sent to Beria on 18 October 1945. The authors of this seven-page document had taken care to weed out erroneous or contradictory information from the Los Alamos spies. The report was so consistent that Sudoplatov thought it represented a chapter of the Smyth report that had been omitted from publication for security reasons. It hardly mattered whether this was true or not. The deliberate omissions in the Smyth report were filled in by espionage. To all intents and purposes, the Soviets now had a complete picture.

---

[3] The Smyth report can be viewed at www.atomicarchive.com. Szilard thought the publication of this report was a big mistake. He believed it gave away the general ideas on which the bomb was based and would make post-war international control all the more difficult. He refused to approve the report's release.

Gunslinger

The first test of post-war atomic diplomacy between America and the Soviet Union came at the first session of the Council of Foreign Ministers in London, which started on 11 September 1945. The council, which included the American, Soviet, British, French and Chinese foreign ministers, had been established at Potsdam to draw up treaties, develop a peace settlement for post-war Germany, and settle outstanding territorial disputes.

Disagreements within the Truman administration had already surfaced a week earlier.

Byrnes was minded to play 'power politics' at the council meeting, to use the fact of the American atomic bomb to leverage concessions from the Soviets who, based on what he had seen at Potsdam, couldn't be trusted to keep any promises they made. Stimson disagreed. He had been working on a memorandum concerning the effects of the bomb on post-war American–Soviet relations in which he warned against taking this line. In this memorandum, which was delivered to Truman on 11 September, he argued that the future development of relations was not merely connected with the bomb, but virtually dominated by it:

> Those relations may be perhaps irretrievably embittered by the way in which we approach the solution of the bomb with Russia. For if we fail to approach them now and merely continue to negotiate with them, having this weapon rather ostentatiously on our hip, their suspicions and their distrust of our purposes and motives will increase.

Stimson was thinking of Byrnes in the role of the ostentatious gunslinger. Rather than foster an atmosphere of deepening mutual distrust, Stimson proposed to enter an arrangement with the Soviets on the control and limitation of the atomic bomb as an instrument of war. This might mean ceasing all work on the further refinement of the bomb and impounding the bombs already stockpiled, provided the Soviets and the British agreed to do likewise. If they could get this far, Stimson suggested, then France and China could be brought into the agreement which could then be taken over by the United Nations.

At a meeting with Stimson the next day, Truman nodded his agreement in principle, but there appears to have been little appetite for Stimson's proposals in Truman's cabinet. By this time, Stimson had submitted his resignation on health grounds.[4] Truman had regretfully accepted.

The Soviets had anticipated that Byrnes might be belligerent, and had decided to counter by visibly playing down the significance of the bomb. Though the bomb was not on the agenda for the Council of Foreign Ministers meeting, it was on everybody's minds. Molotov decided to provoke Byrnes into making his position clear.

At a reception on the third day of the meeting, Byrnes asked Molotov when he was going to stop sightseeing and get down to business. Molotov asked why, did Byrnes have an atomic bomb in his pocket? 'You don't know southerners,' Byrnes replied, 'we carry our artillery in our pocket. If you don't cut out all this stalling and let us get down to work, I'm going to pull an atomic bomb out of my hip pocket and let you have it.' It was no doubt meant as a light-hearted aside, but it was not very subtle and confirmed Molotov's suspicions.

If Byrnes thought the Soviets could be intimidated by the bomb, Molotov was determined to prove him wrong. The London meeting ended on 2 October without agreement on any of its key subjects. Molotov was accused in the British press of recklessly squandering the Allied nations' good will. Journalists referred to him as 'Mr Nyet'.

Duplicate or invent?

The next critical decision for the Soviet atomic scientists was one of strategy. There was obviously no doubt that a Soviet atomic bomb could be built and would be devastatingly effective. Soviet spies right at the heart of the Manhattan Project had provided intimate details of the Fat Man

---

[4] Stimson was 78 on 21 September 1945 and suffered a heart attack the following month. He died in October 1950.

plutonium bomb design. The question that remained was this: should the Soviet scientists simply follow in the footsteps of their counterparts on the Manhattan Project and duplicate the Fat Man design? Or should they develop their own approach to the weapon?

Kurchatov and Khariton reviewed the espionage materials that had been summarised for Beria in October, and concluded that the first Soviet bomb should be a copy of the Fat Man design. 'Given the tension between the Soviet Union and the United States at the time, and the scientists' need to achieve a successful first test,' Khariton and Smirnov later wrote, 'any other decision would have been unacceptable and simply frivolous.'

Not everybody agreed, however. Peter Kapitza had earned the respect of his political masters through his distinguished work as a physicist and his wartime contributions to the development of new methods for producing liquid oxygen. He had received several Stalin prizes and the Order of Lenin. In May 1945 he had been made a Hero of Socialist Labour. Of all the nation's scientists, it was he who had been able to establish the closest rapport with its political leaders, including Stalin himself.

Although Kapitza was not a nuclear physicist, he developed the view that it was wrong for the Soviet programme to attempt simply to copy the Fat Man design. He believed that it was both time-consuming and unnecessary for Soviet scientists to repeat the work that had already been done in America.

Kapitza's relationship with Kurchatov was uneasy. Sudoplatov recalled that Kapitza was 'a marvellous tactician'. He would comment on reports with jokes and anecdotes, and once interrupted a meeting of the Special State Committee to listen to a radio broadcast of a football game.[5] Though the committee members were somewhat startled by the suggestion, the game ended positively and everybody returned to the meeting in a good mood. As the meeting resumed, Kapitza suggested that, to save time, Kurchatov should first consult with him before reporting his results,

---

[5] Presumably this was one of the matches played by Moscow Dynamo during their goodwill tour of Britain in November 1945. Moscow Dynamo played a total of four games, drawing with Chelsea 3–3, beating Cardiff City 10–1, beating Arsenal 4–3 and drawing with Glasgow Rangers 2–2.

allowing time for reflection and the development of their joint recommendations which could then be reported to the committee. It was a move designed to position Kapitza clearly as the scientific head of the project.

Beria and Voznesensky disagreed. Beria, ever distrustful of his scientists, was keen to encourage rivalry between them. He suggested that Kapitza and Kurchatov put forward their separate, possibly contradictory, proposals for review by the committee. Kapitza was incensed. He had already complained about Beria's lack of respect for scientists in a letter he had written to Stalin on 3 October. He now decided to write to Stalin once more, setting out his fundamental concerns about the way that Beria was leading the programme:

> The main deficiencies of our present approach are that it fails to make use of our organisational possibilities and that it is unoriginal. We are trying to repeat everything done by the Americans rather than trying to find our own path. We forget that to follow the American path is not within our means and would take too long ... Comrades Beria, Malenkov and Voznesensky behave in the Special Committee as if they were supermen, particularly Comrade Beria. It is true he has the conductor's baton in his hand. This is fine, but after him, a scientist should play the first violin, for it is the violin that sets the tone of the whole orchestra. Comrade Beria's basic weakness is that as a conductor, he should not only wave the baton, but also understand the score.

Kapitza concluded his letter with a plea. If there was to be no change to the way the Soviet programme was being managed, then he saw no value in his continued participation. If Stalin wasn't prepared to step in and accede to his wishes, then Stalin should release him from the programme. Presumably, Kapitza believed that his past contributions and his position made him relatively invulnerable and provided a sound platform from which he could voice his complaints. But the political game he was playing was very dangerous, as he would soon discover.

On Kapitza's request, Beria was shown the letter. Beria tried to build bridges, but their differences were irreconcilable. Kapitza left the programme on 21 December 1945.

Fishing trip

Kapitza had also been concerned about the effect of the atomic bomb on science and scientists. On 22 October he had written to Bohr, who had now returned to Copenhagen: '[T]he danger exists that scientific discoveries, held in secret, may serve not humanity as a whole, but could be used in the egoistical interests of individual political and national groups.'

He wondered what position scientists should take, and expressed the desire to talk directly with Bohr on the matter. In fact, Bohr had just a few days earlier written to Kapitza expressing very similar sentiments. Their letters had crossed. Bohr remained firm in his conviction that atomic power could not be controlled by political process and had argued in the press for an open exchange of scientific information in an attempt to avoid nuclear proliferation.

So, when Bohr was approached by a former minister in the Danish government, now professor at Copenhagen University, with a request that he meet a Soviet physicist bearing a letter from Kapitza, Bohr agreed. The request had been for a secret meeting, but Bohr said that any meeting with a representative of the Soviet Union should be open, and he asked his son Aage to attend. Bohr also alerted Danish, British and American intelligence. He was calling for an open world, but he was under no illusions.

The visit could, perhaps, be best described as a 'fishing trip'. Mindful of Bohr's open and very visible encouragement of international scientific exchange, Beria had agreed to send a Soviet physicist to visit him in Copenhagen bearing a long list of questions. As Beria explained it to Stalin:

Niels BOHR is famous as a progressive-minded scientist and as a staunch supporter of the international exchange of scientific achievements. This gave us grounds to send to Denmark a group of employees, under the pretence of searching for equipment which the Germans had taken from Soviet scientific establishments, who were to establish contact with Niels BOHR and obtain from him information about the problem of the atomic bomb.

Beria selected Yakov Terletsky for the mission. Terletsky was working as the scientific adviser to Sudoplatov's Department S. He would be joined by Sudoplatov's deputy, Colonel Lev Vasilevsky, and, as Terletsky's English was poor and Vasilevsky spoke only French, an interpreter. Kapitza was obliged to write a letter of introduction for Terletsky, whom he introduced as a 'capable professor of Moscow University' who would 'explain to you the goals of his foreign tour'. Kurchatov and his team compiled a series of questions.

In the event, two meetings took place at Bohr's institute in Copenhagen, on 14 and 16 November 1945. Bohr was at pains to explain that he had not taken part in the construction of the bomb and had not visited a single nuclear installation during his time in America. He provided answers to 22 questions which were carefully recorded and carried back to Kurchatov for evaluation. Nowhere did he provide information that was not already publicly available in the Smyth report.

When Terletsky asked: 'Do you know of any methods of protection from atomic bombs? Does a real possibility of defence from atomic bombs exist?' Bohr replied:

I am sure there is no real method of protection from [the] atomic bomb. Tell me, how can you stop the fission process which has already begun in the bomb which has been dropped from a plane? ... All mankind must understand that with the discovery of atomic energy the fates of all nations have become very closely intertwined. Only international co-operation, the exchange of scientific discoveries, and the internationalisation of scientific achievements, can lead to the elimination of wars, which means the elimination of the very necessity to use the atomic bomb.

Kurchatov's evaluation of Bohr's responses was brusque. Bohr had answered the questions but there was no really new information to be gleaned. Copies of a Russian translation of the Smyth report were already being printed. Thirty thousand copies of the report would be available by the end of January 1946. Kurchatov identified a remark that Bohr had made regarding separation of uranium isotopes and suggested that this be subject to some further study.

On a Russian scale

At 7:30pm on 25 January 1946, Stalin summoned Kurchatov to a meeting. Also present were Beria and Molotov. Kapitza was no longer a member of the Special Committee or its Technical Council, and Stalin wanted once more to impress upon Kurchatov the urgency of his mission and the extent of the support that the Soviet state was prepared to provide. Kurchatov recorded his impressions of their conversation in his diary:

> Viewing the future development of the work Comrade Stalin said that it is not worth spending time and effort on small-scale work, rather, it is necessary to conduct the work broadly, on a Russian scale, and that in this regard the broadest, utmost assistance will be provided.

Stalin stressed that it was unnecessary for Kurchatov to seek out the cheapest paths, and offered to improve the material well-being of the scientists involved with dachas and the promise of prizes for great deeds. They briefly discussed Kapitza, and the utility of Kapitza's work. Kurchatov noted that misgivings were expressed.

One of Stalin's favourite slogans was ДОГНАТЬ И ПЕРЕГНАТЬ, 'dognat i peregnat', meaning 'to catch up and to surpass'. This was a slogan that had been popular since the late 1920s, when Stalin had used it to argue that to ensure survival of the dictatorship of the proletariat it was necessary to catch up with advanced countries and overtake them economically. The slogan now found a new resonance.

At an election speech delivered at the Bolshoi Theatre in Moscow a few days later on 9 February 1946, Stalin proclaimed, to prolonged applause:

> I have no doubt that if we give our scientists proper assistance they will be able in the very near future not only to overtake but even outstrip the achievements of science beyond the borders of our country.

Stalin made no direct reference in his speech to atomic weapons. But thanks to the espionage activities of Hall, Greenglass and particularly Fuchs, the Soviet Union now had an opportunity to catch up with American nuclear technology and then surpass this with the support of Soviet science.

Chapter 19

IRON CURTAIN

September 1945–March 1946

I n late August or early September 1945, Hall told Lona Cohen that he did not share her enthusiasm for life in the Soviet Union. He told her that he thought this was a grim prospect. Although he did not know it, this was an opinion he shared with Igor Gouzenko.

The GRU cipher clerk had arrived in Ottawa for a three-year assignment in June 1943, together with his boss Nikolai Zabotin. On the flight from Moscow, Zabotin told him of his exploits as an artillery officer in the Red Army, mentioning the names of his commanders who, he would then go on to explain, were later shot during the purge. After many such comments, Zabotin exclaimed: 'When I come to think of it, why wasn't I shot, too?' They laughed, and toasted his survival with vodka.

Gouzenko's pregnant wife Svetlana (whom he always called Anna) arrived in October and they had settled into an apartment at 511 Somerset Street. Their son, Andrei, was born shortly afterwards and the family had quickly adjusted to their new life in Canada. This, they discovered, was much more pleasant than their previous austere existence in wartorn totalitarian Russia. 'Candidly, everything about this democratic living seemed good', he later wrote. 'I had been in Canada long enough to appreciate that the free elections were really free, that the press was really free, that the worker was not only free to speak but to strike.'

Their sense of political freedom was coupled with a sense of material well-being. 'The unbelievable supplies of food,' he observed, 'the restaurants, the movies, the wide open stores, the absolute freedom of the people, combined to create the impression of a dream from which I must surely awaken.'

The awakening was to be rude. In September 1944 he was summoned to Zabotin's office and told that he was being recalled to Moscow. This not only threatened to change the Gouzenkos' lives for the worse, but might also mask a darker threat. No reason was given for the recall, less than halfway through his assignment, but if Gouzenko had fallen foul of his superiors in Moscow then his very life might well be at stake.[1]

Zabotin chose to defend him. He sent a message back to Moscow advising that Gouzenko's talents as a cipher clerk were irreplaceable. Moscow agreed to delay the recall. Gouzenko hastened to tell Anna the good news, but she saw that this was only a temporary reprieve. Deep within Gouzenko's mind, a dam burst. 'We won't go back, Anna', he said. 'Andrei deserves his opportunities in this country. You are entitled to live like these Canadian wives. We will pack up and disappear somewhere in Canada or even in the United States. We will change our names. I will take other work. I will ...'

Anna burst into tears. 'I am so glad, so glad, Igor', she sobbed.

'There was no use pointing out the dangers,' Gouzenko later wrote, 'she knew them full well. There was no necessity of stressing absolute secrecy. She knew certain death lay ahead if the least hint of my intended desertion got about.'

When he heard in the spring of 1945 that his replacement would be available within the next few months, he started to put his plan into action.

---

[1] Gouzenko had in fact fallen into disfavour with a Moscow-based GRU colonel who had inspected the GRU residencies in North America and had identified examples of lax security and rule-breaking.

He did not intend to defect empty-handed. He began taking home copies of confidential documents.[2]

At around 8:00pm on 5 September 1945 he left his office at the Soviet embassy on Charlotte Street for the last time. His replacement was due to arrive the next day, and his recall to Moscow was imminent. He was petrified with fear. He made his way first to the offices of the *Ottawa Journal*, the local newspaper, but lost his nerve and hastily retreated. Back home, Anna urged him to return.

He returned later that evening, but the night editor found him shaking, white as a sheet and incoherent. 'It's war. It's war. It's Russia', he mumbled in poor English. Nobody at the *Ottawa Journal* could figure out what Gouzenko wanted. The night editor recommended that he go to the RCMP, at the Ministry of Justice a short walk away. Gouzenko could no longer think straight. He went to the justice building but asked instead to see the Minister of Justice, Louis St Laurent, only to be told to come back the next morning.

He returned the next morning with both Anna, now pregnant with their second child, and young Andrei. Anna carried the stolen documents in her handbag. Gouzenko insisted that he talk only to the Minister of Justice. A trip to the minister's other office on Parliament Hill was in vain. The Gouzenkos returned to the justice building and waited with growing impatience as messages went back and forth.

The Canadian Prime Minister, William Lyon Mackenzie King, was informed. 'It was like a bomb on top of everything and one could not say how serious it might be or where it might lead', King wrote in his diary later that night. King was concerned that a high-profile defection would sour relations with an important wartime ally. 'My own feeling is that the

---

[2] In his biography, Gouzenko claimed that he had marked documents implicating key Soviet spies by turning down their edges and replacing them in the files. When the time came for him to defect, he took the documents he had marked, hid them beneath his shirt and so smuggled them out of the embassy unnoticed. As there were in total some 250 pages of documents, this is somewhat difficult to believe. It is, perhaps, more likely that Gouzenko had been copying the documents and taking them home for some time beforehand.

individual has incurred the displeasure of the Embassy and is really seeking to shield himself', he concluded. The message came back two hours later that Gouzenko should return to the Soviet embassy and return the documents he had stolen.

The Gouzenkos fared no better back at the *Ottawa Journal*. Although Gouzenko was now able to make his intentions clear to a journalist, the newspaper's editors decided against running his story, concerned at what publication might mean for Canadian–Soviet relations. 'Nobody wants to say anything but nice things about Stalin these days', he was told. The journalist suggested they take out naturalisation papers. If they became naturalised Canadian citizens, she suggested, perhaps this would put them out of reach of the Soviets.

They left Andrei with a neighbour and headed for the Canadian Crown Attorney's office on Nicholas Street. They were given naturalisation papers, and told to return the next day to arrange for photographs. Only then did Gouzenko think to ask how long the process would take. 'Oh,' he was told, 'I can't tell you for sure. A few months, perhaps.' Anna burst into tears.

Gouzenko now poured out his story to a sympathetic secretary at the Crown Attorney's office. She promised to help, and called the RCMP. A Mountie came over to the office to talk to Gouzenko but concluded that there was nothing he could do. The secretary then called the RCMP's assistant chief of intelligence, who at first claimed 'we can't touch him', but eventually agreed to see Gouzenko the next morning.

Gouzenko was by now certain to have been missed at the Soviet embassy, and he and Anna were now in real fear for their lives. On returning to their apartment, Gouzenko spotted two men watching from a park bench across the street. He presumed they were NKVD agents. The Gouzenkos entered the building unobserved by a back entrance and collected Andrei from their neighbour. Not long after settling back in their own apartment, at No. 4, 511 Somerset Street, somebody began pounding on the door and shouting Gouzenko's name. Gouzenko recognised the voice. It was Zabotin's chauffeur. They held their breath. Eventually the chauffeur walked away.

Gouzenko now sought help from another neighbour, Harold Main, at No. 5. The Mains were cooling themselves on their balcony. Gouzenko explained his situation and Main, a corporal in the Royal Canadian Air Force, offered to contact the Ottawa police. Two police constables arrived shortly afterwards and agreed to keep the apartment under surveillance. The Gouzenkos were given temporary shelter by another neighbour, at No. 6. The police seemed to know what was happening, and Main thought that they had already been in contact with the RCMP. Later that evening four 'sleazy-looking' Russians led by NKVD agent Vasily Pavlov broke into the Gouzenkos' apartment and started searching for the missing documents.

They were soon confronted by the police constables as Gouzenko watched from across the hallway. Pavlov claimed that the apartment was Soviet property, and that they had permission from the owner to enter it. One of the constables observed wryly that if they had permission to enter the apartment, why had they forced the door? Various exchanges ensued, and Pavlov insisted that the police constables leave. The constables had by now called their inspector to the scene and refused to leave until he arrived. The inspector arrived and continued the interrogation. The Russians eventually backed off, and departed.

A policeman remained with the Gouzenkos in their neighbours' apartment until the next morning, when they were taken into protective custody by the RCMP. In fact, the two men watching the apartment from across the street were RCMP officers, not NKVD agents. The Gouzenkos were moved to a safe house, where Igor's lengthy de-briefing began. He bought the documents, totalling about 250 pages, with him.

It had all been far from straightforward. It had taken Gouzenko two days, but he had now successfully defected to the West.

Codename Alek

The documents that Gouzenko had taken from the Soviet embassy included GRU dossiers on its spies, a series of cables between Ottawa and Moscow, a mailing list and numerous notes prepared by Zabotin and his assistant. Some were handwritten in Russian, some were written in shorthand, making them difficult to decipher. They provided evidence of two spy rings, one operated by Zabotin and the GRU and a second run by Pavlov for the NKVD. Implicated were Canadian Communist Party officials, members of the Canadian government and the Canadian Department of External Affairs, the Canadian armed forces and scientists and engineers. The list of spies exposed by Gouzenko was not limited to Canada. Also implicated were officials in the American State Department, the British High Commission in Ottawa and the British intelligence services. The documents even exposed a Soviet spy ring operating in Switzerland.

One of the most important of the spies identified by Gouzenko was Alan Nunn May, referred to by his codename ALEK in the cables that had gone back and forth between Ottawa and Moscow. This was a major embarrassment to MI5 and the British government. May had made no secret of his Communist sympathies while at Cambridge and it now transpired that he had never been security screened before joining Tube Alloys. May was scheduled to return to Britain on 15 September to take up his lectureship at King's College London. He was put under surveillance during his last few days in Montreal and shadowed by an RCMP officer during his flight to London. On arrival in Britain, surveillance duties were passed to two officers of Scotland Yard's Special Branch, a specialist police unit dealing with counter-terrorism and subversion.

Zabotin had sent a cable to Moscow at the end of July with suggestions for the protocol to be followed to re-establish contact with May in London:

> We have worked out the conditions of a meeting with ALEK in London. ALEK will work in King's College, Strand. It will be possible to find him there through the telephone book. Meetings: October 7.17.27 on the street in front of the British Museum. The time, 11 o'clock in the evening. Identification sign: A newspaper under the left arm. Password: Best regards to Mikel.

The dates referred to 7 October, with fall-back options of 17 and 27 October if the meeting on the 7th failed to occur for whatever reason. Zabotin had received a reply on 22 August recommending that the time be changed to 8:00pm and a slightly more complex series of recognition signals and passwords be adopted. Gouzenko's evidence against May was in itself insufficient to warrant an arrest, so MI5 laid plans to catch May in the act of meeting his Soviet contact.

The meeting never happened, on 7 October or on the fall-back dates. Some time later, May declared that he did not keep the appointment because 'this clandestine procedure was no longer appropriate in view of the official release of information and the possibility of satisfactory international control of atomic energy'.

There is another possible explanation, however. Both MI5 and the British SIS had been alerted to Gouzenko's defection soon after 7 September. Stewart Menzies, the head of the SIS, received information about the case channelled through William Stephenson, the Canadian-born representative of British intelligence in the Western hemisphere.[3] Menzies and his chief of counter-intelligence, 'Kim' Philby, followed the developments very closely. Philby was also an NKVD 'mole', and was sending warnings about Gouzenko to Moscow within weeks of his defection. Philby assessed the evidence against May and, although he reasoned that this was inconclusive, he may have warned May that he was under surveillance.

The Canadian, American and British administrations and their intelligence services were by now embroiled in a debate about what to do next. Opinions varied. King was eager to deal with the matter quietly and diplomatically, confronting the Soviets with evidence of their wrongdoing and asking them politely to stop. Truman favoured keeping the Gouzenko case under wraps for the time being, fearing that a major diplomatic incident would greatly damage efforts to reach an international agreement on the control of atomic energy. Truman and his Undersecretary of State Dean Acheson recommended that May not be arrested by the British unless this became absolutely necessary.

---

[3] Also known by his codename 'Intrepid', one of many possible inspirations for Ian Fleming's most famous creation, James Bond. See Ben Macintyre, *For Your Eyes Only*, Bloomsbury, London, 2008.

Meanwhile Roger Hollis, MI5's chief of counter-intelligence against Communist subversion, appointed by Philby as the principal British intelligence liaison on the Gouzenko case, argued that any action short of arrest would be perceived 'as weakness and the effect of this would be to worsen and not to improve relations'. Hollis feared that May's defection to the Soviet Union was imminent.

To make matters worse, Truman was having to deal with a burgeoning spy scandal of his own. On 6 November, Vassar graduate Elizabeth Bentley had confessed to spying for the Soviet Union and offered to defect. She had acted as a courier between NKVD agent Jacob Golos (with whom she had had an affair) and a number of American government officials. FBI director J. Edgar Hoover asked that no action be taken in the Gouzenko case until the FBI had had the opportunity to analyse and corroborate her evidence.

In the meantime, May continued to behave perfectly normally. He delivered his lectures, moved into rooms at Stafford Terrace in Kensington and lived a relatively quiet life.

UN Atomic Energy Commission

Truman's concern to avoid a major diplomatic incident over the Gouzenko affair betrayed a growing anxiety over the question of the international control of atomic weapons. He had sought advice from Oppenheimer in October 1945, suggesting that his administration first deal with the issue of domestic controls, then international controls. 'The first thing is to define the national problem,' he told Oppenheimer, 'then the international.' Oppenheimer disagreed. He thought it imperative to define and resolve the international problem first.

Oppenheimer went on to voice his own misgivings. 'I feel we have blood on our hands', he remarked. Truman was indignant. The scientists might have discovered atomic energy but it had been his decision to use the bomb against Japan. The last thing he needed was a 'cry baby' scientist. 'Never mind,' he told Oppenheimer, 'It'll all come out in the wash.' But although Truman did not share Oppenheimer's sense of guilt, he had grown increasingly concerned about the indiscriminate destructiveness of the bomb. He had begun to realise that this was a weapon that could never be used again.

Truman met British Prime Minister Clement Attlee and Canadian Prime Minister Mackenzie King in Washington on 11 November. The problem they now wrestled with was one of basic atomic physics. The materials and production processes required to use atomic energy for peaceful, civilian purposes could not easily be disentangled from the materials and production processes required for atomic bombs. The civilian use of atomic energy required the building of nuclear reactors for power generation. However, the spent fuel rods of certain kinds of nuclear reactors were potential sources of plutonium that could be used to produce atomic bombs. While the leaders of the three countries in possession of the knowledge essential to the use of atomic energy accepted the importance of the free interchange of scientific knowledge, they were not convinced that spreading specialised atomic information was wise until appropriate international safeguards had been put in place.

In their joint declaration, issued on 15 November, Truman, Attlee and King acknowledged what Bohr had understood immediately on arriving at Los Alamos in early 1944:

We recognize that the application of recent scientific discoveries to the methods and practice of war has placed at the disposal of mankind means of destruction hitherto unknown, against which there can be no adequate military defence, and in the employment of which no single nation can in fact have a monopoly.

The declaration went on to call for an international commission to be established under the auspices of the United Nations:

In order to attain the most effective means of entirely eliminating the use of atomic energy for destructive purposes and promoting its widest use for industrial and humanitarian purposes, we are of the opinion that at the earliest practicable date a commission should be set up under the United Nations Organization to prepare recommendations for submission to the organization.

In his discussion with Terletsky, Bohr had commented on the meeting in Washington, and expressed hope for a further consultation on international control with the Soviet Union. 'We have to keep in mind,' he told Terletsky, 'that atomic energy, having been discovered, cannot remain the property of one nation, because any country which does not possess this secret can very quickly independently discover it. And what is next? Either reason will win, or a devastating war, resembling the end of mankind.'

Byrnes had learned at the Council of Foreign Ministers meeting in London that the Soviet Union was not about to be cowed by the threat of atomic weapons. A week after the Truman–Attlee–King declaration was issued, Byrnes suggested to Molotov that he host a further, interim, meeting to include the American, British and Soviet foreign ministers, in Moscow in December. Molotov immediately accepted. Relatively little had been agreed at the London conference, so the agenda of the Moscow meeting would be virtually identical. This time, however, Byrnes wanted international control of atomic weapons high on the agenda. Molotov, still minded to be stubborn, moved it to the bottom.

The meeting began on 16 December, and on the question of atomic energy Byrnes found the Soviets to be surprisingly co-operative. Molotov agreed to the establishment of a UN commission on atomic energy, to be proposed at the first session of the UN General Assembly scheduled for January 1946. Molotov insisted that the commission report to the UN Security Council, rather than the General Assembly itself. Byrnes agreed. Molotov probably figured that there was little to be gained either way. And the Soviets could exercise the right of veto over decisions by the Security Council.

On Christmas Eve, the banter that was begun in London continued over dinner at the Kremlin. Molotov proposed a toast to Conant, who had joined the meeting as Byrnes' adviser on atomic energy, and suggested that after a few drinks perhaps they could explore the secrets Conant possessed. Perhaps, Molotov went on, if Conant had a bit of the atomic bomb in his pocket he could bring it out. But as Stalin stood to drink the toast, he said: 'Here's to science and American scientists and what they have accomplished. This is too serious a matter to joke about. We must now work together to see that this great invention is used for peaceful ends.' Molotov's expression did not change.

The interim meeting concluded on 26 December with agreement on the preparation of Allied peace treaties with Italy, Romania, Bulgaria, Hungary and Finland, the establishment of a Far Eastern Commission and an Allied Council for Japan, and the establishment of a UN commission for the control of atomic energy:

> [T]he Commission shall make specific proposals: (a) For extending between all nations the exchange of basic scientific information for peaceful ends; (b) For control of atomic energy to the extent necessary to ensure its use only for peaceful purposes; (c) For the elimination from national armaments of atomic weapons and of all other major weapons adaptable to mass destruction; (d) For effective safeguards by way of inspection and other means to protect complying states against the hazards of violations and evasions.

The UN General Assembly adopted a resolution establishing the UN Atomic Energy Commission on 24 January 1946.

The Acheson–Lilienthal report

Anticipating that the UN resolution would be adopted, in early January Truman called for a concerted effort to formulate American policy and a plan to move the resolution from a statement of intent to political reality. Byrnes established a committee and appointed an initially reluctant Acheson as chairman. Members of the committee included Groves, Bush and Conant.

Acheson complained that he knew nothing about atomic energy, and was prompted to form a Board of Consultants, to be chaired by David Lilienthal. Lilienthal was a lawyer who had in 1933 been appointed by

Roosevelt as one of three directors of the Tennessee Valley Authority, a federally-owned corporation set up to develop the economy of the Tennessee Valley, which had been particularly hard hit by the Depression. Lilienthal had gone on to become the authority's chairman, and was known as 'Mr TVA'. Oppenheimer was the only scientist invited to participate on Lilienthal's board.

Lilienthal was greatly impressed by Oppenheimer. 'He is worth living a lifetime just to know mankind has been able to produce such a being', Lilienthal wrote effusively in his diary. 'We may have to wait another hundred years for the second one to come off the line.' Oppenheimer worked his now legendary charm. 'Everybody genuflected', noted Groves, acerbically. 'Lilienthal got so bad he would consult Oppie on what tie to wear in the morning.'

As the only scientist on the board, Oppenheimer took responsibility for the education of the other consultants. He gave them a crash course on nuclear physics, drawing little stick figures on a blackboard to represent electrons, protons and neutrons. Together the consultants toured the Manhattan Project facilities and interviewed project scientists. They developed arguments and counter-arguments. It seemed to Oppenheimer that simply outlawing atomic weapons and establishing a system of inspections to police compliance was doomed to failure. The entanglement of civilian and military uses of atomic energy ran too deep.

His counter-proposal was to establish an Atomic Development Authority, an international organisation that would take sovereign ownership and responsibility for the entire atomic industry, from end to end, and develop it for peaceful purposes. The authority would own all the world's uranium mines, the means of production and enrichment of uranium, all nuclear reactors and nuclear laboratories. In essence, the proposal was to nationalise the atomic industry, not just within one nation but across all nations, with public ownership vested in a 'world government' represented by the UN.

To make this work, it would be necessary to proliferate atomic technology across the world, establishing uranium mines, production facilities, reactors and laboratories in all nations that could support them. Any nation seeking to breach the international agreement by wresting control of atomic facilities and materials within its territory would be confronted with a harsh reality. All other nations would have atomic facilities and materials of their own. The result would be a form of deterrence. This would not be deterrence derived from the shared mutual threat of atomic weapons, but deterrence derived from the shared mutual threat of the *means to produce* atomic weapons. In this sense, the agreement would be self-policing.

It was an astonishingly bold proposal, containing more than a whiff of scientific and technological socialism. Even more astonishing, Oppenheimer managed to persuade the Board of Consultants to accept it, including Charles Thomas, the Vice President of Monsanto Chemical Company, at the time a $120-million business.

With some relatively minor modifications, Oppenheimer's proposal became the Acheson–Lilienthal report. The report was ready by 7 March 1946, and published on 28 March. It was preceded by the nuclear scientists' own manifesto, *One World or None*, a 'report to the public on the meaning of the atom bomb'. It contained an introduction by Compton and chapters by, among others, Bethe, Edward Condon, Einstein, Philip Morrison, Oppenheimer, Szilard, Urey and Wigner. Bohr provided a foreword. In the opening chapter, Morrison brought the full horror of atomic warfare home to the American public by projecting what he had seen at Hiroshima onto Manhattan:

> The device detonated about half a mile in the air, just above the corner of Third Avenue and East 20th Street, near Gramercy Park ... From the river west to Seventh Avenue, and from the South of Union Square to the middle thirties, the streets were filled with the dead and dying. The old men sitting on the park benches in the square never knew what had happened. They were chiefly charred black on the side toward the bomb. Everywhere in this whole district were men with burning clothing, women with terrible red and blackened burns, and dead children caught while hurrying home to lunch.

It was a portentous vision, designed to spur the public and their political representatives to greater effort on international controls.

The Acheson–Lilienthal report was, as Teller explained in the newly-founded *Bulletin of the Atomic Scientists*, a ray of hope. When Bohr read it he was overjoyed. It embodied the essence of his 'open world' proposals and, in his opinion, offered the best hope for the future.

But other, darker, forces were building which would frustrate Bohr's hopes. Frustrations in which Churchill would yet again play a significant role.

Cold War

If the Soviets were satisfied with the outcome of the December interim meeting of foreign ministers, Truman was not. He had been irritated by Byrnes' failure to keep him informed of the discussions, and did not accept some of the foreign policy decisions that Byrnes had made. To secure Molotov's agreement, Byrnes had accepted that there would be only minor changes to the governments of Romania and Bulgaria, whereas Truman had insisted on more radical changes. In August 1941 British and Soviet troops had invaded Iran in order to ensure the security of supply of Iranian oil to fuel Soviet forces fighting on the Eastern front. At the Tehran conference in November 1943, Roosevelt, Churchill and Stalin had agreed to preserve Iran's sovereignty and independence. In compliance with this agreement, Britain had withdrawn its troops at the end of the war. Soviet troops had remained, however. Byrnes had failed to gain any assurances from Molotov that these troops would be withdrawn.

Truman poured his ire into a letter to his Secretary of State. 'Unless Russia is faced with an iron fist and strong language another war is in the making. Only one language do they understand – "How many divisions have you?" I do not think we should play compromise any longer.' He closed the letter thus: 'I'm tired [of] babying the Soviets.'

Opinion in Washington was beginning to harden. Stalin's speech at the Bolshoi Theatre on 9 February 1946 was in part a reaffirmation of the ideological divide between Communism and monopoly capitalism, and interpreted by some in Washington as little short of a declaration of war. On 22 February George Kennan, the Deputy Chief of Mission in Moscow, drafted a long telegram to Byrnes containing his assessment of the Soviet Union's post-war outlook and its implications for American foreign policy. He was forthright:

[Soviet power is] impervious to logic of reason, and it is highly sensitive to logic of force. For this reason it can easily withdraw – and usually does when strong resistance is encountered at any point. Thus, if the adversary has sufficient force and makes clear his readiness to use it, he rarely has to do so. If situations are properly handled there need be no prestige-engaging showdowns.

Kennan's telegram ran to over 5,000 words. At the time it was the longest telegram in State Department history. Byrnes thought Kennan's analysis 'splendid'.

On 5 March Churchill accepted an honorary degree at Westminster College in Fulton, Missouri. The ceremony took place in the college gymnasium. Truman, a native of Missouri, introduced him. Churchill rose to deliver his acceptance speech, broadcast on radio and by loudspeaker to 40,000 people who had assembled in Fulton to hear what he had to say.

His speech marked the beginnings of a Cold War rhetoric that would last for more than 40 years:

From Stettin in the Baltic to Trieste in the Adriatic an iron curtain has descended across the Continent. Behind that line lie all the capitals of the ancient states of Central and Eastern Europe. Warsaw, Berlin, Prague, Vienna, Budapest, Belgrade, Bucharest and Sofia; all these famous cities and the populations around them lie in what I must call the Soviet sphere, and all are subject, in one form or another, not only to Soviet influence but to a very high and in some cases increasing measure of control from Moscow ...

From what I have seen of our Russian friends and allies during the war, I am convinced that there is nothing they admire so much as strength, and there is nothing for which they have less respect than for weakness, especially military weakness. For that reason the old doctrine of a balance of power is unsound. We cannot afford, if we can help it, to work on narrow margins, offering temptations to a trial of strength.

On the vexed question of the atomic bomb, Churchill had this to say:

It would nevertheless be wrong and imprudent to entrust the secret knowledge or experience of the atomic bomb, which the United States, Great Britain, and Canada now share, to the world organisation,[4] while it is still in its infancy. It would be criminal madness to cast it adrift in this still agitated and un-united world. No one in any country has slept less well in their beds because this knowledge and the method and the raw materials to apply it, are at present largely retained in American hands.

Churchill had warned Britain's House of Commons repeatedly about the threat posed by German rearmament after Hitler had seized power in 1933, and now felt that the Americans needed a similar warning against the threat of Soviet power. He also still clung to the belief that the science and technology of the atomic bomb was a secret that could be kept.

Although Truman later denied it, Churchill had shared the content of his speech with him beforehand, and Truman had approved its tenor.

The climate for the proliferation of atomic weapons was now firmly established.

The arrest of Alan Nunn May

Beria did not take kindly to failures within his intelligence services. Gouzenko had feared for his life. He would later make melodramatic appearances on Canadian television, his head covered by a trademark

---

[4] In other words, the UN.

hood cut with eyeholes.[5] But Stalin forbade his assassination. 'Everyone is admiring the Soviet Union', Stalin said. 'What would they say about us if we did that?' It was nevertheless clear that somebody had to be punished.

Zabotin was recalled to Moscow in December 1945. Together with his wife, he boarded the SS *Alexander Suvarvov* bound for Murmansk, which left port clandestinely by night without complying with Canadian port regulations. Zabotin, his wife and son were sent to a labour camp in Siberia.

Gouzenko's mother died in the Lubyanka prison. Anna's mother, father and sister were sent to prison and her sister's daughter – Anna's niece – was sent to an orphanage.

On 3 February 1946 American journalist and broadcaster Drew Pearson, famous for his syndicated newspaper column, the 'Washington Merry-go-Round', finally broke the Gouzenko story. He informed his nationwide radio audience of the 'recent' defection in Canada of a Soviet spy and of his revelations of other spies in high places in Canada and America. The leak forced King to appoint a Royal Commission to investigate Gouzenko's evidence and initiate proceedings against those accused of spying for the Soviets.

A further broadcast by Pearson on 10 February forced the Royal Commission to carry out raids and make a series of arrests five days later. King now made a public announcement, although he did not mention the Soviet Union. He suspected that the leak had come from Truman's administration. But it was Hoover who had called Pearson on the morning of his first broadcast. Hoover was by now keen to press forward with pursuit of suspected spies in Washington, Harry Dexter White and Alger Hiss, and publicity about the Gouzenko case suited his purposes.

On 15 February, the day of the arrests in Canada, Nunn May was interviewed at the Tube Alloys office at Shell-Mex House in the Strand by Lieutenant Colonel Leonard Burt and Major Reginald Spooner of the

---

[5] A series of television clips featuring Gouzenko can be viewed on the Canadian Broadcasting Company's digital archives website: http://archives.cbc.ca/politics/national_security

Intelligence Corps, attached to the War Office General Staff. The evidence against May was still insufficient to justify his arrest, so Burt needed nothing less than a full confession.

May was told that he was being interviewed in connection with the Canadian Royal Commission investigation. He blanched, but remained composed. Burt mentioned the names Zabotin and Angelov, but May denied having ever heard of them. He denied passing any secret information to unauthorised persons. When Burt asked him directly if he would be prepared to give the authorities all the help he could, May explained that he had heard about the leakage of information in connection with atomic energy for the first time that afternoon, and stated: 'If it means getting any of my late colleagues in Canada into trouble over this, I should feel some reluctance.'

Burt may have sensed an opening: May was holding back information that might implicate his former colleagues. May was released but his home and office at King's College were searched and he remained under close surveillance for the next five days. Burt interviewed him again on 20 February in Savile Row. This time Burt confronted him with information about his dealings with the Soviets in Canada and with the failed meeting with his Soviet contact near the British Museum in London. May probably did not realise that even with this information, the police still had no grounds to arrest him. Instead he broke down and confessed. 'The whole affair was extremely painful to me,' he told Burt, 'and I only embarked on it because I felt this was a contribution I could make to the safety of mankind. I certainly did not do it for gain.'

May was arrested on 4 March by Special Branch Detective Inspector William Whitehead. He was met by Whitehead just as he finished delivering an afternoon lecture at King's College. Not wanting to arrest May in the college grounds, Whitehead informed him that he had a warrant for his arrest and asked him to accompany him to a waiting police car. Whitehead read the warrant in the car. May remained silent.

Half an hour later, at Bow Street police station, May was formally remanded and charged with communicating information contrary to Section 1 of the UK Official Secrets Act, 1911.

The revelation that a member of the Manhattan Project was a Soviet spy sent shockwaves through the community of physicists at Los Alamos. Discussing these events shortly after Nunn May's arrest, Else Placzek remarked that her former husband Hans von Halban had worked at the Montreal laboratory, and she had therefore met May. When pressed to describe him, she remarked that: 'He was just a nice quiet bachelor, very helpful at parties. Just like Klaus here.' Fuchs was visibly discomfited.

One of the suspected GRU spies identified in Gouzenko's documents and arrested by the RCMP was Israel Halperin, professor of mathematics at Queen's University in Kingston, Ontario. Halperin had provided Fuchs with reading material during his incarceration at Sherbrooke in 1940.

Fuchs' name was in Halperin's address book.

Chapter 20

CROSSROADS

November 1945–January 1948

The cordial arrangement on the exchange of atomic secrets between Britain and America embodied in the Quebec agreement and Hyde Park aide-mémoire had been founded on the close relationship between Churchill and Roosevelt. Churchill had told Bohr in May 1944: 'And as for any post-war problems there are none that cannot be amicably settled between me and my friend, President Roosevelt.' But the future is another country. The war was won, but Churchill was no longer Prime Minister, Roosevelt was dead and the Soviet Union was fast transforming from ally into enemy.

It seems that the question of whether or not post-war Britain should strive to become an independent atomic power was never really debated. Britain demanded a seat at the table of the world's most powerful nations, and that meant developing an atomic capability. Within days of the bombing of Hiroshima and Nagasaki, Attlee formed a small cabinet committee to consider Britain's atomic policy. He continued his predecessor's obsession with secrecy on atomic matters by not informing the rest of his cabinet. The committee was known by its 'Gen' number, an identification number assigned to all such ad hoc government bodies whose existence and functions are not publicly reported.

The main focus of Gen 75 was Britain's own peaceful atomic energy programme, the question of international controls and collaboration with the Americans. Although no decision had yet been taken on the need or otherwise for Britain to possess its own atomic weapons, Attlee's informal name for Gen 75 – the 'Atom Bomb Committee' – perhaps belied a decision already made.

Gen 75 needed a body to advise it. The trouble was, there was nobody within the new Labour government with any experience of atomic matters who could chair an Advisory Committee on Atomic Energy. In order to maintain some semblance of continuity, Attlee proposed that John Anderson be appointed. Anderson, who was now an Independent Member of Parliament sitting on the opposition front bench, accepted the appointment. It was an awkward arrangement. Anderson had the status of a minister and access to the Cabinet Office support structures, but felt he was excluded from attending key ministerial meetings.

Attlee and Anderson set off for Washington in November 1945 with at least three missions. The first was to set Britain's stamp on international atomic policy through the Truman–Attlee–King declaration. The second was to perpetuate the cordial arrangement forged in wartime by Churchill and Roosevelt and set post-war Anglo-American atomic collaboration in stone.

The third mission was to resolve an issue with the Quebec agreement, in which Britain had expressly disclaimed 'any interest in these industrial and commercial aspects beyond what may be considered by the President of the United States to be fair and just and in harmony with the economic welfare of the world'. This article, the British wanted to argue, had been superseded by the Hyde Park aide-mémoire of 18 September 1944, which stated: 'Full collaboration between the United States and the British Government in developing tube alloys for military and commercial purposes should continue after the defeat of Japan unless and until terminated by joint agreement'.

But Attlee and Anderson's efforts were not helped by the fact that nobody on the American side seemed to have heard of the Hyde Park aide-mémoire. British representatives of the Combined Policy Committee insisted that the document did exist, but it had been signed in secret and nobody could find it in Roosevelt's files. In the end, the British had to furnish the Americans with a copy.[1]

While Attlee concerned himself with the matter of international policy, Anderson and Groves wrestled with the future of Anglo-American collaboration. Groves was placed in the rather difficult position of drafting a memorandum with Anderson relating to decisions already taken at the White House; decisions of which he was ignorant, taken by people he didn't know. The resulting Groves–Anderson memorandum perpetuated 'full and effective collaboration' in the area of basic scientific research, mirroring the 'full and effective interchange of information and ideas' in the Quebec agreement.

However, as before, co-operation on the more technical aspects of atomic energy development and plant design, construction and operation was to be more closely regulated by the post-war Combined Policy Committee. The Combined Development Trust, which had been established in February 1944, was to secure control and possession of all uranium and thorium deposits in the territories of the USA, Canada, Britain and the British Commonwealth.[2] At the time, nobody seemed to think this three-way carve-up in any way contradictory to the call for international controls.

Agreements between Allied heads of state that are kept secret even from Allied governments are arguably necessary in a time of war. However, formal agreements between heads of state signed by the US President in peacetime had to be ratified by the Senate. Attlee was keen to push for an agreement on collaboration quickly without having to endure the delay and loss of predictability that submission to the Senate would entail. At

---

[1] The Hyde Park aide-mémoire had been misfiled. Groves thought that because the paper referred to 'tube alloys', the clerk who filed it assumed it had something to do with boiler tubes for ships. See Groves, p. 402.

[2] The Combined Development Trust was not a legal entity and was therefore not obliged to make public details of its transactions. It consisted of three American, one Canadian and two British trustees and was set up at the behest of the Americans, possibly to pre-empt any attempt by the Belgian government to play American and British interests off against each other. See Gowing, *Britain and Atomic Energy*, pp. 299–300.

the end of the November meeting, a hastily prepared memorandum was drafted for Truman and Attlee's signatures. It said:

> We desire that there should be full and effective co-operation in the field of atomic energy between the United States, the United Kingdom and Canada. We agree that the Combined Policy Committee and the Combined Development Trust should be continued in a suitable form. We request the Combined Policy Committee to consider and recommend to us appropriate arrangements for this purpose.

With 'full and effective co-operation' thus assured, in December 1945 Attlee appointed Lord Portal as Controller of Production, Atomic Energy, ICI's Christopher Hinton as leader of the effort to produce fissile materials, and John Cockcroft as director of Britain's Atomic Energy Research Establishment (AERE). Cockcroft had served as scientific head of the Montreal project, having taken over the role from Halban in April 1944 in an attempt to raise the profile of the project and to lift the morale of the scientists involved. The AERE, formed on 1 January 1946, was to be located on the site of an RAF airfield at Harwell, about sixteen miles south of Oxford. An independent British atomic energy programme was under way.

Despite these outward expressions of mutual support, Chadwick sensed that the gap between Britain and America was widening. He told Anderson that: 'the cohesive forces which held men of diverse opinions together during the war are rapidly dissolving: any thought of common effort or even common purpose with us or with other peoples is becoming both weaker in strength and rarer to meet.'

Chadwick's senses were to prove finely tuned as, one by one, barriers to 'full and effective co-operation' were erected. At a subsequent meeting of the Combined Policy Committee, Groves pointed out that the secret Truman–Attlee agreement was in violation of Article 102 of the UN Charter, which called for any new international agreement to be reported to the UN Secretariat and openly published. Reporting and publication would mean laying bare the contradictions inherent in the drive for international control of atomic energy through the UN while simultaneously entering into secret agreements behind the back of the fledgling organisation.

As Britain pressed the Americans for full details of atomic energy development, design and production, the precise meaning of 'full and effective' was debated. On 20 April 1946 Truman advised Attlee that he 'considered it inadvisable for the United States to assist the United Kingdom in the construction of atomic energy plants, in view of our stated intentions to press for international control of atomic energy through the United Nations'.

The question of international control was, in the final analysis, neither here nor there. All Anglo-American collaboration on atomic energy was soon to be scuppered by *domestic* legislation sponsored by US Democrat Senator Brien McMahon. The McMahon Bill was a draft 'atomic energy act' for the development and control of atomic energy through the offices of a new civilian US Atomic Energy Commission.[3] It was submitted to Congress in December 1945. On the surface, the draft bill did not appear to pose a threat. It called for 'the dissemination of related technical information with the utmost liberality as freely as may be consistent with the foreign and domestic policies established by the President'.

But the bill that would eventually be enacted would not be quite so liberal with atomic information.

The Baruch plan

Byrnes penned a foreword to the Acheson–Lilienthal report ready for its official release on 28 March 1946, but he didn't much like what the report said. On his recommendation, Truman appointed financier Bernard Baruch to lead the American delegation to the newly-formed UN Atomic Energy Commission. The 75-year-old Baruch had served as a behind-the-scenes adviser to American presidents since Woodrow Wilson in the First

---

[3] It replaced the May–Johnson Bill, which proposed to keep atomic energy matters firmly under military control, and which was slowed by opposition particularly from Met Lab and Oak Ridge scientists. It eventually fell out of favour.

World War. He had amassed a fortune speculating in the sugar market, and his refusal to join a brokerage firm had earned him the nickname 'Lone Wolf on Wall Street'. He was also one of Byrnes' business partners. Both were board members of Newmont Mining Corporation, a company with considerable investments in uranium mines.

It was not hard to guess what the arch-conservative Baruch would make of the Acheson–Lilienthal proposals, or how the Soviets would interpret his appointment. Lilienthal recorded in his diary that: 'We need a man who is young, vigorous, not vain, and who the Russians would feel isn't out simply to put them in a hole, not really caring about international co-operation. Baruch has none of these qualities.' Oppenheimer later said that the day Baruch was appointed 'was the day I gave up hope'.

As expected, Baruch took an instant dislike to the Acheson–Lilienthal report, and insisted that he be allowed to formulate a plan of his own. As he explained to Acheson, he was too old to play 'messenger boy'. He tried to recruit Oppenheimer as a consultant, and Oppenheimer agreed to meet him and three other consultants he had selected – two bankers and the CEO of Newmont. It was unlikely that Oppenheimer would find common ground with Baruch's business cronies. There was no meeting of minds. Though he later regretted it, Oppenheimer refused to co-operate.

Matters came to a head, and a meeting was called for 17 May 1946 at Blair House, the President's official state guest house on Pennsylvania Avenue in Washington. At the meeting, Baruch argued for what would become the main pillars of the Baruch plan. There would be no attempt to nationalise or internationalise uranium mines. There would be no unilateral disarmament. Surely, he argued, it was necessary to retain a stockpile of atomic weapons as a deterrent against any nation found to be in violation of the agreement. If the agreement was to be effective, there could be no Security Council veto. When he stated flatly that there was no provision at all in the Acheson–Lilienthal report for punishment of those in violation, the meeting erupted in fireworks.

Oppenheimer expressed his deep reservations about the Baruch plan to his wife Kitty and to Lilienthal. But Hoover had by now authorised extensive FBI surveillance of Oppenheimer, including phone taps, convinced that he was about to defect to the Soviet Union. Transcripts of his phone conversations were sent to Byrnes.

Baruch delivered his plan to a meeting of the UN Atomic Energy Commission on 14 June. His opening statement was melodramatic: 'We are here to make a choice between the quick and the dead', he said. The plan called for: 'Managerial control or ownership of all atomic energy activities potentially dangerous to world security.'

The manufacture of atomic weapons would stop and existing stockpiles would be dismantled only when 'an adequate system for control of atomic energy, including the renunciation of the bomb as a weapon, has been agreed upon and put into effective operation and condign punishments set up for violations of the rules of control which are to be stigmatized as international crimes'. Nations in violation would suffer 'penalties of as serious a nature as the nations may wish and as immediate and certain in their execution as possible', and for which: 'There must be no veto to protect those who violate their solemn agreements not to develop or use atomic energy for destructive purposes.'

The Soviets simply saw the Baruch plan as an attempt by America to maintain its monopoly on atomic weapons indefinitely. The plan required the Soviet Union to forgo any atomic bomb programme of its own, submit to a powerful international authority (with no right of veto) that could be expected to be strongly influenced if not controlled by America, and yield up whatever uranium deposits could be found beneath Soviet soil. Not surprisingly, it was unacceptable.

Five days later, the Soviets countered with a proposal from Andrei Gromyko, the Soviet representative on the UN Security Council. The Soviet approach was in many ways similar to the convention that had been adopted in 1925 to prohibit the development, production, stockpiling and use of chemical weapons. According to the Soviet proposal, atomic weapons would be banned by international convention. All existing stockpiles would be destroyed within three months of ratification. Signatory states would enact legislation within six months providing for punishment of those in violation. A committee would be established to discuss the exchange of scientific information. A second committee would discuss ways to ensure compliance.

There was no all-powerful international authority in the Soviet proposal. There was no formal provision for inspection and control. Nation-states were expected to fall in line and police themselves. And, most importantly of all, America would be forced to give up its monopoly. It is doubtful that the Soviet policy-makers ever really thought this would be acceptable to the United States. Not surprisingly, it wasn't.

The opportunity to halt what would soon become a madness of atomic weapons proliferation quietly slipped away. The simple truth was that international control appeared to suit nobody.

A somewhat squalid case

Alan Nunn May went on trial at the Old Bailey in London on 1 May 1946, charged with communicating information contrary to the Official Secrets Act. The Attorney General, Hartley Shawcross, opened the case for the prosecution by declaring it: 'quite serious but a somewhat squalid case of a man who having been for some years in the employment of the British Crown in connection with researches which were being made into the problems of atomic energy thought right apparently for reward to communicate to some person whose identity he has refused to divulge information as to the progress which had been made ...'

May had been confronted with the evidence and chose to plead guilty. At stake, therefore, was the magnitude of the punishment.

In his defence, Gerald Gardiner KC sought to downplay the importance of the information that May had transmitted, emphasising that much of this had now been made public in the Smyth report, and argued that the Soviet Union was an ally in the war, not an enemy. Shawcross countered by arguing that the Official Secrets Act is designed to prevent the communication of information to unauthorised persons: 'it might be to your Lordship [Justice Oliver, presiding], it might be to me or to anyone, information which if it eventually got into the hands of persons who were or might become enemies would be useful to them.'

The trial was short. Summing up, Justice Oliver said:

How any man in your position could have had the crass conceit, let alone the wickedness, to arrogate to himself the decision of a matter of this sort, when you yourself had given your written undertaking not to do it and knew it was one of the country's most precious secrets, when you yourself had drawn and were drawing pay for years to keep your own bargain with your country – that you could have done this is a dreadful thing.

May was sentenced to ten years in prison.

A prima facie proof

Fuchs had expected to return to England with the rest of the British mission towards the end of 1945. As a way of saying goodbye to Peierls and his wife, he joined them and Mici Teller on a two-week trip to Mexico City in December 1945 (Teller himself stayed at Los Alamos, pleading a heavy work schedule). Fuchs' Buick broke down on the way.

Feynman had suggested he try to find an academic position in America but, without a hint of irony, Fuchs explained that he owed it to Britain to return. As the independent British atomic programme got under way, there followed a scramble to recruit scientists leaving the Manhattan Project. Both Chadwick and Cockcroft recommended Fuchs for a position. Fuchs was subsequently interviewed in Montreal by British government representatives and offered the position of Head of Theoretical Physics at Harwell. Fuchs was initially cautious, but eventually accepted. Frisch was appointed Head of Nuclear Physics.

With his future thus assured, Fuchs was asked by Norris Bradbury, who had taken over from Oppenheimer as scientific director of the Los Alamos laboratory, if he would be prepared to stay on for a few months more. With so many physicists leaving Los Alamos, and with further atomic bomb tests to prepare for, Bradbury had run out of resources. When 31 scientists reconvened at Los Alamos on 18 April 1946 for a three-day conference on the Super, only seven were on the Los Alamos staff. Teller himself had left Los Alamos for Chicago on 1 February.

Teller and his team had nagged away at the theory of the thermonuclear bomb all through the winter of 1945–46. The problems they encountered were on an altogether different scale from those posed by the theory of the fission bomb. Though the effort was greatly aided by access to the first general purpose electronic computer, ENIAC,[4] the results were still all rather preliminary.

In preparation for the Super conference, the team produced a report entitled *A Prima Facie Proof of the Feasibility of the Super*.[5] Teller took an optimistic view. According to him the prima facie evidence, or the evidence to hand, confirmed that the Super was possible and should be pursued. Fuchs was present throughout the conference and made several contributions, including the suggestion that radiation-induced compression of a deuterium–tritium mixture might increase the chances of initiating fusion. Fuchs would go on to patent this idea with John von Neumann.

A report of the conference was drafted and circulated in May. It concluded:

> It is likely that a super-bomb can be constructed and will work ... The detailed design submitted to the conference was judged on the whole workable. In a few points doubts have arisen concerning certain components of this design ... In each case, it was seen that should the doubts prove well-founded, simple modifications of the design will render the model feasible.

Not everyone was persuaded by Teller's enthusiasm, however. Serber worked with Teller to tone down some of the report's overly-optimistic pronouncements. 'I still thought it was very optimistic,' Serber later wrote, 'but I had no objection to that – I had no desire to throw cold water on Edward's project and was all in favour of his proceeding with it as best he

---

[4] Electrical Numerical Integrator and Calculator, built principally to calculate artillery firing tables. It weighed 30 tons and occupied 680 square feet.
[5] Report LA-551. Access to this report remains restricted to Los Alamos staff.

could, though I really didn't think there was any chance that the weapon would work as it was envisaged.' It made no difference. When a copy of the report arrived at Berkeley, Serber noticed that all the changes he had agreed with Teller had been left out.

The attentions of the Los Alamos physicists were in any case taken up with preparations for further American atomic bomb tests, codenamed Crossroads, in the Pacific at Bikini Atoll in the Marshall Islands. These tests were designed to study the effects of standard Fat Man-type atomic bombs on specific targets, such as ships at sea. A full elucidation of the properties of the basic weapon design was a necessary step towards design improvement. And improvement was a necessary step ahead of any attempt to develop a fusion weapon. Nevertheless, when no immediate action was taken to establish a large-scale project on the Super, Teller was greatly disappointed. He blamed Bradbury.

Fuchs had been party to discussions concerning the Super, work on development of levitated implosion and the composite core, and discussions on plutonium production and processing. Much of this was information that would have been valuable to the Soviets, but Gouzenko's defection and the subsequent arrests in Canada and Britain had resulted in the temporary suspension of espionage activities in America, although it is quite possible that Fuchs passed information to Yatskov via Lona Cohen between October 1945 and June 1946.

On 21 May, physicist Louis Slotin was demonstrating the critical assembly apparatus to a number of colleagues at the Los Alamos laboratory in Parajito canyon, using the same plutonium core that had killed Daghlian. While working on the experiment, Slotin had used the tip of a screwdriver to hold apart two hemispheres of beryllium reflector. This was not normal experimental procedure and Slotin, who had assembled the plutonium core for the Trinity test, was an experienced researcher and should have known better. The screwdriver slipped and the assembly went critical. Slotin received a fatal dose of radiation. He managed to remove the upper hemisphere from the assembly, thereby sparing his colleagues similar fates. Slotin died nine days later, on 30 May.

Fuchs was now asked, together with Philip Morrison, to carry out an investigation into the accident. It was to be his final task at Los Alamos. He left the Hill in June, breaking the security regulations one last time by carrying out a confidential report on deuterium–tritium reactions that he had been asked to take to Chadwick in Washington. He paid a further visit to his sister in Cambridge, where he received an urgent cable from Cockcroft calling him to a meeting of the Harwell Steering Committee scheduled for 1 July. Fuchs flew to Britain from Montreal on 27 June.

Hall's time at Los Alamos was also over. Towards the end of May he was told that he had lost his security clearance. A review of his security file had probably concluded unfavourably, and he was obliged to leave the Hill a few days later. There appeared to be insufficient evidence to bring a case, however, and Hall was given an honourable discharge on 24 June.

Greenglass had turned down a request to continue working at Los Alamos and had been honourably discharged four months earlier, on 29 February. He moved with his wife Ruth to Manhattan and set up a business with his brother-in-law.

Able and Baker

The purpose of the Crossroads test series was to examine the effects of atomic weapons on a 'ghost' fleet of 71 ships that had been assembled and anchored in the Bikini Atoll lagoon. Feeling somewhat excluded up to this point, the US Navy wanted to get in on the atomic act. At issue was whether or not the Navy could withstand the onslaught of atomic weapons, and what this might mean for the future allocation of resources between the Navy and the Army Air Force in post-war American defence budgets.

The test fleet included decommissioned American vessels, such as the aircraft carrier USS *Saratoga* and the battleships USS *Nevada*, *Pennsylvania*, *Arkansas* and *New York*. It also included captured vessels, such as the German cruiser *Prinz Eugen* and the Japanese battleship *Nagato*. Live animals and plants were placed on some ships to test the effects of radiation. The ships held various quantities of fuel and ammunition, to simulate battle conditions.

The first test, codenamed Able, was to be an air drop on the battleship USS *Nevada*, a veteran of Pearl Harbor, painted bright orange to make it easier to identify from the air. There were another 23 vessels anchored within 1,000 yards of the *Nevada*. More distant ships held a variety of instruments to record radiation levels and the effects of the blast.

The tests were a major media event, with over 130 newspaper, magazine and radio correspondents from America, Australia, Britain, Canada, France and China invited to observe from the USS *Appalachian*. Two Soviet observers were also invited.

The Able test took place shortly after 9:00am local time on 1 July 1946. The bomb yield was 23,000 tons of TNT equivalent, but the test was little short of a disaster. The bomb fell a quarter of a mile away from the target. The USS *Gilliam* was sunk instead, and four other vessels were either sunk or severely damaged. On seeing that the target battleship was still afloat, General Joseph 'Vinegar Joe' Stilwell cursed: 'The damned Air Corps has missed the target again.'

If the intention had been to demonstrate the awesome destructiveness of atomic weapons, then the test failed here as well. The observer ship was stationed too far away. On seeing the distant spectacle, one of the Soviet observers, Simon Alexandrov, remarked somewhat contemptuously that it was: 'Not so much.' As the *Economist* reported: 'Dressed in all the trappings of an exaggerated and sometimes frivolous publicity, the first Bikini atom bomb experiment has left rather the impression of a fireworks display which slightly misfired.'

But these would be experiments with long-term legacy effects. American naval personnel entered Bikini Atoll lagoon just seven hours after the explosion. They swam in the lagoon and boarded the target ships that had remained afloat.[6]

A second test, codenamed Baker, was held a little after 9:30 on the evening of 24 July, local time. This test was more successful. The bomb was exploded about 90 feet beneath the surface of the sea with a yield of 23,000

---

[6] The reminiscences of US atomic veterans of the Crossroads tests can be viewed at www.aracnet.com

tons of TNT equivalent, driving a spectacular column of radioactive water and steam high into the air and showering the entire area – and the target ships – with radiation. However, by the time of this second test, interest had waned (or had been deflated by the Able test), and there were fewer to observe and report the results.

The Crossroads tests were not intended to be seen as part of American atomic diplomacy. This was not meant to be a thinly-veiled statement of American atomic supremacy and a further warning to the Soviet Union. But the timing of the tests, hard on the heels of Baruch's overtures to the UN Atomic Energy Commission, was unfortunate. The Truman administration received thousands of letters, some of which pleaded for restraint. One woman from Long Island wrote: 'The United States can not hope to win the confidence of the people of all countries when our endeavours to promote peace are cemented by a display of supreme might.' Oppenheimer refused to observe the tests and questioned their appropriateness 'at a time when our plans for effectively eliminating [atomic weapons] from national armaments are in their earliest beginnings'. An editorial in *Pravda* dismissed the tests as 'common blackmail', which 'fundamentally undermined the belief in the seriousness of American talk about atomic disarmament'.

A third test was called off, on Groves' suggestion. Relations with the Soviet Union were now at an all-time low. Los Alamos needed to concentrate its efforts on building the American stockpile of atomic weapons.

Born secret

By the time the McMahon Bill had passed through both Houses of Congress, it had undergone a dramatic transformation. Gone was the call for a liberal dissemination of technical information. In its place, a section entitled 'Control of Information' introduced the term 'restricted data':

The term 'restricted data' as used in this section means all data concerning the manufacture or utilisation of atomic weapons, the production of fissionable material, or the use of fissionable material in the production of power, but shall not include any data which the [US Atomic

Energy] Commission from time to time determines may be published without adversely affecting the common defence and security.

No doubt mindful of the fallout from the Gouzenko affair and Nunn May's espionage activities, in cases where the offence was clearly intended to be damaging to US interests, the punishment for communicating restricted data was established to be death or life imprisonment. Where no such intent could be proved, the lesser punishment was to be a fine of not more than $20,000 or imprisonment for no more than twenty years, or both.

In America this was (and remains today) an unprecedented restriction of free speech:

The phrase 'all data' included every suggestion, speculation, scenario, or rumor – past, present, or future, regardless of its source, or even of its accuracy – unless it was declassified. All such data were born secret and belonged to the government. If you related a dream about nuclear weapons, you were breaking the law.

Truman signed the Atomic Energy Act on 1 August 1946. It came into force on 1 January 1947. It ended all prospects for 'full and effective' Anglo-American co-operation.

In Britain, another secret cabinet committee, again known only by its 'Gen' number – Gen 163 – took the decision to produce atomic weapons independently of America. The British sentiment behind this decision was encapsulated by a comment made by Foreign Secretary Ernest Bevin:

We've *got* to have this. I don't mind for myself, but I don't want any other Foreign Secretary of this country to be talked at, or to, by the Secretary of State in the United States as I just have in my discussions with Mr Byrnes. We've got to have this thing over here, whatever it costs. We've got to have the bloody Union Jack on top of it.

Britain now wanted its own deterrent.

A complicated cock and bull story

The Atomic Energy Act called for the establishment of a civilian US Atomic Energy Commission (AEC), a purely national organisation tasked with the management of domestic atomic energy matters. Responsibility for the Manhattan Project laboratories and factories passed to the AEC from the Manhattan Engineer District in January 1947.

Truman appointed Lilienthal as chairman of the AEC. The Atomic Energy Act acknowledged the need for an advisory committee on scientific and technical matters, called the General Advisory Committee (GAC). Although Truman had grown wary of Oppenheimer, the 'cry baby' scientist, it was inevitable that the former scientific director of Los Alamos would be called to serve on the GAC. Oppenheimer was duly appointed, along with Rabi, Seaborg, Fermi, Conant and others. Oppenheimer, delayed by bad weather, arrived late for the GAC's first formal meeting in January 1947, to discover that in his absence he had been elected chairman.

After the euphoria of the Acheson–Lilienthal report and the debacle of the Baruch plan, Oppenheimer had become somewhat withdrawn. He had left Los Alamos to return to teaching at the California Institute of Technology in Pasadena. But teaching had lost its appeal. His mind was barely on the task, and his telephone wouldn't stop ringing as politician after politician sought his views on atomic energy. He seemed to be forever on a plane, bound for Washington, Los Angeles or San Francisco.

One of the recently appointed Atomic Energy Commissioners was Lewis Strauss, a self-made millionaire businessman who had worked in the navy during the war. Strauss was also a trustee of the Institute for Advanced Study in Princeton. Towards the end of 1946 Strauss offered Oppenheimer the position of director of the institute. Oppenheimer thought long and hard about it, and eventually agreed to move from the West Coast. At least at Princeton he would be nearer to Washington.

The intransigence of the Soviet Union, which Oppenheimer had seen at first hand, convinced him that no agreement on the international control of atomic energy was likely in the near future. He told Hans Bethe that he had 'given up all hope that the Russians would agree to a plan'. He saw the

Soviet counter-proposals to ban the atomic bomb as a device designed to 'deprive us immediately of the one weapon which would stop the Russians from going into Western Europe'.

Though hardly the stance of a 'hawk', Oppenheimer's transformation from 'leftwandering' idealist to Cold War realist was complete.

But Oppenheimer was to find it impossible to put his past indiscretions behind him. The FBI had continued to poke around in the murky depths of the 'Chevalier incident'. Because of the allegations contained in his FBI file, Chevalier himself had found it impossible to obtain the necessary clearance for war-related work and had stayed on in New York as a freelance translator and writer. He returned to teaching in Berkeley in the spring of 1945 before being asked to provide translation services at the Nuremberg war crimes tribunal. On returning to Berkeley once more in May 1946 he found that he had been denied tenure.

In June, FBI agents had simultaneously, but separately, interviewed Chevalier and Eltenton, the agents cross-checking their stories by phone. At one stage, Chevalier's interviewer pulled a file towards him and said: 'I have here three affidavits from three scientists on the atomic bomb project. Each of them testifies that you approached him on three separate occasions for the purpose of obtaining secret information on the atomic bomb on behalf of Russian agents.' Chevalier was nonplussed. He thought this was all a joke at first, then realised he had no alternative but to relate the conversations he had had with Eltenton and Oppenheimer. The FBI agents didn't seem all that interested in his story.

Chevalier had an opportunity to compare notes with Eltenton a few months later, and realised that they had both been interviewed by the FBI at the same time; Chevalier in San Francisco and Eltenton across the bay in Oakland. Then the opportunity came to raise the matter directly with Oppenheimer during a cocktail party at the Oppenheimers' home on Eagle Hill. Oppenheimer suggested they talk outside.

'I had to report that conversation, you know', Oppenheimer said.

'Yes,' Chevalier said, 'but what about those alleged approaches to three scientists, and the supposed repeated attempts to get secret information?'

Oppenheimer gave no reply. Chevalier saw that his friend was extremely nervous and tense. When Kitty called a second time for Oppenheimer to attend to their guests, he lost his temper and 'let loose with a flood of foul language, called Kitty vile names and told her to mind her goddamn business'.

Oppenheimer himself was interviewed by the FBI on 5 September 1946, a little over three years since his ill-fated (and recorded) conversation with Pash and Johnson. He now conceded that in attempting to protect Chevalier he had fabricated a 'complicated cock and bull story' about Eltenton's approach to three scientists. If this was indeed the truth, and it is difficult to understand Oppenheimer's motives otherwise, then he might have felt that confessing the lie would be the end of the matter.

The Atomic Energy Act obliged the FBI to review the security clearances of all involved and removed any obstacles to an open and thorough investigation of Oppenheimer's past activities. The level of surveillance was increased and Oppenheimer's colleagues were questioned about his loyalty. Lawrence vouched for him once again, declaring that Oppenheimer 'had a rash and is now immune', even though the personal distance between the two physicists was growing.

Hoover sent a summary of Oppenheimer's weighty FBI file to the AEC in early March 1947. Although Strauss was visibly shaken by what he read, he told Oppenheimer shortly afterwards that he saw nothing in it that would stand in the way of his appointment as director of the Institute for Advanced Study. The Oppenheimers arrived in Princeton in July.

Oppenheimer was given a 'Q' clearance for his work with the GAC the following month.

Modus vivendi

The Atomic Energy Act had ended the prospect for a 'full and effective' collaboration between British and American nuclear scientists but the commitment to share raw materials, managed by the Combined Development Trust, remained. The Groves–Anderson memorandum had not changed the basic premise of the wartime agreements. This meant that Britain could, in principle, lay claim to half the production of uranium ore from the Belgian Congo. Britain had no immediate use for the uranium ore, but having taken the decision that Britain should become an independent atomic power, Attlee had decided to stockpile the ore for future use in Britain's own bomb project. He duly staked his country's claim.

This gave Lilienthal some cause for concern. When, as chairman of the AEC, he had taken control of Los Alamos in January 1947 he had found that there were far fewer atomic bombs in the American arsenal than Truman had been led to believe. In fact, the number of actual bombs was zero. While there were many bomb cores, these hadn't been assembled into bombs that could be deployed at short notice. 'I was shocked when I found out', Lilienthal commented. 'Actually we had one [bomb] that was probably operable when I first went off to Los Alamos; one that had a good chance of being operable.' Building a stockpile meant assembling all the pieces that existed, and procuring sufficient raw materials for additional bombs. Lilienthal and his commissioners calculated that America needed the entire free world supply of uranium ore. They could not afford to share materials with the British.

But then a solution offered itself. Post-war Britain had almost run out of money. If British citizens thought that the VE Day celebrations signalled the beginning of the end of wartime austerity, they were soon to be disappointed. Rationing actually bit deeper in the months and years after the war as Britain, though victorious, sank slowly to its knees.

Byrnes had resigned from Truman's cabinet in early 1947. His relationship with Truman had become inevitably strained by what Truman perceived as Byrnes' tendency to set foreign policy without consultation. Byrnes' successor as Secretary of State was General George C. Marshall. In a Harvard commencement address on 5 June, Marshall had outlined an aid package designed to restore the shattered post-war economies of Europe:[7]

---

[7] The Marshall Plan involved loans and technical assistance to the European allies totalling some $13 billion over four years (equivalent to roughly $130 billion in 2006).

It is logical that the United States should do whatever it is able to do to assist in the return of normal economic health in the world, without which there can be no political stability and no assured peace. Our policy is directed not against any country or doctrine but against hunger, poverty, desperation and chaos. Its purpose should be the revival of a working economy in the world so as to permit the emergence of political and social conditions in which free institutions can exist.

It was now suggested that Attlee might be willing to give up his claims to the uranium ore or at least sell Britain's share to America in exchange for Marshall Plan aid. This, of course, was little short of blackmail, and the strategy was passionately debated in meetings at the Pentagon through September and November 1947. Eventually a cable went to the American ambassador in London spelling out the terms: 'further aid to Britain … should be conditioned on Britain's meeting our terms with respect to the allocation of atomic raw materials.'

The British government had little choice but to comply. After some tough negotiations Britain agreed to relinquish both its veto on American use of atomic weapons and its rights to the uranium ore for at least the next two years. Britain further agreed to supply two-thirds of the country's stockpile of ore to America. Britain had, in effect, delayed its own atomic bomb programme by several years.

The agreement, a modus vivendi, was signed at a rather muted ceremony on 7 January 1948. It contravened the US Atomic Energy Act and Article 102 of the UN Charter. It was consequently kept secret, even from Congress.

Donald Maclean, appointed co-secretary of the Combined Policy Committee in Washington in February 1947, had participated fully in the negotiations.

Chapter 21

ARZAMAS-16

April 1946–June 1948

T he espionage materials provided by Hall, Greenglass and especially
Fuchs undoubtedly delivered an enormous advantage to the Soviet
programme. They held solutions to many of the scientific and technical problems that the Manhattan Project physicists had encountered.[1]
But these solutions still had to be checked by Soviet physicists through
meticulous experiments and calculations. Knowing that it could be done
was one thing. Knowing how it could be done was another. Actually doing
it was something else entirely. No Soviet physicist was going to risk testing a weapon without first gaining the practical experience needed to be
certain that the test would be a success.

Yuli Khariton had adopted a motto: 'We have to know ten times more
than we are doing.'

An entire atomic industry had to be built. Isotope separation facilities, nuclear reactors, plutonium production facilities and their associated
laboratories had to be constructed, as well as a weapons laboratory for

---

[1] However, it is worth noting that some of the problems and their potential solutions had
been identified – at least in outline – by Soviet physicists before the espionage materials
were available.

assembling and testing the bomb. There could be no doubt that the Soviet programme would be a major undertaking.

Before anything else could be done the Soviet scientists desperately needed to find sources of uranium. The uranium liberated from occupied Germany was critical in helping to kick-start the Soviet programme, but a lot more would be needed if Russia was to become a sustainable atomic power. Between them, Britain and the United States had cornered about 97 per cent of the world market for uranium and about 65 per cent of the market for thorium, a source of the fissionable isotope U-233. Large-scale exploration in Central Asia was immediately begun and ore production got under way at several mines, including one in Taboshary, near Tashkent, where uranium deposits had been discovered before the war. In the meantime, operations were restarted at mines in the Soviet zone of occupation in eastern Germany.

Work on an experimental nuclear reactor had begun in 1943 but had been constrained by shortages of uranium and purified graphite. By the end of 1945 quantities of purified graphite were becoming available from the Elektrostal plant which lay some 45 miles to the south-east of Moscow. The equipment that had been salvaged from the Auer company's Oranienburg plant was reassembled at the Elekrostal plant and Nikolaus Riehl now supervised uranium production and processing there. Progress was slow but, aided by information in the Smyth report, by the summer of 1946 Riehl's group was providing its first few tons of uranium metal.

Work on the construction of isotope separation facilities began early in 1946. A gaseous diffusion plant was to be built in the central Ural mountains, near Neviansk about 30 miles north of Sverdlovsk. An electromagnetic separation plant was to be built close by, in the northern Urals at Severnaia Tura. The facilities were given the codenames Sverdlovsk-44 and Sverdlovsk-45, respectively.[2] Kikoin and Artsimovich were appointed as scientific directors, and work on separation methods was supported by research teams led by German émigrés, including Ardenne.

---

[2] Each secret atomic facility was assigned a codename based on the name of a nearby city and the last digits of a post office box number.

Certain aspects of the Soviet programme were reorganised on 9 April 1946. It was recognised that work on the high explosives required for implosion could not be safely accommodated at a laboratory situated so near to Moscow. Kurchatov proposed setting up a weapons laboratory in a more remote area – the Soviet equivalent of Los Alamos.

Beria agreed. Sector No. 6 of Laboratory No. 2 subsequently became a distinct entity, Design Bureau No. 11 (*konstruktorskoe biuro*-11, or KB-11), assigned the task of designing and manufacturing prototype atomic weapons. Beria appointed General Pavel Zernov, Deputy People's Commissar of the Tank Industry, to head it. Yuli Khariton was appointed chief designer and scientific director. Just as Groves and Oppenheimer had searched for a location for Site Y towards the end of 1942, so Zernov and Khariton now sought a location for their new weapons laboratory. Khariton described what they found:

Finally, after a long search, on 2 April 1946 Pavel Mikhailovich Zernov and I arrived in the small town of Sarov, where St Seraphim had once worshipped. Here there was a small factory which, during the war, had produced munitions, including shells for 'Katiusha' rocket-launchers. All around were impenetrable woodlands. There was plenty of space and a lack of population, and we were thus able to carry out the necessary explosions.

Sarov is located about 250 miles east of Moscow, on the border between Gorky Oblast and the former Autonomous Soviet Socialist Republic of Mordovia. At the time it had a population of a few thousand. In the centre of the town were the remains of an Orthodox monastery which had been closed down in 1927. The first nuclear laboratories were set up in the former monks' quarters. The facility was known by several codenames – KB-11, Base 112, Site 550, *Privolzhskaya Kontora* (the 'Volga Office'), Installation No. 558, Kremlev, Moscow, Centre 300, and Arzamas-75. However, its most enduring codename was to be Arzamas-16, after the town of Arzamas some 40 miles to the north of Sarov. It was known colloquially as 'Los Arzamas'.

The laboratory was given some tight deadlines. The 'technical assignment' – a summary of the technical requirements for RDS-1, the Soviet version of the Fat Man plutonium bomb, and RDS-2, a U-235 weapon based on the gun method – was to be ready by 1 July 1946. The acronym RDS was invented by Makhnev. It stood for *Reaktivnyi Dvigatel Stalina*, or 'Stalin's Rocket Engine'. The designs for both RDS-1 and RDS-2 were to be complete by 1 July 1947. RDS-1 had to be ready for testing by 1 January 1948, RDS-2 by 1 June 1948.

Kharition and his team developed a scale model of the RDS-1 implosion bomb – a structure of nested metal shells fourteen inches in diameter – and sent it to Beria and Stalin for them to examine. Kharition provided the technical assignment shortly afterwards, on 25 July.

The will of Soviet man

Work on *Fizicheskii*-1 (F-1), the Soviet Union's first pilot nuclear reactor, was now accelerated. The reactor was constructed at Laboratory No. 2 on the outskirts of Moscow, in a special building which housed a pit about twenty feet in depth. Details of the first Chicago reactor had been published in an appendix to the Smyth report, but the dimensions of F-1 were subsequently identified to be very close to those of the experimental reactor Hanford 305, suggesting that design details might have been obtained by espionage at Hanford or the Met Lab in Chicago, where the design was developed.

Preparations for the construction of F-2, an industrial-scale plutonium production reactor, had already begun at Cheliabinsk-40, about ten miles east of Kyshtym and 50 miles north-west of Cheliabinsk in the Urals. Herbert Hoover had helped to establish copper mining and smelting at Kyshtym before the October Revolution.

Kurchatov supervised the reactor programme through the months August to October. As the Met Lab physicists had done, he constructed small-scale sub-critical assemblies to test the extent of neutron multiplication and make the measurements necessary to predict the amounts of uranium and graphite that would be required to reach criticality. Assembly of F-1 subsequently began on 15 November, one layer at a time. Kurchatov had estimated that the reactor would reach criticality with 76 layers, and pressed every last ounce of available uranium into service.

But, just as the Met Lab physicists had discovered four years previously, the extrapolation from small-scale models tended to overestimate the quantity of uranium required. As layer 61 was put in place on 24 December, it became obvious that one further layer would be sufficient to tip the reactor over the threshold of criticality.

At 2:00pm on 25 December, the three cadmium control rods were inserted and layer 62 was added. Kurchatov arrived to supervise the next step as the building was cleared of all non-essential personnel. Those who remained were silent. Only the clicking of the neutron counters could be heard. There followed a series of experiments in which the control rods were partially withdrawn and the neutron counters monitored to confirm that everything was behaving as expected.

At 6:00pm, F-1 went critical. 'Well, we have reached it', Kurchatov observed. It was the Soviet programme's first major success. The physicists congratulated each other, as Kurchatov declared: 'Atomic energy has now been subordinated to the will of Soviet man.'

Beria arrived to inspect F-1 a few days later. The physicists went through their routines once more and declared that the reactor was operational. But, apart from the noise of the neutron counters, there was nothing to hear and there was certainly nothing to see. Beria was immediately suspicious. 'Is that all?', he said. 'Nothing more?' He asked if he could go into the reactor room and take a closer look. When Kurchatov insisted it would be too dangerous, Beria grew even more suspicious.

Stalin received a report on the success of F-1 on 28 December 1946:

In the first days of work of the uranium–graphite pile (December 25–26–27), we have already obtained the first nuclear chain reaction to be launched in the USSR on a semi-industrial scale. It is now possible to regulate the functioning of the pile in the necessary range and to control the run of the nuclear chain reaction.

Stalin received members of the Special Committee and the scientists who had taken part in the successful F-1 project at a formal session at the Kremlin on 9 January 1947. It was the first, and last, time that Stalin agreed to receive reports directly from his atomic scientists.

Book-breaker

The Japanese surrender and the ending of hostilities meant that teams of Army Security Agency (ASA)[3] cryptanalysts assigned to work on German and Japanese codes were freed to work on the Soviet message traffic. The breakthroughs secured by Hallock and Phillips at Arlington Hall had now rendered the Soviet messages vulnerable. It had become possible partly to strip away the one-time pad cipher to reveal the underlying, 'plain' code groups contained in hundreds of messages that had gone back and forth between Moscow and Soviet embassies, consulates and trade organisations in America.

As the ranks of cryptanalysts swelled to between 50 and 75, that vulnerability increased. One particularly gifted cryptanalyst, Samuel Chew, made use of the highly predictable patterns in messages detailing scheduled shipments of Lend-Lease aid from American ports.

There remained the problem of breaking the code itself. This could only be done either by acquiring a copy of the code book or recreating it from a painstaking analysis of the plain code groups that had now been revealed.

In early 1946 Meredith Gardner joined the Russian project. Gardner was an accomplished linguist who had taught languages at universities in Texas and Wisconsin before the war. He was able to read German, Spanish, French, Sanskrit and Lithuanian and had studied Old High German, Middle High German and Old Church Slavonic. He had astonished his Arlington Hall colleagues by learning Japanese in three months. He now proceeded to learn Russian.

---

[3] The Army Security Agency was the 1945 successor to the Army Signals Intelligence Service.

Gardner was the 'book-breaker'. His task was to use his language skills, identify patterns in the code groups and so, step by step, recreate the code book that the Soviet cipher clerks had used before enciphering their messages using one-time pads. This was a task that called for infinite patience and a certain type of personality. Gardner fitted the bill: 'tall, gangling, reserved, obviously intelligent, and extremely reluctant to discuss much about his work', was how FBI counter-intelligence agent Robert Lamphere described him.

The materials provided by Gouzenko were presumably not of direct assistance to Gardner, but he did have some clues in the form of obsolete codebooks. A partially burned codebook had been recovered from a battlefield in Finland, and Donovan had purchased a copy from the Finns for the OSS. This had been passed to the ASA. Such material served as a basis for imagining how newer codebooks might be structured.

In the summer of 1946, Gardner finally began to read portions of some of the messages dated two years previously. He saw enough to convince him that some of these messages related to Soviet espionage.

Some of the messages inevitably referred to English-language names or places, and the Soviet cipher clerks had used a 'spell table' to code letters from the Latin alphabet. Gardner managed to reconstruct this spell table and on 20 December he broke into a message from New York to Moscow dated 2 December 1944. He read a list of names:

Hans BETHE, Niels BOHR, Enrico FERMI, John NEWMAN, Bruno ROSSI, George KISTIAKOWSKY, Emilio SEGRE, G.I. TAYLOR, William PENNEY, Arthur COMPTON, Ernest LAWRENCE, Harold UREY, Hans STANARM, Edward TELLER, Percy BRIDGEMAN, Werner EISENBERG, STRASSMAN.

It was a list of Manhattan Project scientists, to which Heisenberg ('Eisenberg') and Fritz Strassman's names had for some reason been appended. Although Gardner did not know it yet, it was the list taken from the report that Hall had passed to Kurnakov and Yatskov in October 1944.

Gouzenko's defection had unmasked Nunn May, exposing Soviet espionage against the Manhattan Project from the distant Montreal laboratory. Here, it seemed, was evidence of atomic espionage from within America itself.

I hope your baby will be born soon

Fuchs may have rationalised his decision to return to Britain on the basis of some curious sense of loyalty, but the Britain he returned to at the end of June 1946 was bleak and inhospitable. The Attlee government decided to put bread on ration in July, provoking anger from the opposition Conservatives and protests from the Master Bakers' Federation and the British Housewives' League. Bread had never been rationed during the war, and the decision to ration it now was taken as a sign of national impoverishment.[4] To make matters worse, the UK winter of 1946–47 was one of the most severe on record, with heavy snow falls and temperatures falling to minus 20° Celsius.

Despite these hardships, Fuchs settled into a reasonably comfortable existence at Harwell. There was a small nucleus of Los Alamos physicists at the British research establishment, and a sense of community not unlike that which had developed on the Hill. The scientists also shared a sense of idealism about the possibilities for the peaceful uses of atomic power. Fuchs made some new friends. Cockcroft decorated the site with lawns and flower beds.

Fuchs suspended his espionage activities. At their last meeting he had given Gold instructions for his Soviet contact back in Britain, but he chose not to follow up on those instructions and let the contact lapse. Perhaps he was still somewhat unnerved by the revelations of the Nunn May trial or perhaps he suspected that he was under surveillance. In fact, several expatriate scientists working on sensitive projects were put under surveillance

---

[4] It was introduced primarily to force the pace of negotiations with America regarding Marshall Plan aid and food for the British occupation zone of Germany. Rationing bread demonstrated the dire financial circumstances with which post-war Britain was struggling.

by British intelligence for a time. Fuchs got on with his job in his typically dry, reserved way.

When Peierls and his wife decided to escape the vicissitudes of the severe British winter and take a skiing holiday in Switzerland, Fuchs gratefully accepted an invitation to join them. When he returned to Britain two weeks later, he had resolved to continue to spy for the Soviet Union.

As his contact with Soviet agents was now broken, he thought to re-establish communication through Jurgen Kuczynski, but Kuczynski was now back in Germany, working in the Russian zone of occupation. Another German Communist émigré, Johanna Klopstech, put him in touch with Soviet intelligence. Klopstech was not an agent, but was deemed a 'reliable person' by Russian intelligence. Fuchs was instructed to meet his new contact at the Nag's Head pub in Wood Green, north London, on 27 September 1947. He was to carry with him a copy of *Tribune* magazine, and look for a man carrying a red book.

The man was Alexander Feklisov, the NKVD agent who had managed the Rosenberg network and who had worked alongside Yatskov in New York. Feklisov watched from across the street as Fuchs entered the pub and then followed him inside. Feklisov was able to identify Fuchs from a photograph, but Fuchs did not know what his contact looked like. However, Fuchs spotted the red book and walked up to a panel carrying framed photographs of famous British boxers.

'I think the best British heavyweight of all time is Bruce Woodcock', Fuchs declared.

'Oh no, Tommy Farr is certainly the best', Feklisov responded.

With these code phrases correctly exchanged, they left the pub separately. Once outside, Feklisov caught up with Fuchs, introduced himself as 'Eugene' and gave him a series of questions that he had been asked to put by Moscow Centre, including one concerning the possibility of building the hydrogen bomb, or Super. Fuchs promised to have the answers ready for their next meeting.

Feklisov set the protocols for future meetings, at the Spotted Horse pub in Putney High Street and outside Kew Gardens tube station. They were to meet once every two or three months.

'I'm very happy to be with you again', Fuchs said, towards the end of their meeting. 'I hope your baby is born soon!'

Feklisov didn't understand. 'Which baby?' he asked.

'Your bomb. From the questions you're asking me I estimate it will take you one to two years. The Americans and our own research scientists think it'll take you seven or eight years. They're very wrong and I'm delighted.'

Fuchs then handed over a bulky package containing important information on the production of plutonium which he had acquired after he had returned to Britain.

'Thank you', Feklisov said simply.

'My pleasure', said Fuchs. 'I shall always be indebted to you.'

Fuchs advised Feklisov of the British decision to embark on an independent bomb programme. Britain's first atomic weapon was to be a plutonium bomb, and William Penney had been appointed to head weapons development at the Ministry of Defence's Fort Halstead site in Kent. At the time, not even Fuchs' Harwell colleagues were aware of this decision.

When the Combined Policy Committee convened a three-day meeting in Washington in November 1947 to determine what kind of atomic information could be declassified, it was Fuchs who represented British interests. He sat in the meeting alongside the British co-secretary, Donald Maclean, presumably neither knowing that the other was a Soviet spy. One participant later recalled a certain sense of exasperation with Fuchs' rather conservative assessment of which information could be safely declassified, and which should remain secret.

Life in Sarov

Soviet physicists began to arrive in numbers at Arzamas-16 in the spring of 1947. Veniamin Tsukerman, an expert in the flash radiography of explosions, recruited to the project by Khariton, described his arrival in May:

We had arrived in what was for us a new world. Everything was unexpected: the thick forest, the beautiful, centuries-old pines, the monastery on the high river bank with its cathedrals and white bell tower. And, in sharp contrast, the grey columns of prisoners who went through the village in the morning and in the evening.

The Autonomous Republic of Mordovia was a region of prison camps. Because of the sensitive nature of the installation being built in Sarov, no political prisoners were used. The prisoners who constructed it were *ukazniki*, people found guilty of infringing the many decrees (called *ukazy*) which applied outside the normal criminal code. For those prisoners whose sentences unhelpfully ended during the construction phase, Beria's solution was simple. He extended their sentences. After their release, they were sent to the far eastern corner of the Soviet Union, as far away as possible from the facility they had helped to build. The 'grey columns of prisoners' were rarely mentioned by the scientists now assembling at Arzamas-16 but, as Khariton observed, they regularly intruded on the scientists' consciousness.

Although the scientists were notionally free, in truth they too were cosseted prisoners. The Soviet atomic industry was being developed as a network of highly secret facilities within a 'Closed Administrative and Territorial Formation' (known by its Russian initials as a ZATO). It was eventually to become known as the 'White Archipelago'. Conditions were at least better than those prevailing in the prison camps that formed the notorious 'Gulag Archipelago'. The American and émigré scientists and their families that had gathered on the Hill in 1943 complained bitterly about the 'concentration camp' conditions behind the razor-wire-topped fences. But at Arzamas-16 the oppressive conditions were greatly amplified by the thinly-veiled threat of extreme punishment in the event of failure. This was, after all, a project led by Stalin's most feared executioner.

Physicist Lev Altshuler described the effect this had on him:

It was not merely a regime, it was a way of life, which defined people's behaviour, thoughts and spiritual condition. I often dreamt the same dream, from which I would awaken in a cold sweat. I dreamt that I was in Moscow, walking down the street carrying top secret and extremely top secret documents in my briefcase. I was killed because I could not explain why I had them.

But it is a characteristic of the specifically Russian human condition under Stalin that amid the oppressive secrecy and fear there was also great enthusiasm, art, romance and humour. The immense suffering that the country had endured in the war was a raw collective memory. The scientists perceived the threat of an American attack using atomic weapons as very real, and now worked hard to restore the balance of atomic power. As Tsukerman observed:

> We worked without heed for ourselves, with huge enthusiasm, mobilising all our spiritual and physical strength. The working day for senior researchers lasted from twelve to fourteen hours. Zernov and Khariton worked even longer hours. There were practically no days off, nor was there any leave; permission to travel on business was granted comparatively rarely.

The scientists and their families would entertain themselves with ancient gramophones. Tsukerman, slowly going blind from a rare form of pigmentary retinitis, would play foxtrots, tangos and waltzes on a mahogany piano that he had brought with him from Moscow. There were competitions, and parties, and picnics, skiing trips and practical jokes. Yakov Zeldovich and Vitaly Alexandrovich clubbed together to buy a Harley Davidson motorcycle and sidecar. Zeldovich would always drive it. Alexandrovich would always repair it.

The summer of 1947 was hot. Tsukerman felt the heat, both literally and figuratively, as the scientific divisions at Arzamas-16 grew and came up to full strength.

Codenames

By July 1947 the ASA was becoming increasingly alarmed at the information that was being revealed by Soviet message traffic decrypted by the code-breakers at Arlington Hall. The messages contained countless code-names, many clearly referring to Soviet agents.

The codenames used by the Soviet cipher clerks were not necessarily intended as a security measure to disguise the identity of the agents, locations or sources of information, but were used rather to reduce the amount of coding that would be required.[5] This much was clear from the information that Gouzenko had been able to provide. An agent's codename would often be given in a message shortly after recruitment and assigned to his or her real name. Sometimes a certain tongue-in-cheek logic would be used in the selection. A Communist was referred to as a FELLOWCOUNTRYMAN. Trotskyists and Zionists were referred to as POLECATS and RATS. The FBI was referred to disparagingly as KHATA, or 'the Hut'. San Francisco was BABYLON. Washington was CARTHAGE.

Among the messages now being partially deciphered were many references to an agent with the codename ANTENNA, which had subsequently been changed to LIBERAL. A message from New York to VIKTOR (Pavel Fitin) in Moscow dated 27 November 1944 had revealed the following:

Your No. 5356. Information on LIBERAL's wife. Surname that of her husband, first name ETHEL, 29 years old. Married five years. Finished secondary school. A FELLOWCOUNTRYMAN since 1938. Sufficiently well developed politically. Knows about her husband's work and the role of METR and NIL. In view of delicate health does not work. Is characterised positively and as a devoted person.

The cryptanalysts were starting to reveal information that required proper investigation if the identities of the agents were to be discovered. The ASA was not equipped to undertake such detective work, so in September 1947 Carter Clarke, now a general in the army's intelligence organisation G-2,

---

[5] These codenames should not be confused with the agents' cover names. For example, Harry Gold's cover name was Raymond, and this was the only name by which Fuchs knew him during the time of their meetings in New York and New Mexico. Gold's codenames as used by Soviet cipher clerks were later revealed to be GOOSE and ARNO. Gold probably never knew his codenames.

made contact with S. Wesley Reynolds, G-2's FBI liaison. Clarke briefed the FBI on the breakthroughs that had been achieved in cracking the Soviet messages. Over the years the project would go by many names: Jade, Bride, Drug and, finally, Venona.

FBI special agent Robert Lamphere was assigned to the project in October and made the first of what would become regular pilgrimages to Arlington Hall every two or three weeks. He found Gardner to be distant and reserved at first, but as they worked together they started to become friends.

Gardner warmed enough to ask Lamphere in early 1948 if he could obtain the plain texts (that is, the uncoded and unenciphered texts) of some Soviet trade messages from 1944. Lamphere was not optimistic, but by return post received from the FBI's New York field office a mass of material, stacked seven or eight inches high, most of it in Russian. Some commentators have speculated that the material was the result of a 'black bag' job, an authorised FBI burglary of Soviet premises involving the photography of sensitive documents.

Lamphere immediately took the material to Gardner. When he returned on his next pilgrimage, two weeks later, he found Gardner in a highly excited state: 'In his shy way he explained that we'd hit the jackpot. He now had the plain texts of some very important material.'[6]

Shortly afterwards, Gardner began giving Lamphere some messages that had been completely decoded. Lamphere remembered Gardner's slight smiles of pleasure as he would reach for his own steadily growing version of the codebook, and write a word in Russian alongside one of the code groups.

Grand jury

The spy networks that had been stitched together in America with such impunity by Soviet agents were now starting slowly to unravel.

---

[6] Note that the official NSA history of the Venona project makes no reference to this material, though Lamphere distinctly remembered passing it to Gardner.

Elizabeth Bentley secretly testified before a federal grand jury in New York in the spring of 1947. The grand jury investigation started as a wide-ranging inquiry into Soviet espionage, particularly in US government agencies such as the State Department, the Treasury Department and the OSS. The investigation would eventually focus on the case against Alger Hiss, a former official in the State Department.

Bentley described how someone she knew only as 'Julius' had approached her Soviet contact and lover Jacob Golos with the offer of industrial secrets from a group of Communist engineers he had assembled. She identified other Americans who had acted as spies and couriers, including industrial chemist Abraham Brothman. When Brothman was interviewed by the FBI, he named Harry Gold.

Gold had obtained a job at Brothman's small commercial chemical laboratory on Long Island in May 1946. They had first met five years previously, Gold acting as courier to Brothman's industrial spy. Both were obviously aware of the other's espionage activities, although Brothman was unaware that Gold had also couriered atomic secrets for the Soviets.

Yatskov knew that Brothman was under surveillance, and had warned Gold to cut the link. When at his last meeting with Gold towards the end of 1946 he discovered that Gold was now working for Brothman, he accused Gold of ruining eleven years of espionage work. Yatskov left the meeting hurriedly. His cover had been blown by Gouzenko, and shortly after this last meeting with Gold he left America for a new posting at the Soviet embassy in Paris.

Both Brothman and Gold were called before the grand jury towards the end of July 1947. Together they had concocted a detailed cover story. Gold was able to persuade the jury that they were both innocent bystanders, and with no corroboration of Bentley's testimony, neither Brothman nor Gold were called to face charges.

A relieved Gold walked free. But he had now acquired a thick FBI file which identified him as a suspected Soviet courier.

Annushka

With F-1 up and running satisfactorily, Kurchatov turned his attention to the industrial-scale reactor being constructed at Cheliabinsk-40. It was an area of great beauty, nestled among mountains, forests and lakes. By the end of 1947 Cheliabinsk-40 was already a large city built by the forced labour of as many as 70,000 prisoners, drawn from twelve different camps. The prisoners worked in stages, one group starting construction work, others continuing, and yet others completing. When the prisoners were discharged, none could say precisely what it was they had been helping to build.

Kurchatov now travelled to the site with Vannikov to oversee the final preparations. Both stayed in a railway carriage parked next to the site and settled in for the long, harsh mountain winter.

The plutonium production reactor was called Installation-A or Annushka, meaning Little Anna. It was assembled in a pit blasted 60 feet deep, over which a large building had been constructed. Work began on the reactor assembly in March 1948. In a speech to his engineers, Kurchatov quoted from Pushkin's *The Bronze Horseman*, in which Peter the Great founds a great city on the banks of the Neva 'to spite our arrogant neighbour'. The neighbour in question was Sweden. 'We still have enough arrogant neighbours', Kurchatov said.

MVD[7] General Zavenyagin, Pervukhin and other senior officials would visit often, entering the reactor room through a special manhole that workers called 'the General's manhole'. By May, the reactor assembly was complete.

Kurchatov supervised the first run to 'dry' criticality on 8 June 1948. Over the next few days the Soviet scientists introduced water cooling, cautiously added more uranium and ran the reactor up to higher and higher

---

[7] The People's Commisariats were redesignated as Ministries in March 1946. The NKVD became the Ministerstvo Vnutrennikh Del (MVD), the Soviet Ministry of Internal Affairs. The subordinate NKGB became the MGB.

power outputs. The reactor reached its designed output of 100,000 kilowatts on 19 June.

The scientists soon discovered for themselves some of the technical problems associated with nuclear reactors. When the cans containing the uranium fuel swelled and became lodged in their channels, Beria's watchful representatives claimed sabotage. Kurchatov was able to explain that with this technology the scientists were in largely unexplored territory. Some changes to the reactor design were made and the problems were solved.[8]

Eniwetok

The United States initiated further atmospheric atom bomb tests in April and May 1948, at Eniwetok Atoll in the Marshall Islands. These tests, designated Sandstone, were developed to investigate the levitated core and composite plutonium/enriched uranium core designs. Sandstone was the first test series to be managed by the new US Atomic Energy Commission, with the military providing a supporting role. Their purpose was scientific, rather than military.

Little Boy and Fat Man had exploded over Hiroshima and Nagasaki with efficiencies of 1.4 per cent and 14 per cent respectively, and the Los Alamos scientists were now bent on raising this efficiency and perfecting the technology. This was not so much about perfecting the bomb's ability to kill people (although this would become a preoccupation of later generations of bomb designers), it was about developing more efficient bombs that required smaller cores, thereby extending the US stockpile simply through the ability to make more weapons using the same amount of fissile material.

The first test – X-ray – was held at 6:17am local time on 15 April. It was, like the Trinity test, a 'tower shot', with the bomb mounted at the top of

---

[8] The Soviet reactor programme may have been aided by intelligence passed by Melita Norwood, an NKVD spy recruited in Britain in 1934. Norwood, the 'Bolshevik of Bexleyheath', was exposed in 1999 and died in 2005. However, recent claims that her espionage accelerated the Soviet bomb programme by five years are a gross exaggeration. See the *Sunday Telegraph*, 31 August 2008.

a 200-foot tower on Engebi Island. The new design yielded about 37,000 tons of TNT equivalent, with a utilisation efficiency of 35 per cent for the plutonium in the composite core and 25 per cent for the enriched uranium. Another tower shot, Yoke, followed at 6:09am on 1 May on Aomon Island. This was another levitated, composite core probably containing more fissile material. Although this produced an explosive force of 49,000 tons, almost four times the size of the bomb that had destroyed Hiroshima, it was regarded as less efficient. At 6:04am on 15 May on Runit Island a final tower shot, Zebra, with a levitated, enriched uranium core, yielded 18,000 tons. Although of lower total yield, the Zebra device was more efficient than that used for Yoke.

The tests demonstrated conclusively the efficiency of levitated implosion over the solid core compression that had been used in the Fat Man design. They also demonstrated the efficiency of implosion compared to the gun method that had been used in Little Boy. The results enabled a 75 per cent improvement in efficiency: bombs could now be made with less than half the amount of plutonium used in Fat Man and a tenth of the enriched uranium used in Little Boy. At a stroke, the US stockpile of atomic weapons had been extended by 63 per cent.

The Soviets, uninvited this time, watched from a warship stationed some twenty miles away.

Acknowledging the failure of the first, faltering steps towards international control of atomic weapons, the United Nations Atomic Energy Commission was wound up on 17 May 1948.

The Berlin blockade

The democratically elected government of Czechoslovakia had bid for Marshall Plan aid in July 1947. In post-war Eastern Europe, this was the region's only democracy, a coalition government led by Prime Minister Klement Gottwald, who was also leader of the Czech Communist Party. The government consisted in part of Communist representatives but was not dominated by them.

But Stalin was having none of this. Sensing an attempt to push Western democracy into the Soviet sphere of influence, he leaned on Gottwald and the bid for aid was withdrawn. By 25 February 1948, with the Red Army waiting on the country's borders, a Soviet-supported coup had displaced all but one of the non-Communist ministers from important posts in the Czech government. The last was Foreign Minister Jan Masaryk. On 10 March Masaryk was found dead in the courtyard of the Foreign Ministry. Czech President Edvard Beneš refused to sign the new, post-coup constitution of 9 May and resigned in June. He was replaced by Gottwald. Beneš died three months later.

The Communist coup in Czechoslovakia, so recently liberated from Nazi dictatorship, sent shockwaves around the West. The opinions of advisers who had only recently anticipated that there would be no war with the Soviet Union for years to come were dramatically re-evaluated. The American military lobbied for more funding and a reinstatement of the draft. Britain signed the Treaty of Brussels, aligning the country with France, Belgium, the Netherlands and Luxembourg and, in September 1948, creating the Western Union Defence Organisation, a military alliance designed to confront the forces of the Eastern European bloc. Negotiations were begun that were eventually to result in the formation of the North Atlantic Treaty Organisation (NATO) in April 1949.

To many, another war now seemed inevitable. And then the Cold War took an even more ominous turn.

Occupied Berlin had been organised into four zones, city-wide echoes of the American, British, French and Soviet occupation zones of Germany itself. But the city was buried deep in the heart of the Soviet zone, 100 miles from the nearest border (with the British zone), and therefore extremely vulnerable to a Soviet threat.

In February, the Americans and the British had proposed to create a new German currency, to replace the greatly devalued Reichsmark, and to be backed by Marshall Plan aid. Their aim was to stifle a new threatened wave of hyperinflation and undermine a black market in which American cigarettes were the principal currency. Not surprisingly, the Soviets refused to co-operate. Stalin preferred to see Germany remain economically weak.

The Americans, British and French continued with the currency plan in secret.

On 12 June 1948, the Soviets announced that the main autobahn linking Berlin to the border was to be temporarily closed for repairs. All road traffic into and out of Berlin was halted three days later. On 21 June, the day that the new Deutsche Mark was introduced in the American, British and French occupation zones, all barge traffic into the city was stopped. Three days later, on 24 June, all rail traffic was stopped due to 'technical difficulties'.

Access to the Western zones in Berlin had never been a right, governed by a formal agreement between the former Allies. The Soviets argued that the Western powers had no legal claims to such access, and on 25 June declared that they would not provide food to the Western sectors of the city. General Lucius D. Clay, military governor of the US occupation zone, declared it: 'one of the most ruthless efforts in modern times to use mass starvation for political purposes.'

The population of the Western sectors of Berlin numbered two and a half million. It was estimated that there was food to last about 35 days and coal to last 45 days. Clay argued that it was important for the Western powers to remain in Berlin at all costs, using force if necessary. Yet, as the Soviets tightened the screw by cutting the supply of electricity, Truman became increasingly concerned that the wrong response could precipitate all-out war.

One of the Truman's advisers urged that the AEC hand over its arsenal of atomic weapons to the military. But, despite the fact that the Soviet Union did not have atomic weapons of its own, Truman appreciated that any decision to use the weapon was one that could not be taken lightly:

> I don't think we ought to use this thing unless we absolutely have to. It is a terrible thing to order the use of something like that, that is so terribly destructive, destructive beyond anything we have ever had. You have got to understand that this isn't a military weapon ... It is used to wipe out women and children and unarmed people, and not for military uses ... This is no time to be juggling an atom bomb around.

Right of access to Berlin by road, barge or rail may not have been agreed with the Soviet Union, but a written agreement of 30 November 1945 did provide for three twenty-mile-wide air corridors between Berlin and the borders with the British and American zones. It was estimated that to keep the population of Berlin alive, they would need at least 1,700 calories a day, which translated into an airlift of over 1,500 tons of foodstuffs a day. The population would also need nearly 3,500 tons of coal and gasoline a day.

The Berlin airlift, Operation Vittles, began on 25 June, modestly at first as more and more transport planes were drafted. By the second week, cargoes were averaging 1,000 tons per day.

The world stood by and watched anxiously.

Chapter 22

JOE-1

June 1948–January 1950

S talin was taking a gamble, one he had judged to be of low risk. He did not believe that the Berlin blockade would provoke a military reaction from the United States, despite that nation's atomic monopoly. He could not envisage the Truman administration sanctioning the use of the bomb against Soviet targets just to resolve a dispute over the fate of a single city.

In one sense at least, the threat of atomic weapons was largely empty, at least for the time being. The bomb might be an effective deterrent against large-scale acts of aggression, but against small-scale acts of political confrontation or limited, local wars it offered no deterrent at all. No nation, least of all one that perceived itself as leader of the world for the common welfare, would countenance such massive retaliation, out of all proportion to the act of provocation.

Stalin fully expected the Berlin airlift to fail, leaving the Americans, British and French with no alternative but to withdraw from a city deep in the heart of Soviet-occupied East Germany.

The blockade tested the American government's nerve. As the airlift got under way, attention turned to what the blockade might presage. If, as many senior military figures now expected, a Soviet invasion of Western

Europe was only a matter of time, then America had to prepare itself properly for such an eventuality.

But America had quickly demobilised after the war, scaling back conventional armed forces and relying on the threat of atomic weapons as the ultimate deterrent against foreign aggression. Five months after Japan's surrender, about three million Army Air Force personnel had returned to civilian status, with the air force inevitably losing its most experienced air and ground crews. The Soviet Union, in contrast, had not demobilised. Intelligence assessments for the American Joint Chiefs of Staff concluded that the Red Army and allied Eastern European forces could quickly overrun most of continental Europe. The only thing holding them back, it was believed, was the threat of retaliation with atomic weapons. America therefore had to ensure that its strategic atomic strike force was combat-ready.

The strike force was far from combat-ready, however. The air force faced three substantial problems. First, atomic weapons were owned by the civilian AEC, not the military: '[T]he military services didn't own a single one', LeMay remarked years later. 'These bombs were too horrible and too dangerous to entrust to the military. They were under lock and key of the Atomic Energy Commission. I didn't have them, and that worried me a little bit to start with.'

It was estimated that if it became necessary to launch an attack on invading Soviet forces in Europe, then atomic weapons-capable B-29s of the 509th Composite Group, now part of the Strategic Air Command (SAC), would need five to six days to prepare and depart from their base at Roswell, New Mexico, fly to an AEC location to load the weapons, and then fly on to a forward base in readiness to attack. By the time the B-29s had crossed the Atlantic, a war in Europe might already be over, with Soviet territories already well defended against unescorted bombers.

The second problem was the size of the atomic arsenal. The successful Sandstone tests had shown that it would be possible for the stockpile to be built up quite quickly. An estimated stockpile of thirteen bombs in 1947 had grown, and towards the end of 1948 the AEC possessed about 50. SAC had 60 bombers that had been modified to carry them. Many senior military figures believed this was far from sufficient to provide an effective deterrent.

But perhaps the overwhelming problem was not how many bombs America possessed or who owned them. What worried America's military leaders was the question of whether, in time of war, the bombs could be delivered accurately to their targets. Any atomic bombing mission over Soviet cities would likely involve targeting by radar, at night, from altitudes above 25,000 feet. It was clear that under these circumstances SAC personnel could not guarantee delivery of an atom bomb within one or two miles of the designated target.

Before the Berlin airlift began, the SAC bombing crews were encouraged to improve their navigational and bombing accuracies through an annual competition. Each crew had to drop six bombs from 25,000 feet – three visually and three by radar. The results were greatly disappointing, with circular-error averages ranging from over 1,000 feet to almost 3,000 feet.

LeMay took command of SAC on 19 October 1948 and ordered a major shake-up. He organised a combat exercise against an American practice target – Dayton, Ohio – designed to be as realistic as possible. Crews were issued with photographs of the target that were ten years old, on the basis that reconnaissance photographs of Soviet cities were of a similar vintage. Neither the crews nor the aircraft were used to flying at high altitude. The crews were insufficiently trained to target using radar. And the weather was bad. The results were disastrous. Of the 150 crews that flew the mission, none completed it as directed. Few crews even managed to find Dayton, let alone target the city accurately. LeMay called it the 'darkest night in American military aviation history'.

Fortunately for LeMay, SAC was not required to go to war just yet. The Berlin airlift was a humanitarian operation of massive proportions, and it was successful. American, British and French civilian and military aircraft were used to airlift cargoes ranging from containers of coal to small packets of sweets dropped with tiny, individual parachutes for the children.

By January 1949 Stalin realised that he would be unable to starve and freeze the Berlin population into submission and force the Western powers out of the city. Economic sanctions that had been imposed on Soviet East Germany had reduced imports into the country by almost half, and were starting to take their toll. Secret negotiations to end the crisis began in February. On 12 May the blockade was lifted and rail traffic once more flowed into Berlin.

The airlift nevertheless continued until September to build up supplies in case the Soviets blocked the routes again. By the time the airlift ended, nearly 280,000 flights had been made and over two million tons of coal, food and other essential supplies had been delivered.

Sloika

Although the decision had been taken to base RDS-1 on the Fat Man design, experimental work on a specifically Soviet atom bomb design had begun in the spring of 1948. The work was carried out by a small group of physicists at the Institute for Chemical Physics in Moscow under the guidance of Yakov Zeldovich, who spent most of his time at Arzamas-16. The Soviet physicists believed the resulting design to be much more progressive than the American original, half the size but twice as powerful, to a large extent vindicating the position that Kapitza had argued.

But Zeldovich was soon to become caught up with another problem. At their second meeting in Golders Green in London on 13 March 1948, Fuchs had passed to Feklisov a detailed report about the latest work on what was to become known as the 'classical' Super, the original Teller design for the hydrogen bomb. Although the report still lacked many of the calculations that could have confirmed the feasibility of the weapon, it nevertheless electrified the Soviets.

A translation was sent to Stalin, Molotov and Beria on 20 April. Three days later Beria ordered Kurchatov, Khariton and Vannikov to undertake a thorough study of the espionage materials and develop proposals for a parallel effort on a 'Soviet Super'. A resolution to supplement the working plan of KB-11 to include the Super – codenamed RDS-6 – was adopted by the Special State Committee on 10 June.

Zeldovich, who had been conducting independent research on the possibility of thermonuclear fusion, was now asked to investigate the Super design that had been delivered to Soviet intelligence by Fuchs. Meanwhile, a second, parallel investigation was initiated under the leadership of Igor Tamm at the Physics Institute of the Soviet Academy of Sciences. Each research group was aware of the other's existence, but only Zeldovich had sight of Fuchs' report. Tamm recruited some of the Soviet Union's most talented young physicists: Semyon Belenky, Vitaly Ginzburg, Yuri Romanov and Andrei Sakharov.

Tamm approached Belenky and 27-year-old Sakharov in a rather furtive manner after a Friday seminar at the institute. He explained what they had been asked to do. 'Our task would be to investigate the possibility of building a hydrogen bomb and, specifically, to verify and refine the calculations produced by Yakov Zeldovich's group at the Institute of Chemical Physics', Sakharov later explained. Nobody had sought his consent to work on thermonuclear weapons, and Sakharov felt he had no real choice in the matter. But it was an opportunity to do physics in an area he regarded as a genuine theoretician's paradise.

With guards mounted outside their office doors, and armed with new German-made calculators, the young physicists set to work with great enthusiasm, as though 'possessed by a true war psychology':

> I understood, of course, the terrifying, inhuman nature of the weapons we were building. But the recent war had also been an exercise in barbarity; and although I hadn't fought in that conflict, I regarded myself as a soldier in this new scientific war. (Kurchatov himself said we were 'soldiers', and this was no idle remark.)

Sakharov was particularly impressive. 'I envy Andrei Sakharov', Zeldovich said. 'My brain is built to work like a well-maintained computer. But a computer only works if it is pre-programmed. Sakharov's brain writes its own programmes.'

Sakharov spent two months studying Zeldovich's reports and improving his knowledge of gas dynamics. He suspected that the original Super design being investigated by Zeldovich's group had been inspired by espionage, and he quickly identified its flaws.

By the end of the summer he had devised an alternative design, which in his memoirs he calls the 'First Idea'. This was the 'Sloika' or 'layer cake', consisting of alternating layers of deuterium and tritium and U-238. The basic idea is that a core plutonium bomb creates the temperatures and pressures necessary to ignite fusion of the deuterium and tritium nuclei. The fast neutrons released by the fusion reactions then initiate fission in the U-238 layer. This layer serves both to confine the layer of lighter elements and, on fissioning, provides compression of the fusion fuel which further enhances the yield of thermonuclear energy.

The compression achieved by the ionised nuclei from the U-238 layer became known as 'sakharisation', at once crediting the scientist responsible for the invention of the process and providing a pun – 'sakhar' being the Russian for sugar.

The Soviet design took a further step forward in December 1948, when Ginzburg suggested that the deuterium and tritium fuel in the bomb be replaced by the chemical compound lithium-6 deuteride. Sakharov calls this the 'Second Idea'. It has the advantage that at room temperatures lithium deuteride is a non-radioactive, chalk-like solid substance, thereby avoiding the problems associated with handling gaseous mixtures of deuterium and tritium or the refrigeration apparatus required to turn these to liquid.

On absorption of a neutron the rare isotope Li-6, accounting for about 7.5 per cent of naturally-occurring lithium, fissions to produce tritium and helium nuclei. So, triggering a plutonium fission bomb in the presence of lithium-6 deuteride produces tritium and deuterium nuclei *in situ* at the temperatures and pressures required for these light nuclei to fuse together, releasing much greater quantities of energy in a thermonuclear explosion.

It was immediately seen as a much more promising design.[1] It was subsequently agreed that Tamm's group would concentrate on Sakharov's

---

[1] In fact, both the First and Second Ideas had been anticipated by Teller. During a period of consultancy work at Los Alamos in the summer of 1946, Teller had worked with the

'Sloika' proposal, and that Zeldovich's group would support this work while pursuing further investigations of the original design.

In early 1949 Tamm and Sakharov were summoned to Vannikov's spacious office in Moscow. Vannikov explained that Sakharov was to be transferred to Arzamas-16 to work with Khariton: 'It's necessary for the project', he said. But Tamm was extremely reluctant to let Sakharov go. He argued that Sakharov promised much in key scientific fields and to limit him to applied research would be a great mistake, not in the country's best interests.

When the direct line from the Kremlin rang, Vannikov answered. He tensed. It was Beria. 'Yes, I understand', he said. 'Yes sir, I'll tell them.' He hung up. 'I've just been talking with Lavrenty Pavlovich [Beria]', he explained to Sakharov. 'He is *asking* you to accept our request.'

Beria did not often ask nicely, and he did not ask twice. 'There was nothing left to say', wrote Sakharov. 'Things seem to have taken a serious turn', said Tamm.

Emergency War Plan 1-49

By January 1949 SAC possessed more than 120 aircraft capable of delivering atomic weapons, consisting of B-29s and B-50s that had been modified for air-to-air refuelling. There were now six bomb-assembly teams, with a further team in training. Flight crews were being drilled relentlessly, and navigation and targeting skills were improving.

LeMay drew up his first war plan. The result, the SAC Emergency War Plan 1-49, was delivered in March 1949. It encapsulated all the lessons LeMay had learned from his experience firebombing Japanese cities. These were simply summarised: hit fast, and hit hard. He called on SAC 'to increase its capability to such an extent that it would be possible to deliver

---

head of the Theoretical Division, Robert Richtmyer, and had developed a very similar design which they had called the 'Alarm Clock' (thereby hoping to 're-awaken' interest in the Super at Los Alamos). The design was not pursued.

the entire stockpile of atomic bombs, if made available, in a single massive attack'.

At the time this meant striking 70 Soviet cities with 133 atomic bombs, targeting urban industrial centres, government offices, the oil industry, transport networks and power stations. Of course, not all the bombs were of the latest, higher-yield, design. But a conservative assumption that each bomb would yield a Nagasaki-scale 20,000 tons of TNT equivalent implies a plan to hit the Soviet Union with an explosive force totalling three million tons (three megatons) of TNT. It was estimated that there would be nearly three million civilian deaths, and four million casualties.

Such was the calculus of atomic war. To countenance such a plan required a moral compass set spinning by nothing less than one of Lawrence's giant magnets. But as LeMay later pointed out, before leaping to moral judgements it helped first to establish a proper sense of perspective:

Incidentally, everybody bemoans the fact that we dropped the atomic bomb and killed a lot of people at Hiroshima and Nagasaki. That I guess was immoral; but nobody says anything about the incendiary attacks on every industrial city of Japan, and the first attack on Tokyo killed more people than the atomic bomb did. Apparently, that was all right ...

Towards the end of the war, LeMay had ordered the firebombing of 63 Japanese cities, which had resulted in the deaths of two and a half million civilians. As far as LeMay was concerned, all that atomic weapons had done was make the process so much more efficient.

America was not at war, however, and this was still just a plan. Nevertheless, the temptation to launch a pre-emptive strike must have been very real. A noon SAC reconnaissance exercise over Vladivostok was met with no resistance. 'We practically mapped the place up there with no resistance at all', LeMay said later. 'We could have launched bombing attacks, planned and executed just as well, at that time. So I don't think I am exaggerating when I say we could have delivered the stockpile had we wanted to do it, with practically no losses.'

LeMay's plan was judged sufficient to precipitate a collapse of the Soviet Union, or at least sufficient to destroy the Soviet capability for offensive operations. As if this was not enough, the AEC confidently predicted that it could make a stockpile of 400 atomic weapons available by the end of 1950.

A perfect spy

Fuchs was enjoying his job as division head at Harwell and, perhaps for the first time in his life, he was making good friends. There were even rumours of an affair. He had acquired an MG sports car, somewhat more reliable than the dilapidated Buick he had driven while at Los Alamos. He continued to gain respect for his work. Over dinner at a restaurant in Abingdon in September 1948 Oppenheimer had offered him a position at the Institute for Advanced Study in Princeton. He had politely declined.

He had been the perfect spy. But beneath the veneer of calm and quiet authority Fuchs was now in turmoil. He was having some serious doubts about what he had done. He was quickly losing the ability to compartmentalise the two very different sides of his life. At the same time he had become all too aware that he was betraying the trust of his friends. Worse, he was betraying secrets to a regime whose true character was now beginning to emerge into the wider public consciousness, as Moscow crushed its Eastern European satellite states into submission. His political world was changing, dramatically.

'I then realized that the combination of the three ideas which had made me what I was, were wrong,' Fuchs later wrote, 'in fact that every single one of them were wrong, that there were certain standards of moral behaviour which are in you and that you cannot disregard.'

Fuchs started to drink heavily. He grew increasingly nervous during visits to Harwell by his father, Emil, fearing that he would let slip his son's early Communist affiliations. He began mentally to prepare himself for a final end to his espionage activities. At a meeting with Feklisov in February 1949, sitting on a park bench in Putney Bridge Park near the Spotted Horse pub, Fuchs described his plans for the future.

'I'd like to help the Soviet Union until it is able to test its atomic bomb. Then I want to go home to East Germany where I have friends. There I can get married and work in peace and quiet. That's my dream', he told Feklisov, with a smile.

Fuchs went on to tell of recent meetings he had had with members of his family: his father Emil during a visit to Abingdon, Kristel in Cambridge, Massachusetts, his brother Gerhard, suffering from tuberculosis, in Davos, Switzerland. Feklisov saw an opportunity to provide some small reward for Fuchs' efforts in support of the Soviet cause.

'Klaus, I know you're not working for the money and want none for yourself', Feklisov said. 'But we do wish to help you with your daily financial problems. I hope you will not feel offended if I offer this small token of our gratitude.'

Fuchs hesitated, but accepted the envelope. 'Thank you,' he said, 'I don't need the money, but I do appreciate your offer. I'll send a money order to my brother right away.'

One of Fuchs' close friends at Harwell was Henry Arnold, the AERE Security Officer. For reasons of his own, Arnold had decided that if there was a Soviet spy operating at Harwell, then Fuchs would be high on his list of suspects.

Fuchs arranged to meet Feklisov again in early April. It would be their last meeting.

First Lightning

Kurchatov's brother Boris had been the first Soviet scientist to separate plutonium from spent fuel slugs taken from the F-1 reactor. A full-scale separation plant, called Installation B, was constructed alongside F-2 at Cheliabinsk-40. It was ready in December 1948. A third installation, designed to purify the plutonium further and convert it to metallic form, was not quite ready by the time the first plutonium nitrate solutions were becoming available in early 1949, and a temporary workshop was set up. By April 1949 Cheliabinsk-40 was producing pure plutonium dioxide.

Shortly thereafter, sufficient plutonium metal was available for the Soviet Union's first bomb test.

Just as the two hemispheres of the first bomb core were being nickel-plated prior to sending them to Arzamas-16 for criticality tests, Pervukhin and a number of Soviet generals arrived to inspect them. They wanted to know how the scientists could be certain it really was plutonium and not just a lump of iron disguised to look like plutonium. Anatoly Alexandrov, responsible for plutonium separation at the plant, assured them it was the genuine article and pointed out that as a result of its radioactivity, the hemispheres were warm to the touch. However, this did not seem to convince them. They pointed out that it was easy just to heat a piece of iron. In frustration, Alexandrov suggested that they could sit with the hemispheres until the next morning and check that they remained warm. The generals went away.

A state commission had been formed to monitor the test, with Beria as chairman and Kurchatov as the scientific supervisor. In May Kurchatov left for the test site. This was Semipalatinsk-21, a small settlement some 50 miles north-west of Semipalatinsk along the Irtysh river in Kazakhstan. In August the scientists and observers began to gather.

Essential hardware and test instruments were transported to Kazakhstan by train. The Soviet scientists travelled quickly, stopping only to change locomotives and check the cargo. At the stations where they did stop, they were surprised to find the platforms deserted. At one such stop, Zeldovich and a number of young scientists disembarked to play volleyball on the empty platform. Zavenyagin ordered them back on the train: 'They're supposed to be serious people', he grumbled. 'They're on a responsible mission and they behave like a bunch of eighteen-year-old kids.'

On 28 August Beria, Kurchatov, Khariton, Flerov, Zavenyagin and Zernov watched as RDS-1 was assembled according to Khariton's strict instructions. The first Soviet test, codenamed First Lightning, was to take place at 6:00 the following morning atop a specially constructed tower over 100 feet high. The scientists were all too aware of the problems that had beset the Trinity test and even now there could be no absolute guarantee that the bomb would work. Few got any sleep that night. To make matters worse, the weather was poor.

At 2:00am the assembled bomb was moved into the freight elevator which would lift it into position. The intention had been to send the bomb up the tower unaccompanied, but when Beria raised his eyebrows Zernov quickly climbed in alongside. Flerov and a small team of scientists then ascended the tower to mount and check the detonators. Flerov was the last to return.

As dawn broke the rain subsided. Although the sky remained overcast, there was sufficient visibility for the optical instruments that had been set up to monitor the explosion. Kurchatov delayed the test by an hour. A number of structures had been built near the tower – one-storey wooden buildings, four-storey brick houses, bridges, tunnels and water towers. Locomotives, carriages, tanks and artillery had also been placed in the area. Animals were put in open pens so that the initial effects of radiation exposure could be examined.

The tower was not visible from the command bunker, but Kurchatov opened the glass-panelled door on the opposite side so that the scientists and generals gathered within could watch the flash reflected from the distant hills. There would be time enough to close the door before the shockwave struck. Beria remained deeply sceptical: 'Nothing will come of it, Igor', he growled at Kurchatov.

The countdown was completed at 7:00am on 29 August:

An explosion. A bright flash of light. A column of flame, dragging clouds of dust and sand with it, formed the 'foot' of an atomic mushroom. Kurchatov said only two words: 'It worked' ... What remarkable words these are: 'It had worked! It had worked!' Physicists and engineers, mechanics and workmen, thousands of Soviet people who had worked on the atomic problem, had not let the country down. The Soviet Union had become the second atomic power. The nuclear balance had been restored.

Flerov, bathed in reflected light, closed the door before it was shattered by the shockwave. Beria rushed to hug Khariton, but the only thing Khariton felt was relief. The bomb yielded about 20,000 tons of TNT equivalent, virtually identical to the yields of the Fat Man design used for the Trinity test and against Nagasaki. Had it not worked, the Soviet physicists would all have been shot.[2]

Beria rushed to inform Stalin. Woken from sleep, an angry Stalin told him that he already knew. The next day Beria and Kurchatov submitted a hand-written report: 'We hereby report to you, Comrade Stalin, that a large team of Soviet scientists, designers, engineers, managers and industrial workers has succeeded, after four years of hard work, in fulfilling the task that you set before them and had made the Soviet atomic bomb.'

On 3 September airborne radioactive fallout from the Soviet test was detected by instruments on board an American WB-29 (a B-29 modified for weather reconnaissance missions), flying a few miles east of the Kamchatka Peninsula. During the following week scientists tracked the radioactive air mass as it passed across America. The British were alerted on 9 September as the air mass crossed the Atlantic. By 14 September there was little room left for doubt. Scientists at Tracerlab in Berkeley, a private radiological laboratory, placed the time of the explosion – dubbed 'Joe-1' – at 6:00am on 29 August. They were out by only one hour.

Prime suspect

Gardner and Lamphere had been hard at work putting the jigsaw puzzle of evidence together. What the Venona decrypts told them was frightening: atomic secrets had haemorrhaged from the Manhattan Project.

---

[2] Though possibly apocryphal, the story that Beria handed out honours to his atomic scientists in inverse proportion to their likely fate in the event of failure reflects the mood. Those who would have been shot were awarded the highest honour: they became Heroes of Socialist Labour. Those who would have received maximum prison sentences were awarded the Order of Lenin.

In mid-September 1949 Lamphere had found a startling piece of information in a recently-decrypted message that had been sent to Moscow Centre from New York on 15 June 1944.[3] The message stated:[4]

[1 group unrecovered] received from REST the third part of report MSN-12 Efferent Fluctuation in a stream [37 groups unrecoverable]. Diffusion method – work on his speciality. R[EST] expressed doubt about the possibility of remaining in the COUNTRY without arousing suspicion.

This was a partial summary of theoretical work on gaseous diffusion. The message went on to describe the strife between the ISLANDERS (the British mission) and the TOWNSMEN (the Americans) working on ENORMOZ. The message told Lamphere that spies had been at work not only in Canada, but also in America, at the very heart of the Manhattan Project.

The message also revealed the existence of a spy with the codename REST, which had been changed to CHARLES in another cable dated 5 October 1944. A message dated 16 November 1944 provided further details about REST/CHARLES:

On ARNO's last visit to CHARLES' sister it became known that CHARLES has not left for the ISLAND but is at Camp No. 2. He flew to Chicago and telephoned his sister. He named the state where the camp is and promised to come on leave for Christmas. He [ARNO] is taking steps to establish liaison with CHARLES while he is on leave.

---

[3] In his 1986 book *The FBI–KGB War: A Special Agent's Story*, Lamphere is not specific about the dates or content of decrypted messages. However, the decrypts were made public by the NSA between July 1995 and September 1997, and it is therefore now possible to identify them.

[4] A number of the code groups in this message remained unrecovered or unrecoverable (i.e. they could not be read or decoded).

Camp No. 2 was the codename for Los Alamos. On 27 February 1945 Moscow Centre had sent a message back to New York with a long list of questions concerning CHARLES. Moscow wanted to know what CHARLES had been doing since August 1944 and the purpose of his trip to Chicago. The message also assigned CHARLES' sister the codename ANT.

In another message, dated 10 April 1945, Moscow Centre advised ANTON (Leonid Kvasnikov) in New York of the value of CHARLES' information:

CHARL'Z's information 2/57 on the atomic bomb (henceforth 'BAL...') is of great value. Apart from the data on the atomic mass of the nuclear explosive and on the details of the explosive method of actuating 'BAL...' it contains information received for the first time from you about the electromagnetic method of separation of ENORMOZ. We wish in addition to establish the following: 1. For what kind of fission – by means of fast or slow neutrons – [35 groups unrecovered] [281 groups unrecoverable].

The partially recovered codename 'BAL...' in this message is most likely 'balloon'. Whoever REST/CHARLES was, it seemed that his information had been very highly regarded by the Soviets. This was betrayal on a massive scale.

Lamphere narrowed the list of suspects down to just one man. Fuchs had been the author of the theoretical paper on gaseous diffusion mentioned in the June 1944 message. From Fuchs' personnel file Lamphere learned that he had started at Los Alamos in August 1944 and that he had a sister – Kristel – living in Cambridge, Massachusetts. The file showed that Fuchs had made the visit to Chicago that was detailed in the message traffic. It showed that he had visited Cambridge while on leave from Los Alamos during February 1945.

There was some further evidence of possible culpability from other sources. A Gestapo file that had fallen into the hands of the Allies in Kiel towards the end of the war stated that Fuchs had been a Communist since 1934. Fuchs' name was also found in the address book of Israel Halperin.

Lamphere was convinced. Fuchs was his prime suspect. He opened a case file, initiated an investigation and wrote to alert British intelligence on 22 September.

Quantum jump

On the morning of 23 September 1949 Truman stated publicly that: 'We have evidence that within recent weeks an atomic explosion occurred in the USSR.' America may have had over 100 atomic bombs in its stockpile, but it now no longer had a monopoly on the technology.

Oppenheimer and other members of the GAC who met later the same day were shocked by the news. While some, like Rabi, believed that this brought closer the prospect of an atomic war, Oppenheimer was more sanguine. He advised Teller to 'Keep your shirt on'.

Speculation mounted as to how the Soviets had managed to build the bomb so quickly, years before they were predicted to have the capability. But for many in senior positions in the AEC and the Truman administration, the Soviet atom bomb was now a political and military reality, demanding a political and military response. Inevitably, attention returned to the Super.

Lewis Strauss had no doubt that every effort should now be made to accelerate development of a super-bomb:

It seems to me that the time has now come for a quantum-jump in our planning (to borrow a metaphor from our scientist friends) – that is to say, that we should now make an intensive effort to get ahead with the super. By intensive effort, I am thinking of a commitment in talent and money comparable, if necessary, to that which produced the first atomic weapon. That is the way to stay ahead.

The time had also come to tell the President. Truman still did not know that the Super was possible.

The so-called hydrogen or super bomb

Teller had left Los Alamos in 1946. But he had watched, seething, as the Communist Party gradually took control of his native Hungary. In the elections of May 1949, candidates from the Hungarian Workers' Party, formed by the 'merger' of the Communist Party and Social Democratic Party in 1947, had stood unopposed. The People's Republic of Hungary was declared shortly afterwards and a period of repressive Stalinist rule under the authoritarian leader Mátyás Rákosi began. Soviet-style purges and show trials soon followed. Teller was persuaded to return to Los Alamos. He resumed work on the Super in July.

On 21 October Oppenheimer noted in a letter to Conant that the Super is 'not very different from what it was when we first spoke of it more than seven years ago: a weapon of unknown design, cost, deliverability and military value.' Oppenheimer's position on the morality of the Super was at this stage somewhat ambiguous. 'We have always known it had to be done,' he continued in his letter, 'and it does have to be done ... But that we become committed to it as the way to save the country and the peace appears to me to be full of dangers.'

By the time of the next GAC meeting on 29 October Oppenheimer's views were still not fixed, but Conant was utterly opposed to the development of a thermonuclear bomb on moral grounds. Although each member of the committee argued somewhat differently, there appeared to be a strong consensus regarding the conclusion: there were better things America could do than embark on a major programme to build a thermonuclear bomb. Oppenheimer, chairman of the GAC, aligned himself with the majority view.

This view was detailed in a report dated 30 October. Having summarised the technical problems associated with the production of thermonuclear weapons, the moral objections were put forward in a majority annex (so called because this was signed by the majority of GAC members, including Conant and Oppenheimer):

If super bombs will work at all, there is no inherent limit in the destructive power that may be attained with them. Therefore, a super bomb might become a weapon of genocide ... We believe a super bomb should never be produced. Mankind would be far better off not to have a demonstration of the feasibility of such a weapon, until the present climate of world opinion changes ... In determining not to proceed to develop the super bomb, we see a unique opportunity of providing by example some limitations on the totality of war and thus of limiting the fear and arousing the hopes of mankind.

It was an instinctive reaction to a morally repulsive weapon. If different moral codes indeed prevail during wartime compared with peacetime, then building an insanely destructive weapon without the justification of war was doubly reprehensible. In a minority annex to the report, Fermi and Rabi went further: 'Necessarily such a weapon goes far beyond any military objective and enters the range of very great natural catastrophes. By its very nature it cannot be confined to a military objective but becomes a weapon which in practical effect is almost one of genocide ... It is necessarily an evil thing considered in any light.'

Emotions were running high. The GAC report and recommendations of the AEC commissioners were passed to Truman. The commissioners themselves were split on the issue, with Strauss in favour of proceeding with the Super but with Lilienthal and others opposed. Strauss argued his position in a memorandum to the President dated 25 November 1949, in which he concluded: 'In sum, I believe that the President should direct the Atomic Energy Commission to proceed with all possible expedition to develop the thermonuclear weapon.'

For a time, Oppenheimer may have thought that his views on the Super would prevail but, in truth, there were few supporters of his position in the Truman administration. Truman's response to the split within the AEC was to appoint another study group consisting of Lilienthal, Acheson (now Secretary of State) and Secretary of Defense Louis Johnson. Lilienthal's position had not changed. Johnson was in favour. The decision was therefore down to Acheson.

But Acheson was an astute politician, firmly in tune with the mood that now prevailed in the administration. This was a mood summarised in the response to the GAC report by the Joint Chiefs of Staff in January 1950, sent to Johnson, in which the Joint Chiefs advanced counter-arguments that we would hear again and again in subsequent years:

It would be foolhardy altruism for the United States voluntarily to weaken its capability by such a renunciation. Public renunciation by the United States of the super bomb development might be interpreted as the first step in unilateral renunciation of the use of all atomic weapons, a course which would inevitably be followed by major international realignments to the disadvantage of the United States ... the security of the entire Western Hemisphere would be jeopardized.

Oppenheimer and George Kennan, whose 'long telegram' had helped to establish the rhetoric of the Cold War in 1946, made representations to Acheson in a final attempt to stop the Super. But it was already too late. At a scientific conference organised by the American Physical Society on 29 January 1950, Teller asked Oppenheimer if he would be prepared to join the Super programme. 'Certainly not', was Oppenheimer's curt reply.

Although Acheson agreed with many of Lilienthal's arguments, he had decided that domestic politics would demand a crash programme on the Super. A project to develop the thermonuclear bomb would be necessary to ensure Truman's survival as president. At a meeting in the Oval Office on 31 January, Lilienthal began to put forward the objections. Truman interrupted him.

'Can the Russians do it?' he asked.

They all nodded.

'In that case, we have no choice. We'll go ahead', Truman stated flatly.

Later that evening, Truman delivered a radio broadcast:

It is part of my responsibility as Commander in Chief of the Armed Forces to see to it that our country is able to defend itself against any possible aggressor. Accordingly, I have directed the Atomic Energy

Commission to continue work on all forms of atomic weapons, including the so-called hydrogen or super bomb. Like all other work in the field of atomic weapons, it is being and will be carried forward on a basis consistent with the overall objectives of our program for peace and security.

The decision had been made, and the world had now been advised. The Super would be built. But the scientists at Los Alamos didn't really have the first idea how to do it.

Confession at the War Office

Lamphere had a problem. Neither the FBI nor British intelligence knew if Fuchs was still actively spying for the Soviet Union. Using the evidence of the Venona decrypts would betray the existence of the project and the extent to which American counter-intelligence had managed to decode the Soviet spies' secret messages. And yet Fuchs could not be interrogated without a good reason.

Fuchs supplied the reason himself. His father had now moved to Leipzig, in East Germany, and Fuchs had asked his good friend Henry Arnold if this compromised him in any way. It may have been that Fuchs was looking for an excuse to leave Harwell and start a new life away from atomic secrets and espionage. But Arnold saw it as an opportunity to question him.

Arnold advised Fuchs that someone from the security service wanted to talk to him about his father's move and its implications for the work that Fuchs was doing at Harwell. William Skardon, an experienced and skilful MI5 interrogator, drove to Harwell on 21 December 1949. Lamphere likened Skardon to the fictional television detective Columbo: 'complete with dishevelled appearance and an intellect that was sometimes hidden until the moment came to use it to point out incongruities in a suspect's story.' Arnold took Skardon along to Fuchs' office, then left them to it.

Skardon got Fuchs to talk about himself and his career. Fuchs' own paroxysms of doubt meant that he was now less reserved in talking about his student politics. But after an hour and a quarter of gentle, rambling discussion, Skardon made his move. Fuchs was describing his work on gaseous diffusion as part of the British mission in New York when Skardon interrupted. 'Were you not in touch with a Soviet official or a Soviet representative while you were in New York?' he asked. 'And did you not pass on information to that person about your work?'

Fuchs mumbled that he didn't think so, but his lack of reaction convinced Skardon of his guilt. The interrogation proceeded through the afternoon. On his return to MI5 headquarters, Skardon reported that he believed Fuchs was guilty, but that if left to think it over he would likely confess voluntarily.

Fuchs travelled to Birmingham for Christmas with Peierls and his wife. Skardon interrogated him again on his return to Harwell on 30 December, and again two weeks later.

On 22 January 1950 Fuchs asked Arnold to meet him for lunch at a local pub, and they agreed to meet the next day. Unusually, Fuchs talked openly about his politics and claimed to Arnold that he now disagreed with Communism as practised in the Soviet Union. He told Arnold that there was something further he needed to tell Skardon. Arnold asked him directly if he had passed information to the Soviets. Fuchs confessed that he had.

Skardon was back at Harwell the next morning, but Fuchs was still unwilling to make a full confession. They drove out to the Crown and Thistle, a hotel on the river near Abingdon, where they had lunch. By the time they returned, Fuchs had come to a decision. 'What do you want to know?' Fuchs asked. Skardon wanted to know when Fuchs had started passing atomic secrets and how long it had been going on. 'I started in 1942 and had my last meeting last year', Fuchs replied. Skardon was shocked. He had thought this was all about a single instance of espionage in New York. He now began to realise that this was much, much more serious than anybody had realised.

Fuchs travelled to London on 27 January and wrote out a full confession at the War Office. When Skardon asked him to be more specific about the information he had passed on to the Soviets, Fuchs refused. He claimed that Skardon didn't have the appropriate security clearance.

In subsequent discussions with Michael Perrin, who had been deputy director of the Tube Alloys project and was security cleared, Fuchs admitted to passing on basic information about the design of the Super.

MI5 does not have the powers to make arrests. On 2 February, Commander Leonard Burt, head of the Special Branch and by now entirely familiar with the business of arresting atomic spies, took Fuchs into custody at Perrin's office in Shell-Mex House on the Strand. Fuchs was charged with communicating information that might be useful to an enemy in violation of the Official Secrets Act.

Fuchs had been in denial, thinking that his confession would have no further repercussions and that he would be left to continue his work at Harwell. Now his world caved in.

Oppenheimer was sitting at an oyster bar in New York's Grand Central Station when he found out about Fuchs' betrayal two days later from the *New York Times*. His face crumpled as he read the lurid headlines.

The pattern of nuclear escalation

A new chapter in the arms race had begun, with a certain crushing inevitability. The tragedy of the time is that, to a large extent, Truman's assessment of the Soviet position was correct. If the Soviets could do it, they would.

'The Soviet government,' wrote Sakharov years later, 'already understood the potential of the new weapon, and nothing could have dissuaded them from going forward with its development. Any US move toward abandoning or suspending work on a thermonuclear weapon would have been perceived either as a cunning, deceitful manoeuvre or as evidence of stupidity or weakness. In any case, the Soviet reaction would have been the same: to avoid a possible trap, and to exploit the adversary's folly at the earliest opportunity.'

In America, atomic and thermonuclear weapons would consume $5.5 trillion. They would also consume countless roubles, pounds sterling, yuan, francs and possibly rupees.

All for weapons that were simply too dreadful to use.

Epilogue

MUTUAL ASSURED DESTRUCTION

Despite the vision and the far-seeing wisdom of our wartime heads of state, the physicists felt a peculiarly intimate responsibility for suggesting, for supporting, and in the end, in large measure, for achieving the realization of atomic weapons. Nor can we forget that these weapons, as they were in fact used, dramatized so mercilessly the inhumanity and evil of modern war. In some sort of crude sense which no vulgarity, no humor, no overstatement can quite extinguish, the physicists have known sin; and this is a knowledge which they cannot lose.

These were Oppenheimer's words, delivered during a lecture at MIT in November 1947. His message seems clear but what, precisely, was the nature of the sin that the physicists had come to know?

The scientific path to Hiroshima and Nagasaki can be traced directly back to Frisch and Meitner, sitting on a tree trunk in the snow-covered woods in Kungälv, scrambling for pieces of paper on which to scribble their calculations. Was the very discovery of nuclear fission a sin? The Frisch–Peierls memorandum on critical mass? The discovery of plutonium? The first self-sustaining nuclear chain reaction?

No. By its very nature, a scientific fact is amoral: it is not right or wrong in a moral sense; not good or evil. Like a stone or a tree, it just *is*. This goes for the scientific fact of an explosive nuclear chain reaction. Obviously, it is *people* who are right or wrong in a moral sense, people who are good or evil. And yet, while it is true that many of the greatest physicists of a generation had found themselves drawn inexorably into a project to build the world's most dreadful weapon of war, it requires a gross twist of logic to describe their discoveries and their participation as sinful.

This was a weapon which Frisch and Peierls had rightly judged in 1940 to be 'practically irresistible' at a time when the world was threatened by the darkest evil. This confluence of historical events meant that there was a startling inevitability about the chain that led from the discovery of fission to the development of the bomb, to the use of the bomb against Japan, to the development of a Soviet weapon.

The scientific fact of nuclear fission set the physicists firmly on the stage of world events. As C.P. Snow put it: 'With the discovery of fission, physicists became, almost overnight, the most important military resource a nation-state could call upon.' The physicists became contaminated with the political and military decision-making process at the end of a long and profoundly immoral war, a war marked by its relentless, unparalleled barbarity. By virtue of their scientific expertise, the physicists became ravelled in decisions about who was to live, and who was to die. These were people more used to sitting in quiet judgement on the results of laboratory experiments and the veracity of abstract scientific theories. The physicists had acquired, as Oppenheimer proclaimed, a 'peculiarly intimate responsibility' for making or at least participating in fateful decisions against a background of determinedly ambiguous political and military facts. These were decisions for which they were poorly prepared.

Today, we look back on the atomic bombing of Hiroshima and Nagasaki from the comfort of more than 60 years of relative peace, a peace if not free from war then at least free from conflagration on the scale of world war. Many look back with horror, or with a sense of deep shame. Of course, the bombings were a tragedy. But at the time the evidence of Nazi and Japanese brutality made more palatable 'lesser' evils, such as the firebombing and atomic bombing of German and Japanese cities. The atomic bombing of Hiroshima and Nagasaki was just one more catastrophe on top of a long list of utter catastrophes, one that brought the war to an end with such a powerful exclamation that it burned into the consciousness of all who lived then, and all who have lived since.

If we want to understand what this particular catastrophe meant to ordinary citizens who had survived nearly six years of mechanised slaughter, we could, perhaps, do worse than ask an innocent young girl. In April 1945 she was just eight years old. There she sits, the glimmer of a Pathé newsreel flickering across her face in a dark, smoke-filled cinema:[1]

Our greatest fear finally materialised in mid-April 1945, when Allied troops entered Buchenwald and Belsen concentration camps. And this was only the beginning of the nightmare. Here was man's inhumanity to man exposed as never before. To young eyes the mounds of pale, twisted corpses seemed like mountains. They had once been beautiful people.

With Victory in Europe we had a few days of jubilation and that was that. It was victory but it wasn't the end. The rationing and shortages remained. Too many of our servicemen were still prisoners of the Japanese. We saw their pictures – they reminded us of the horrors of Belsen; the difference was that they were still walking, still being beaten and starved. On 6 August 1945 the Americans dropped an atom bomb on the Japanese city of Hiroshima and I didn't know anyone who wasn't glad. Three days later another bomb was dropped on Nagasaki. On the 14 August the war was over.

Standing at the pond of Auschwitz concentration camp in the early 1970s, Polish-born mathematician and biologist Jacob Bronowski made a plea to the cameras filming a BBC television documentary series. He repeated words once spoken by Oliver Cromwell: 'I beseech you, in the bowels of Christ, think it possible you may be mistaken.' In *The Ascent of Man*, he wrote:

---

[1] A British Pathé newsreel dated 30 April 1945 features Mavis Tate, MP, visiting Buchenwald. The newsreel shows the corpses piled at Buchenwald and Belsen, and ranks of German civilians digging mass graves to bury them. See www.britishpathe.com

It is said that science will dehumanise people and turn them into numbers. That is false, tragically false. Look for yourself. This is the concentration camp and crematorium at Auschwitz. This is where people were turned into numbers. Into this pond were flushed the ashes of some four million people. And that was not done by gas. It was done by arrogance. It was done by dogma. It was done by ignorance. When people believe that they have absolute knowledge, with no test in reality, this is how they behave. This is what men do when they aspire to the knowledge of the gods.

When Bronowski, not long returned from Hiroshima, overheard someone say in Szilard's presence that it was the tragedy of scientists that their discoveries had been used for destruction, Szilard replied that it was not the tragedy of scientists: 'it is the tragedy of mankind.' This may be true, but the triumph of the scientists was no less tragic for all that. In helping to defeat the evils of arrogance, ignorance and dogma they had unleashed a primordial force upon the world, the threat of which would endure long after the perpetrators of evil were gone. Through their efforts they had helped to put the world in an even greater jeopardy.

The world was not about to thank them for what they had done.

Witch-hunt

Of course, the great dictators of the twentieth century had no monopoly on arrogance, ignorance and dogma. These human failings can birth, grow and thrive also in free, democratic societies, as Oppenheimer would discover to his cost. Julius and Ethel Rosenberg would pay the ultimate penalty.

The House of Representatives Un-American Activities Committee (HUAC) had been established originally as a special committee in 1934 and was then constituted four years later as a special investigating committee. It was charged with the investigation of Nazi propaganda, German-American involvement in Nazi activities, and the Ku Klux Klan. Having failed to gather sufficient information on Klan activities, the HUAC turned its attention to the American Communist Party. It became a standing committee in 1945. Its investigation of the American movie industry began two years later and led ultimately to the blacklisting of some 300 artists, including Richard Attenborough, Leonard Bernstein, Charlie Chaplin, Dashiell Hammett, Lillian Hellman, Paul Robeson and Orson Welles.

In 1949 HUAC turned its attention to Rad Lab physicist Joe Weinberg. The evidence assembled by the FBI was compelling but there had been no warrant for the bug in Steve Nelson's home that had caught Weinberg betraying atomic secrets. This evidence was therefore inadmissible in court. In an attempt to trigger a confession, in April 1949 both Weinberg and Nelson were called to a HUAC hearing where they met face to face. Weinberg denied ever having met Nelson before.

Knowing full well that Weinberg had committed perjury, the HUAC lawyers issued subpoenas to Lomanitz, Friedman and Bohm in an attempt to force the truth. Lomanitz had been pursued relentlessly around the country by the FBI, hounded out of a succession of jobs when the FBI exposed his Communist past to his employers. Bohm, in contrast, had moved to an academic position at Princeton University and was building the beginnings of a promising career in theoretical physics.

Einstein advised Bohm to refuse to testify, suggesting that he 'may have to sit for a while', meaning that the penalty for his silence might be a short prison term. Lomanitz and Bohm met in Princeton to discuss the impending hearings. Encountering Oppenheimer in the street, they explained what was happening. 'Oh my God,' Oppenheimer exclaimed, 'All is lost. There is an FBI man on the Un-American Activities Committee.' Oppenheimer too had been served with a subpoena, and was aware that one member of the HUAC was a former FBI agent who had investigated the Rad Lab during the war.

Bohm chose to testify. When asked on 25 May 1949 if he had ever been a member of the Young Communist League he refused to answer, pleading freedom of assembly and association under the First Amendment. When asked if he knew Steve Nelson he again refused to answer, pleading the right of refusal to self-incriminate under the Fifth Amendment. When asked if he had ever been affiliated with any political party or association, he told the hearing: 'I would say definitely that I voted the Democratic ticket.'

Bohm refused to divulge names at this hearing and at a subsequent hearing on 10 June. Princeton University expressed support and declared him a 'thorough American'.

Oppenheimer testified on 7 June and skilfully negotiated his way around the questions. When asked about the 'Chevalier incident', he gave the version of events that he had reported to the FBI in September 1946. He was not asked to elaborate the 'cock and bull' story he had given to Pash and Johnson in Berkeley three years earlier.

Oppenheimer's brother Frank had been denounced in 1947. The headlines of the *Times Herald* had screamed: 'US atom scientist's brother exposed as Communist who worked on A-bomb.' When questioned by HUAC about his brother's membership of the Communist Party, Oppenheimer replied: 'I will answer, if asked, but I beg you not to ask me these questions.' Oppenheimer, lauded as the 'father of the atom bomb', was not yet the target. The counsel for HUAC withdrew the question.

At the end of Oppenheimer's testimony one HUAC member, Congressman Richard M. Nixon, voiced his appreciation: 'I think we all have been tremendously impressed with him and are mighty happy we have him in the position he had in our program.'

But events over the next twelve months would build anti-Communist sentiment to unprecedented heights. Earlier, in August 1948, Whitaker Chambers, a former GRU agent and editorial staff member of *Time* magazine, had revealed in testimony to HUAC the names of highly-placed Communists in the Truman administration – Alger Hiss in the State Department and Harry Dexter White at the Treasury.[2] Hiss was subsequently convicted on two counts of perjury in January 1950 and sentenced to two concurrent five-year sentences. As the 'red scare' now gathered momentum, Truman's Democrat administration was targeted particularly by Republicans for its lax security and apparent unconcern. Even after the

[2] The Venona decryptions implicate both Hiss and White as espionage agents, although the evidence against Hiss is somewhat less conclusive.

conviction of Hiss, Truman himself dismissed the allegation that there were Soviet spies in the White House as a 'red herring'.

Republican Senator Joseph McCarthy saw an opportunity to make political hay. In a Lincoln Day speech in Wheeling, West Virginia, in February 1950, he declared: 'I have here in my hand a list of 205 that were known to the Secretary of State as being members of the Communist Party and who nevertheless are still working and shaping the policy of the State Department.'[3] McCarthy touched a nerve left raw by a series of post-war international events that had spooked the American public. He was quickly overwhelmed by media attention.

'Senator McCarthy's crusade, which was to last for the next several years, was always anathema to me', wrote FBI special agent Lamphere some years later. 'McCarthy's approach and tactics hurt the anti-Communist cause and turned many liberals against legitimate efforts to curtail Communist activities in the United States ... McCarthy's star chamber proceedings, his lies and overstatements hurt our counterintelligence efforts.'

But the unfolding events lent credibility to McCarthy's claims of Communist spies in high places. Fuchs, convicted at the Old Bailey on 1 March 1950, had been given the maximum sentence of fourteen years' imprisonment. He had initially resisted naming his espionage contacts and, in any case, knew them only by their cover names. However, Fuchs agreed to co-operate when, during a visit to Wormwood Scrubs in May, Lamphere dropped vague but dark hints about the future safety in America of his sister Kristel. When it became clear that Gold was the FBI's prime target, Fuchs admitted on 22 May that Gold had 'very likely' been his contact, Raymond. Gold was arrested that same day. A search of his Philadelphia apartment had uncovered the map of Santa Fe that Fuchs had given him. He had previously denied ever having been to New Mexico. He slumped

---

[3] There is no recording of McCarthy's speech and the figure of 205 was soon amended to 57. In John Frankenheimer's 1962 movie *The Manchurian Candidate*, McCarthy's lack of precision was parodied through the fictional character Senator John Iselin (played by James Gregory), who latched onto Heinz's '57 varieties' as a number he could more easily remember.

in his chair and declared: 'I am the man to whom Fuchs gave the information.'

Gold was the weak link in the espionage network. From Gold the trail led to Greenglass, already a suspect based on the Venona decryptions. He was arrested in June. From Greenglass the trail led to Julius and Ethel Rosenberg, arrested a month later. In the meantime, North Korean forces had on 25 June attacked the Republic of Korea along the Ongjin Peninsula, intent on reunifying Korea under a Communist flag. On 5 July, the US Army's 24th Infantry Division was fighting and losing its first battle at Osan in South Korea. The Cold War had escalated to a limited 'hot' conflict. A real, large-scale war loomed, ominously.

HUAC gained no new evidence from the testimonies of Bohm and Lomanitz, yet it had concluded in September 1949 that Weinberg, Lomanitz and Bohm had been members of a Communist cell that had passed atomic secrets to the Soviets. Bohm was arrested on 4 December 1950 and charged with contempt of court. He was subsequently bailed. This time the Princeton University administrators were not so supportive. Concerned for the continuation of financial support from the university's wealthy benefactors, Princeton president Harold Dodds suspended Bohm from his post for the duration of the trial.

Bohm was brought to trial on 31 May 1951. He was acquitted. Lomanitz was also acquitted, as was Weinberg a few years later. Princeton did not renew Bohm's contract when it expired in June 1951. Einstein tried to bring him to the Institute for Advanced Study, declaring that if anyone could create a radical new quantum theory then it would be Bohm. But Oppenheimer vetoed the move. Bohm left America for exile in Brazil in October 1951. He carried with him a copy of his new book *Quantum Theory*. The thought processes sparked by the writing of this book would lead eventually to some remarkable discoveries, and would help set a path towards some of the most profound experiments in modern quantum physics.

Technically sweet

When Truman had announced on 31 January 1950 the intention to build a hydrogen bomb, the scientists at Los Alamos hadn't the faintest idea how to do it. Despite Teller's constant nagging and his report declaring a prima facie proof of the weapon's feasibility, the general consensus among scientists was rather different. George Gamow would demonstrate the practical problems of the classical Super design by trying – and failing – to ignite a piece of petrified wood by setting fire to a small cotton wool ball sat alongside. The cotton wool, representing the fission bomb, would flare up and burn out quickly, leaving unaffected the wood, representing the deuterium/tritium fuel for the thermonuclear bomb. 'That is where we are just now in the development of the hydrogen bomb', he would say.

Teller had by now returned to Los Alamos from Chicago but continued to be unpredictable, erratic and irascible. He flitted from one obsession to the next, unable to fix on any one idea and run with it. His frustrations continued to build along with his paranoia, as Oppenheimer and the GAC and Bradbury at Los Alamos continued to put obstacles in his path.

The breakthrough came from Ulam, staring intently out of a window one afternoon in late January 1951. The problems they had been experiencing were related to the difficulty of triggering the fusion reactions simultaneously or near-simultaneously with fission. At issue was the fact that much of the energy from the fission device would be carried away rapidly by radiation, much as Gamow's cotton wool ball burnt itself out without igniting the wood. Ulam now realised that by staging the sequence – more clearly separating the 'primary' fission device from the 'secondary' thermonuclear device – it would be possible to make use of the massive flux of neutrons from the primary to compress and heat the fuel in the secondary sufficient to trigger thermonuclear fusion reactions.

The idea quickly evolved. Teller realised that compression could be achieved by the X-ray radiation from the primary fission device which, as high-speed photons, would bombard the deuterium/tritium fuel a lot sooner than the heavier, much more sluggish neutrons. In itself, radiation compression was not new – Fuchs and von Neumann had filed a patent application for this idea in 1946 – but Teller had hitherto dismissed it as irrelevant. Now, with radiation compression combined in a two-stage design, he realised that he had made a mistake, one that was 'simple, great and stupid'.[4]

When Oppenheimer reviewed the two-stage Teller–Ulam design at a GAC meeting in June 1951, he declared it 'technically sweet' and supported further work to build the H-bomb. This time there was no debate about the morality of the weapon. It seems that, once again, fear of the potential for 'enemy' scientists to make similar breakthroughs led the physicists to set aside any moral misgivings. Bethe, resolutely opposed to the development of the H-bomb, recognised that the new design changed everything. Soviet physicists were quite clearly capable of coming up with a similar design and, with the Korean war threatening to escalate way beyond a regional conflict, he felt that the Americans should strive to be the first to build such a weapon. History was repeating itself. Once again, a major discovery in nuclear physics was made beneath the storm clouds of a threatening war. Bethe returned with some reluctance to Los Alamos that summer.

Teller, who had pushed relentlessly for development of the Super since the summer school at Berkeley in 1942, had finally won the argument by showing, with Ulam, how it could be done. He should have been satisfied, but he was not. He offered to stay on at Los Alamos and assume managerial responsibility for the H-bomb programme, but was told that he could continue only as an assistant director or consultant. To have put the peripatetic Teller in charge would have been a calamity. When Bradbury appointed physicist Marshall Holloway to head the H-bomb programme, Teller resigned from Los Alamos. He continued to carp from the sidelines as he lobbied for the establishment of a second nuclear weapons laboratory.

After some ups and downs, the design of the first thermonuclear bomb was fixed in March 1952. This was the sequence of physical events. The fission primary would explode at the tip of a 20-foot long cylinder weighing

---

[4] It should be noted that this simple description of the discovery process was disputed by Teller, who, in his memoirs, did not credit Ulam with any kind of breakthrough. Norris Bradbury later argued that the invention of the H-bomb was a team effort, with no single individual deserving right of paternity.

82 tons. X-rays from the fission explosion would be channelled down the cylinder and around the outside of the secondary device positioned along the cylinder's long axis. A lining of polyethylene would be turned to plasma, radiating more X-rays towards the centre of the secondary, the radiation pressure collapsing and concentrating the cold, liquid deuterium it contained. Squeezing the deuterium from the outside in would implode a fission 'sparkplug' made of plutonium and suspended down the centre of the secondary. This would trigger a second fission explosion boosted by high-energy neutrons from fusion reactions. Squeezing the deuterium now also from the inside out and heating it with X-rays would trigger a series of nuclear fusion reactions. These reactions would all liberate nuclear energy but the greatest source of explosive force would come from high-energy neutron fissioning of U-238 nuclei from the uranium 'pusher' which would line the outside of the secondary device.

The bomb, designated Mike, was tested at Eniwetok Atoll on 1 November 1952, part of the Ivy test series. Its yield was 10.4 million tons of TNT equivalent, about 1,000 times more powerful than the Hiroshima atomic bomb. Within seconds of ignition it had produced a fireball that spread three miles in diameter. It vaporised the tiny island of Elugelab, leaving a circular crater more than a mile wide and 200 feet deep.

For what it was worth, America had re-established its technical lead in nuclear weapons.

Scorpions in a bottle

Stalin died on 5 March 1953 of a cerebral haemorrhage. Molotov later revealed in his memoirs that Beria claimed to have poisoned him, although the precise circumstances of Stalin's death may never be known. Georgei Malenkov was appointed as chairman of the Council of Ministers. Beria himself was arrested by Marshal Zhukov on 26 June. Towards the end of July the Party Central Committee issued a letter declaring Beria a 'bourgeois degenerate' and detailing his horrific crimes. But it was not opprobrium over Beria's reign of terror that had led to his arrest. As part of his strategy to seize power after Stalin's death he had initiated a relaxation of Soviet policy which had backfired in East Germany. He was arrested because he had gone too far and was judged a liability. He was executed on 23 December.

Stalin was gone but it was business as usual. On 12 August 1953 the Soviet Union tested RDS-6s, or Joe-4, a 'layer cake' thermonuclear device which made use of fusion in lithium deuteride to enhance the yield of a fission bomb. It yielded 400,000 tons. Vyacheslav Malyshev, who had taken over from Beria as head of the Soviet atomic programme, declared after the test that he had just received a call from Malenkov: 'He congratulates everyone who helped build the hydrogen bomb – the scientists, the engineers, the workmen – on their wonderful success. Georgei Maximilianovich [Malenkov] requested me to congratulate and embrace Sakharov in particular for his exceptional contributions to the cause of peace.'

The nature and scale of nuclear weapons escalation was now set. In a speech to the Council on Foreign Relations in New York on 17 February 1953, Oppenheimer had made this comparison:

The trouble then is just this: During this period the atomic clock ticks faster and faster; we may anticipate a state of affairs in which two Great Powers will each be in a position to put an end to the civilisation and life of the other, though not without risking its own. We may be likened to two scorpions in a bottle, each capable of killing the other, but only at the risk of his own life.

Although the Soviet 'layer cake' was in many ways inferior to the two-stage Teller–Ulam design tested the previous November, the Soviet physicists were indeed not that far behind, at least in their understanding of the theory. Having pushed various 'exotic' variations on the design of RDS-6s, in early 1954 Sakharov abandoned these in favour of a two-stage radiation compression design – essentially the Teller–Ulam configuration – that had been developed independently by several theoreticians at Arzamas-16. In his memoirs Sakharov called it the 'Third Idea'.

Meanwhile, with support from Lawrence, Teller had arm-twisted the AEC into setting up a second thermonuclear weapons laboratory, at a cost of nearly $12 million, on the site of a former air base at Livermore in California. Teller's popularity rating at Los Alamos fell to a new low as a bitter rivalry developed.

Despite this rivalry, the Livermore laboratory and Los Alamos combined their efforts in the Bravo test on Bikini, part of the Castle series, on 1 March 1954. Bravo was a much smaller lithium deuteride-fuelled thermonuclear device, weighing in at nearly twelve tons and therefore much easier to 'weaponise' than Mike. The rare Li-6 isotope had been enriched in the fuel from 7.5 to 40 per cent.

An error in the measurement of the rate of a nuclear reaction involving the dominant Li-7 isotope led the physicists to underestimate its potential yield. Bravo should have yielded five million tons. It actually yielded three times this figure; at fifteen million tons it was the largest nuclear weapon ever tested by the United States. Its fireball measured nearly four miles across as it vaporised three islands and threw radioactive debris over nearly 50,000 square miles, exposing the task force personnel thought to be at a safe distance out at sea and the crew of a Japanese fishing boat. The Japanese fishermen arrived at port suffering radiation sickness similar to the initial survivors of Hiroshima and Nagasaki. The inhabitants of the islands of Rongelap and Ailinginae had to be hastily evacuated.

The Soviet Union tested its first two-stage, lithium deuteride-fuelled, megaton-yield weapon on 22 November 1955. Designated RDS-37, and dropped from an aeroplane rather than triggered atop a tower, its theoretical yield was three megatons but it had deliberately been limited to little more than half this to reduce the risks from fallout. The shockwave cracked the ceiling of the observation bunker, raining loose plaster on Zavenyagin's head. It also collapsed a trench in which a platoon of soldiers was sheltering, killing one young soldier in his first year of service. There were further fatalities among the civilian population living close to the Semipalatinsk test site. Another collapsed shelter had killed a two-year-old girl.

'For my part, I experienced a range of contradictory sentiments,' Sakharov wrote, 'perhaps chief among them a fear that this newly released force could slip out of control and lead to unimaginable disasters. The accident reports, and especially the deaths of the little girl and the soldier, heightened my sense of foreboding. I did not hold myself personally responsible for their deaths, but I could not escape a feeling of complicity.'

The pattern of escalation was repeated again and again throughout the Cold War. In truth, the Soviets did catch up in terms of the science and technology, but they never did surpass the potential destructive force of the American nuclear arsenal. But this had long since ceased to be the point. The larger American scorpion carried the greater sting, but the Soviet scorpion's smaller sting was still deadly. The American policy of 'massive retaliation' did not change the simple fact that in nuclear war destruction remained mutual, and assured.

The inexorable growth of the American nuclear arsenal simply increased the American military's ability to 'bounce rubble'.

In the matter of J. Robert Oppenheimer

Oppenheimer had used the position he had acquired on the world stage as the 'father of the atom bomb' to make political pronouncements and exert influence over American nuclear policy. But what some perceived to be clear-headed and reasoned arguments, others saw as misguided liberal sermonising. And although Oppenheimer retained the ability to charm, he had lost none of his character flaws. His casual arrogance, his thinly-disguised disdain for those he thought intellectually inferior or wrong-headed, and his acidic put-downs had helped create enemies where friends might have proved more useful.

Chief among his enemies was Lewis Strauss, who had resigned as AEC commissioner in protest over delays to the Super programme. When Eisenhower, elected as US President on 4 November 1952, appointed Strauss as chairman of the AEC in January 1953, Strauss began a campaign to discredit Oppenheimer and remove him completely from his position of influence. By this time Oppenheimer was no longer chairman of the GAC. He had stepped down in 1952, largely in frustration, but had been persuaded to continue as a consultant, extending his Q-clearance for at least another year.

Strauss had contributed to Eisenhower's election campaign fund and now moved to sow seeds of doubt in the President's mind regarding Oppenheimer's fitness even for his lesser role as consultant. Strauss also encouraged William L. Borden, a young member of the Joint Committee on Atomic Energy[5] and already deeply suspicious of Oppenheimer, to examine the evidence that had sat festering in Oppenheimer's weighty FBI file for more than a decade. On 7 November 1953 Borden, on his 'own personal initiative and responsibility', wrote a letter to Hoover in which he asserted, 'based upon years of study, of the available classified evidence, that more probably than not J. Robert Oppenheimer is an agent of the Soviet Union'. The letter went on to declare that: '[H]e has since acted under a Soviet directive in influencing United States military, atomic energy, intelligence and diplomatic policy.'

In truth, Borden had no new evidence on which to base his denunciation. The most serious allegation against Oppenheimer's loyalty to America remained his attempted obfuscation regarding the 'Chevalier incident'. This was an incident that had been examined several times in numerous security reviews and, while all who knew of it regarded it as very unfortunate, it had not previously been seen as sufficiently serious to deny security clearance. The real issue was the perceived loss of American technical leadership on the H-bomb, for which Oppenheimer was regarded as worthy of blame. Oppenheimer's obstinacy, amplified by his extensive influence, was seen by Borden to be motivated not by rational scientific or moral judgements, but by directive from Moscow. There was absolutely no evidence for this.

Strauss knew he would get only one chance to bring Oppenheimer down and wanted to wait for precisely the right moment to spring his trap, but events forced his hand. Borden's letter was forwarded to Eisenhower who, fearing that a failure to act would expose his administration to damaging allegations of incompetence by McCarthy, secretly suspended Oppenheimer's security clearance on 3 December. Hoover had already

---

[5] The JCAE was a congressional committee formed in 1946 which, among other duties, oversaw the activities of the AEC.

moved to divert McCarthy from the Oppenheimer case, concerned that the Wisconsin senator would bungle it.

Oppenheimer was told of the decision to suspend his clearance on 21 December. Two days later he requested a formal hearing so that he could clear his name.

This was to be a hearing of the AEC's Personnel Security Board. As such it had no precedent and no formal basis in law and, as AEC chairman, Strauss was free to set it up in whatever way he saw fit. He proceeded to stack the deck as firmly and as unfairly against Oppenheimer as he could, through a series of manoeuvres worthy of the Soviet Politburo.

Strauss handpicked the members of the Security Board that would sit in judgement. He ensured that the members of this 'jury' had full access to all the evidence assembled by the FBI, which they reviewed in the presence of the prosecuting attorney, Roger Robb. Strauss had selected Robb on the strength of his reputation: he had the cross-examination skills of a Rottweiler.

Oppenheimer was once again under 'technical surveillance' by the FBI, meaning that his phones were tapped and his offices bugged. Strauss ensured that the surveillance continued throughout the hearing and that the prosecuting attorney had access to the results, including taped conversations between Oppenheimer and his defence attorney, Lloyd Garrison. In his turn, Garrison was denied access to the FBI evidence against his client. When the AEC finally relented and agreed to expedite his Q-clearance, he was told that this couldn't be extended to the other members of the defence team. Garrison made a profound mistake, withdrawing the request for clearance in the belief that it would be impossible for the defence team to work together if only one member had access to the relevant information. By the time he changed his mind it was too late. Garrison never gained access to the FBI files and several times during the hearing the defence team was obliged to leave the room.

The hearing began on 12 April 1954, in room 2022 at the AEC's headquarters, Building T-3 on the corner of Sixteenth and Constitution in Washington, DC. It began by reading into the record a list of charges levelled by AEC general manager Kenneth D. Nichols, Groves' former aide on the Manhattan Project, and Oppenheimer's personal, lengthy, rebuttal. Two days later, Robb was relentlessly pursuing Oppenheimer on the circumstances surrounding the Chevalier incident.

'Now let us get back to your interview with Colonel Pash. Did you tell Pash the truth about this thing?' Robb asked.

'No', Oppenheimer answered.

'You lied to him?'

'Yes.'

Robb now probed the details of the story Oppenheimer had given Pash and Johnson that August day in Berkeley, nearly eleven years before. He asked if Oppenheimer had said that Chevalier, then unnamed, had approached three people.

'Probably', Oppenheimer replied.

'Why did you do that, Doctor?'

'Because I was an idiot.'

Robb later told a reporter that at this stage in the proceedings Oppenheimer was visibly struggling with his testimony, wringing his hands between his knees as he sat, under oath, in the witness-box. Robb proceeded to read from the transcript of the recorded conversation, advising Oppenheimer that 'for your information, I might say we have a record of your voice'. The nature of Oppenheimer's 'cock and bull story' was fully exposed. Robb forced him to admit that he had told 'not one lie ... but a whole fabrication and tissue of lies'.

Robb then turned his attention to the night that Oppenheimer had spent with Jean Tatlock. He first established that Oppenheimer had no grounds for believing that in 1943 Tatlock was no longer a Communist, before levelling an accusation.

'You spent the night with her, didn't you?' he said.

'Yes', Oppenheimer replied.

'That is when you were working on a secret war project?'

'Yes.'

'Did you think that consistent with good security?'

Oppenheimer's reply was shot through with the sound of defeat: 'It was, as a matter of fact', he said. 'Not a word – it was not good practice.'

In his testimony the next day, Groves explained that while he hadn't liked some of the things that Oppenheimer had done, it had not been his job as head of the Manhattan Project to like everything his subordinates did. He felt that Oppenheimer had made a mistake regarding the Chevalier incident through a misplaced desire to protect a friend, but he also felt that he had eventually got what he needed and had decided not to make an issue of it. However, under Robb's cross-examination he was obliged to admit that under the terms of the 1946 Atomic Energy Act, 'I would not clear Dr Oppenheimer today if I were a member of the Commission on the basis of this interpretation'.

Character witnesses were brought before the hearing to testify for Oppenheimer's integrity and loyalty. Among them were Bethe, Conant, Fermi, Kennan, Lilienthal and Rabi. Vannevar Bush questioned the entire basis for the hearing, declaring: '[H]ere is a man who is being pilloried because he had strong opinions [about the H-bomb], and had the temerity to express them.' He concluded his testimony with the statement: 'I think this board or no board should ever sit on a question in this country of whether a man should serve his country or not because he expressed strong opinions. If you want to try that case, try me ...'

Although some, like von Neumann, had disagreed with Oppenheimer's position on the H-bomb programme, he did not doubt Oppenheimer's loyalty to America. Others, including Wendell Latimer and Kenneth Pitzer, testified against Oppenheimer. A turning point in the hearing was reached when Teller took the stand on 28 April.

Robb asked Teller about Oppenheimer's loyalty. Teller replied that he did not doubt his loyalty.

'Now, a question which is a corollary of that', Robb continued. 'Do you or do you not believe that Dr Oppenheimer is a security risk?'

'In a great number of cases,' Teller replied, 'I have seen Dr Oppenheimer act – I understood that Dr Oppenheimer acted – in a way which for me was exceedingly hard to understand. I thoroughly disagreed with him in numerous issues and his actions frankly appeared to me confused and complicated. To this extent I feel that I would like to see the vital interests of this country in hands which I understand better and therefore trust more.

In this very limited sense I would like to express a feeling that I would feel personally more secure if public matters would rest in other hands.'

Under cross-examination later that afternoon, Teller was asked by Gordon Gray, chairman of the Security Board, if granting clearance to Oppenheimer would endanger American defence and security. Teller delivered this indictment:

'I believe, and that is merely a question of belief and there is no expertness, no real information behind it, that Dr Oppenheimer's character is such that he would not knowingly and willingly do anything that is designed to endanger the safety of this country. To the extent, therefore, that your question is directed toward intent, I would say I do not see any reason to deny clearance. If it is a question of wisdom and judgement, as demonstrated by actions since 1945, then I would say one would be wiser not to grant clearance ...'

There lay the rub. Oppenheimer was to be indicted for his stubborn refusal to bow to political pressure and sanction the development of a weapon that he had believed to be both unnecessary and technically unfeasible. As Teller left the hearing room, he walked past Oppenheimer. He extended his hand and said, 'I'm sorry'. Oppenheimer shook hands and replied: 'After what you've just said, I don't know what you mean.'

The Personnel Security Board voted two-to-one to deny clearance and their verdict was returned on 23 May. Oppenheimer had violated no laws or security regulations, but with regard to the H-bomb he was guilty of poor judgement, of 'conduct ... sufficiently disturbing as to raise a doubt'.

Oppenheimer lost his Q-clearance just one day before it was due to expire. He was visibly aged by the proceedings. As an influential advocate of clear-thinking atomic policy and international arms control, he was finished.

If it could be called a victory, then it was surely hollow. Within the physics community Teller became a pariah. After the hearing, Strauss tried to block Oppenheimer's reappointment in October 1954 as director of the Institute for Advanced Study. He failed. When Strauss sought appointment as Secretary of Commerce in the Eisenhower administration in 1959, the Senate voted marginally against. Accusations of an abuse of power while chairman of the AEC were factors in the Senate decision.

TEODOR KhOLL

Fuchs was tried at the Old Bailey in 1950 for transmitting atomic secrets in breach of the Official Secrets Act. He had initially imagined that he was facing the death penalty. But despite the judge declaring in his summary that his crime was 'thinly differentiated from high treason', this was not the crime for which he was tried. Instead he received the maximum penalty of fourteen years' imprisonment.

Gold was sentenced to 30 years in prison in December. Greenglass was sentenced to fifteen years on 6 April 1951. Julius and Ethel Rosenberg, also tried for espionage, not treason, nevertheless received the death penalty on 5 April for a crime declared by the judge to be 'worse than murder'. The Rosenberg case became something of an international cause célèbre, sparking mass protests and vigils. Intent on martyr status, the Rosenbergs refused to confess or co-operate, and both went to the electric chair on 19 June 1953. Lamphere, who had been heavily involved in providing evidence for the case, experienced mixed emotions: 'I felt, not satisfaction, but defeat. I knew the Rosenbergs were guilty, but that did not lessen my sense of grim responsibility for their deaths.'

By coincidence, Theodore Hall and his wife Joan were on their way to a dinner party on the evening of the executions and drove past Sing Sing prison in New York, where the executions were scheduled to take place. Hall, too, was experiencing mixed feelings. After leaving Los Alamos he had moved to Chicago, were he had worked for a time with Teller. He had met Joan Krakover in 1946 and they had married on 25 June 1947. Hall's friend and one-time espionage contact Sax had also moved to Chicago and Joan was made aware of their past activities in support of the Soviet cause. Hall and Joan were drawn increasingly into political activism, and joined the American Communist Party in late 1947. They joined in full knowledge of how such affiliation was likely to be received in Cold War America. It also meant that Hall could not continue with his clandestine activities, as there was obviously nothing clandestine about Communist Party membership. He wrote to Sax, by now also married and living back in New York, that it was time for him to cut ties with the Soviet spy network.

But it was not to last. Sax had retained his Soviet contacts and they now argued that Hall was still desperately needed. Despite the tacit agreement he had made with his wife, Hall decided to return to espionage. Without the requisite security clearance, Hall no longer enjoyed access to atomic secrets. However, it seems possible that he gleaned at least some important secrets about the production of polonium-210, an isotope used in the initiators of fission weapons, from other scientists working at Hanford. By August 1949 Joan was four months pregnant with their first child, and Hall once again wanted to terminate his espionage activities. They both met Morris and Lona Cohen in a New York City park to discuss the matter. The Halls were not persuaded. When Truman announced on 31 January 1950 that the Soviets had tested Joe-1, Hall figured that his work was already done. He told Sax to advise his Soviet contacts that his career as a Soviet spy was over.

It was to prove a timely decision. During the spring of 1950, Meredith Gardner at Arlington Hall was assembling a report based on a partially decrypted NKGB cable dated 12 November 1944. It pointed to a Los Alamos spy named Theodore Hall:

BEK visited Theodore Hall [TEODOR KhOLL], 19 years old, the son of a furrier. He is a graduate of HARVARD University. As a talented physicist he was taken on for government work ... According to BEK's account Hall has an exceptionally keen mind and a broad outlook, and is politically developed. At the present time H. is in charge of a group at 'CAMP-2' [SANTA-FE]. H. handed over to BEK a report about the CAMP and named the key personnel employed on ENORMOUS. He decided to do this on the advice of his colleague SAVILLE SAX [SAVIL SAKS], a GYMNAST living in TYRE ... We consider it expedient to maintain liaison with H. [1 group unidentified] through S. and not to bring in anybody else. MAY has no objection to this ...

BEK was the codename for Sergei Kurnakov. GYMNAST probably meant the Young Communist League, and TYRE was the codename for New York City. MAY was Stepan Apresyan, the Soviet vice consul in New York.

The cable had been sent before the codenames for Hall and Sax had been assigned. Of all the messages decrypted in the Venona project, this was one of the most clear and unambiguous in its identification of atomic spies.

The Hall case was passed to FBI agent Robert McQueen. Gardner made the connection between Hall and the spy codenamed MLAD, and a cable dated 23 January 1945 indicated that MLAD (whose codename had by then changed to YOUNG) had been called up into the army but was left to work at Los Alamos. The timing coincided precisely with Hall's personnel record.

This was damning evidence, but the Venona decrypts themselves were inadmissible. McQueen needed to gather evidence that Hall was still an active spy or he needed to secure a confession. It was not going to be plain sailing. Both Hall and Sax were now highly politically active, extremely unusual behaviour for supposedly active spies. The FBI had yet to make a connection with Morris and Lona Cohen, but this husband-and-wife spy team had quietly left America and made their way to the Soviet Union, arriving in Moscow in November 1950.

Hall and Sax were separately picked up by the FBI for interrogation on 16 March 1951. However, both had anticipated this eventuality and had prepared for it. Although McQueen's suspicions were even further aroused by Hall's coolness during three hours of questioning, the case went no further forward. Intense surveillance turned up no new evidence. McQueen was taken off the case in late 1951, and it slipped down the FBI's list of priorities.

Despite his coolness during interrogation, Hall was churning inside. In mid-1952 he was again approached by Soviet intelligence. He and Joan moved to New York in the autumn of that year. It is not known what, if any, secrets Hall might have transmitted to the Soviets in this third and final phase of his espionage career, but as the Rosenbergs faced execution he offered an extraordinary exchange to his Soviet controller. He offered to show that the Rosenbergs were not wholly responsible for atomic espionage by giving himself up and confessing his own wartime role. 'I would have done it,' he later told a friend, 'I felt that strongly about it. But he [Hall's Soviet controller] felt it wasn't a good idea at all, and so it came to nothing.' The Soviets had already decided that the Rosenbergs were expendable.

After the news of the explosion of the first Soviet thermonuclear device in 1953, and the realisation that Joan was expecting their second child, Hall decided to break off contact with Soviet intelligence for the last time. His Soviet controller thanked him for everything he had done.

Hall went on to build a reputation as an academic biophysicist, and moved to the Cavendish Laboratory in Cambridge in July 1962, bringing Joan and their three daughters with him to England. Although they remained nervous that Hall's espionage activities could still be exposed (especially after the arrest and trial of Morris and Lona Cohen, caught masquerading as antiquarian booksellers Peter and Helen Kroger in London in March 1961),[6] as time went on the possibility receded. It was only with the July 1995 release of the Venona decrypt of the 12 November 1944 cable that Hall's wartime role as a Soviet spy was finally revealed. Hall would never be called to account in court, but the publicity surrounding his case eventually forced him to explain his actions. Hall died in November 1999. In April 2003 Joan published a long memoir, in which she concluded:

But in 1995, when the whole story came out, I was glad. For us personally the danger was clearly past, and Ted now had a chance to tie together the two ends of his life in a coherent way. This gave him more confidence that in the end he had done something worthwhile in addition to his scientific work – something he had paid for with long years of anxiety and constraint. Ted's efforts and those of Fuchs, and of others who tried in legal ways to contain the nuclear threat, probably helped to stave off a violent crisis for several decades during which the 'socialist'

---

[6] The Cohens were exchanged in 1969 for Gerald Brooke, a British teacher who had been arrested in Moscow for anti-Soviet activity. They had served just nine years of their twenty-year prison sentence. They returned to teach spycraft to a new generation of Soviet intelligence agents.

nations tried to control the arms race. There is good reason to believe that the delay gave humanity an extended time of hope – a time during which our children and grandchildren, with millions of others, grew up to take the reins of struggle into their hands.

Unusual suspects

Nobody was immune from suspicion. Harwell physicist Bruno Pontecorvo disappeared in the summer of 1950 while on holiday with his family, eventually to reappear in Moscow. He later told a Russian journalist that he had defected because he feared arrest for his wartime espionage activities, although the precise details of this espionage are unclear.

The Venona decrypts revealed more codenames, names of spies clearly connected either centrally or peripherally with the Manhattan Project. The FBI compiled substantial dossiers on other physicists, such as Bethe, Fermi, Peierls, Lawrence, Serber and Szilard. Some of the codenames remain to this day unidentified; names such as QUANTUM and VOGEL/PERS.

A cable sent to Moscow on 21 June 1943 revealed that QUANTUM (KVANT in the decrypts) had visited the Soviet embassy in Washington carrying with him secret documents:

On 14 June a meeting took place with 'KVANT' in CARTHAGE … KVANT declared that he is convinced of the value of the materials and therefore expects from us a similar recompense for his labour – in the form of a financial reward.

QUANTUM had clearly established contact before the 14 June meeting. He had already been assigned a codename and was able to gain an audience with a high-ranking Soviet diplomat, most probably Andrei Gromyko. Whoever he was, he handed over information about gaseous diffusion, for which he received $300. The FBI was never able to identify him.

Peierls, whose wife Genia was Russian and an admitted former Communist, was also a suspect. After Fuchs' arrest, Peierls was able to figure this much out for himself: 'I later learned that, in the course of tracing the source of leaks from Los Alamos, the evidence indicated at one stage that a theoretician in the British Group was responsible, which pointed to Fuchs and me. I must therefore have been under great suspicion for a time, but at no stage was I made to feel it.'

But Peierls did experience problems with his security clearance in the early 1950s. When his clearance was eventually withdrawn in 1957 he resigned from his post as consultant to the AERE. Whatever suspicions about him prevailed, these did not prevent him from receiving a knighthood in 1968.

In 1999, after the release of the Venona decrypts, *Spectator* journalist Nicholas Farrell quoted 'British security service sources' to accuse Peierls of being the Soviet agent VOGEL/PERS. This seems certain to be an accusation without much foundation[7] but may reflect the degree of suspicion that led to Peierls' loss of clearance.

Mikhail Gorbachev's *glasnost* and the subsequent formal ending of the Cold War had former Soviet spymasters scrambling to tell their stories. For those who had been involved in atomic espionage, it was an opportunity to gain recognition for the contribution they had made to the Soviet programme. Howevever, Pavel and Anatoly Sudoplatov's *Special Tasks*, first published in 1994, muddied the waters considerably by hurling fantastic accusations against leading Manhattan Project physicists. I will not dignify these accusations by repeating them here. Suffice to say that many of these have since been refuted by Russian historians.

The apologetic thesis

In 1956, Berlin-born writer and journalist Robert Jungk published *Brighter than a Thousand Suns*, the first comprehensive popular history of the development of atomic and thermonuclear weapons. It was written as a

---

[7] In their detailed study of the Venona decrypts, Haynes and Klehr associate VOGEL with CAMP-1 (Oak Ridge) and declare: 'All that can be said about [VOGEL] with certainty is that he had access to technical information about the atomic bomb project; very probably he was an engineer or scientist of some sort, but even that is a guess.' See Haynes and Klehr, p. 314.

'personal' history, based on interviews with many of the physicists who had been directly involved up to that time.

Jungk's thesis regarding the German atomic programme was clear. Drawing on his interviews particularly with Weizsäcker and Heisenberg, he painted a stark contrast between the behaviour of German physicists under Nazi rule and that of Allied physicists working in the relative peace and freedom of New Mexico:

> It seems paradoxical that the German nuclear physicists, living under a sabre-rattling dictatorship, obeyed the voice of conscience and attempted to prevent the construction of atom bombs, while their professional colleagues in the democracies, who had no coercion to fear, with very few exceptions concentrated their whole energies on production of the new weapon.

Jungk's words echo those of Weizsäcker at Farm Hall, rationalising the reasons for the German physicists' failure to build even a working nuclear reactor. Weizsäcker had sown the seeds of the *Lesart* that claimed the German physicists could have built atomic weapons if they had really wanted to, but deliberately refrained from doing so on moral grounds. By the time of publication of Jungk's book, Heisenberg too had bought into the *Lesart*. He had published articles in the scientific press which argued along similar lines.

It was a thesis which contrasted sharply with Goudsmit's, set out in his memoir of his wartime involvement in the Alsos mission, which had been published nearly ten years previously in 1947. In *Alsos*, Goudsmit argued that the German physicists had failed largely because they were incompetent, and claimed that 'science under fascism was not, and in all probability could never be, the equal of science in a democracy'.

Historian Mark Walker has called these opposing interpretations the 'apologetic' and 'polemic' theses. The apologetic thesis is framed in the Farm Hall *Lesart*, expounded by Jungk, elaborated by journalist Thomas Powers in his more recent book *Heisenberg's War* into a story implying conscious sabotage by the German physicists, and treated sympathetically in Michael Frayn's award-winning play *Copenhagen*. It asserts that the German physicists did what they had to do to prevent Hitler from gaining the ultimate weapon. It is an apologetic thesis because, as Walker explains, it is an 'apologia for being willing to work on the economic and military applications of nuclear fission for the National Socialist government during World War II, in other words, for being apolitical, irresponsible, and, some might add, amoral'.

Jungk's book provoked Bohr to draft the unsent letters recalling the circumstances and substance of Heisenberg's visit to Copenhagen in September 1941, quoted in Chapter 4. Controversy has raged ever since and, despite the release of the Farm Hall transcripts in 1993, the Bohr letters in 2001 and Heisenberg's correspondence with his wife Elisabeth in 2003, it seems likely to continue in the absence of further compelling historical documentation that could settle the matter one way or the other. It is doubtful that such documentation exists. In seeking a conclusion to this story, I believe Walker is the clearest guide:

> Why were myths and legends of active resistance against Hitler created and propagated after the war? Obviously because something is being repressed. Scientific work, exactly like any other occupation, can be politicized. Scientists in general are morally neither superior nor inferior to the general public. Finally sometimes – for example under National Socialism during World War II – there are neither simple answers nor simple questions.

Jungk eventually came to understand that he had been misled, even betrayed, by the German physicists whom he had interviewed. He felt that he had been used to promulgate 'eine Legende'.

Thirteen days

By 1962, the American arsenal of over 27,000 strategic and non-strategic nuclear warheads outnumbered the Soviet arsenal by more than eight to one. American foreign policy remained resolutely belligerent, with politicians now wearing the bloated arsenal even more ostentatiously on their hips. The Soviet test of RDS-220, the *Tsar Bomba*, in October 1961 had demonstrated a capability to build thermonuclear weapons with ever-greater yields (at 50 million tons, the three-stage *Tsar Bomba* was the largest nuclear weapon ever tested),[8] but it did not address the imbalance in numbers.

This superiority was understood by America's military leaders, but it did not engender any real sense of security. Still dissatisfied, the new American administration under recently-elected President John F. Kennedy turned its attention from size to proximity. In 1961 America deployed fifteen Jupiter intermediate-range ballistic missiles at a base in Izmir in Turkey, close to the Soviet Union's southern border and targeted at western Soviet cities, including Moscow. If these had been launched, Muscovites would have had sixteen minutes' warning of impending doom.

So, when Cuba's Fidel Castro sought support from the Soviet Union to defend against what he perceived to be an imminent American invasion of his new socialist republic, Nikita Khrushchev saw an opportunity to kill two birds with one stone. By installing missiles on Cuba, Khrushchev reasoned that he could forestall any hasty American intervention against the fledgling revolutionary socialist regime and at the same time restore the nuclear balance, at least in terms of proximity. From Cuba, the Soviet Union could strike all the major cities in the eastern US, including Washington, New York and Philadelphia, with virtually no warning.

The Soviets constructed nine missile sites, with about 40 launchers. Medium-range R-12 (known to NATO forces as SS-4) missiles arrived in September 1962. Despite growing evidence of the presence of missiles on Cuba, the Kennedy administration was in denial. Years later, LeMay explained what happened next: 'The administration would come back and say, "there is no evidence that there are missiles in Cuba". Finally they gave the mission to SAC to overfly Cuba with our U-2s, and they found the

---

[8] Its theoretical yield was 100 megatons, but had been limited to half this to reduce fallout. The *Tsar Bomba* was developed for demonstration purposes only – it was never 'weaponised'.

missiles.' Photographic evidence of the construction of a base for medium-range missiles was gathered on 14 October and presented to Kennedy two days later.

Kennedy assembled an Executive Committee of the National Security Council (ExComm) to determine the response. The Joint Chiefs of Staff urged Kennedy to sanction an invasion. LeMay, now Air Force Chief of Staff, believed that this was an opportunity to eliminate the missile threat and drive the Communists out of Cuba. He felt certain that American nuclear superiority would prevent the Soviets from retaliating. Kennedy was not so sure. He believed that if the Soviets didn't strike back in Cuba, they would strike in West Berlin.

The ExComm agreed to a naval blockade against offensive weapons – a quarantine. Kennedy made a televised announcement on 22 October and the US armed forces defence readiness condition (DEFCON) was moved from 5 to 3, a state of high alert. Khrushchev declared the quarantine an act of piracy and stated that the Soviet Union would take measures to protect its rights. Work on the Cuban missile sites continued around the clock. Cuban forces prepared to repel an invasion, to be supported if necessary by Soviet troops stationed on the island.

When the quarantine came into effect on 24 October, SAC was moved to DEFCON 2. The number of American nuclear weapons now made ready for use by planes already in the air or submarines already at sea reached nearly 3,000, and included many thermonuclear weapons. They promised a total destructive force of *7,000 million tons* of TNT. If they had been used, conservative estimates put the total likely death toll at 100 million people. SAC commander Thomas Power made sure the Soviets knew what was coming. He broadcast the new state of readiness to all SAC pilots in plain English, rather than in code.

During these tense moments, accidents were perhaps inevitable. Unauthorised missile launches, erroneous signals for planes to scramble, unauthorised U-2 flyovers and a 'phantom' missile launch from Cuba which turned out to be a computer test tape all contributed to the general fraying of nerves.

Meanwhile, negotiations continued. Via Soviet ambassador Anatoly Dobrynin, Attorney General Robert Kennedy secretly signalled his brother's willingness to consider a trade. If the Soviets agreed to remove their missiles from Cuba, then America would remove its missiles from Turkey. When Khrushchev formalised this offer the next day, 27 October, the ExComm judged it unacceptable, as removal of the missiles from Turkey would undermine NATO, the Turkish government (which wanted to keep the missiles) and Kennedy's presidency. The ExComm decided to respond instead to an earlier proposal, to link withdrawal of the missiles from Cuba with American assurances of non-aggression against Castro's republic.

Robert Kennedy's secret offer of a trade was shared with a trusted inner circle of ExComm members. It was recognised that this was a real opportunity – possibly the only opportunity – to resolve the crisis and avert disaster. The inner circle agreed that Robert Kennedy should go back to Dobrynin with the proposal for a separate, and secret, agreement to withdraw missiles from Turkey. Messages went back and forth, some through the 'back channels', via ABC News reporter John Scali and Alexander Fomin, a cover name for Alexander Feklisov, the former spymaster of the Rosenberg network and of Fuchs in England, now promoted to the position of Soviet *rezident* in Washington.

Meanwhile, the tension had continued to grow. Kennedy had promised instant retribution if any American plane was shot down over Cuban airspace by Soviet forces (the Cubans did not possess surface-to-air missiles of their own). When a U-2 spy-plane was shot down, Kennedy assumed it had been a mistake and held back from a military response. He remained patient when a further four reconnaissance planes were fired upon.

Khrushchev's response to the public and the private offer was received in Washington at 9:00am on 28 October. He acknowledged the previous message and expressed his gratitude. He explained that he had ordered the missile bases to be dismantled. There hadn't been time to advise Castro of the decision. He heard about it on the radio and was absolutely furious.

The world, which had been collectively holding its breath for thirteen days, breathed a great sigh of relief. A few months later the Jupiter missiles were quietly withdrawn from Turkey.

Trading on fear

The Cuban missile crisis brought the world to the brink of disaster. It was a salutary lesson. Nuclear war was unthinkable, yet, somehow, the posturing and brinkmanship of the world's superpowers had not only made it thinkable, they had very nearly made it happen. Fortunately for the world, Kennedy and Khrushchev stepped back from the brink. American forces were never again placed at DEFCON 2. Soviet forces were never placed on a comparable level of nuclear alert.

Some have argued that these insane weapons not only kept the peace, but they also contributed to the eventual economic collapse of the Soviet Union in the late 1980s, creating an opportunity for at least some form of democracy to emerge. This may be so, but at what risk?

Deterrence was already in place and working quite effectively with weapons capable of Hiroshima-scale devastation. Despite growing paranoia bordering on hysteria concerning Soviet weapons capabilities, the Soviet Union was *never* ahead in the arms race. Perhaps the arms race afforded just too good an opportunity to build careers and profits for those – scientists, technologists, businessmen, politicians, generals – involved in sponsoring and perpetuating the endless cycle of nuclear escalation.

The opportunists certainly knew how to trade on fear. When Viktor Berlenko, a Soviet pilot with the 513th Fighter Regiment, defected with his MiG 25 Foxbat on 6 September 1976, the aircraft was examined in detail by the Foreign Technology Division of the US Air Force. The investigators were astonished to discover that it was full of obsolete valve technology. This simply didn't fit the natural presumption that the Soviets were far ahead in terms of technology development.

The answer was quickly forthcoming. In an exchange of battlefield nuclear weapons, the electromagnetic pulse resulting from a nuclear explosion would completely disable aircraft filled with more sophisticated transistor technology. Indeed, the Soviets *were* ahead of the game. Better double the research budget ...

The Spanish philosopher, poet and novelist George Santayana once wrote that those who cannot remember the past are condemned to repeat it. The threat of global annihilation may have diminished, but the lessons of history are indeed too easily forgotten as we continue to live in the shadow of the fear inspired by The Bomb.

In his book *Doomsday Men*, Peter Smith describes his experiences at an anti-nuclear protest demonstration gathering in Trafalgar Square, London, in April 2004, ready to march to Britain's Atomic Weapons Research Establishment at Aldermaston. Damon Albarn, front-man of the Brit-pop band Blur, railed at the fact that fewer than 1,000 protesters had turned up. But it is not hard to understand this apathy. In 1987, America began to dismantle its stockpile of 24,000 nuclear warheads. By the end of 2007 America had passed the half-way point towards its commitment under the Strategic Offensive Reduction Treaty (also known as the Moscow Treaty), signed five years earlier by President George W. Bush and Russian Federation President Vladimir Putin. As of January 2008, the US stockpile contained an estimated 5,400 nuclear weapons, of which about 4,000 were operational. At the same time, the Russian Federation arsenal consisted of about 9,000 weapons, with about 5,200 operational. The prospect of Armageddon is considerably less that it was at the height of the Cold War, so why protest?

And yet the danger is far from past. The Moscow Treaty will reduce, but not eliminate, the American and Russian arsenals. Even if the Treaty objectives are met at the end of December 2012, the combined American and Russian arsenals will still likely represent an explosive potential in excess of one billion tons, or 80,000 Hiroshimas. Proliferation continues, inexorably. On 9 October 2006 North Korea – the last bastion of Stalinism – joined the world's elite nuclear club. In January 2007 the *Bulletin of the Atomic Scientists* advanced its 'Doomsday Clock' by two minutes. It remains at five minutes to midnight, the closest it has been to midnight since 1984. In August 2007 an accidental breach of protocols left six nuclear warheads sitting unattended in a B-52 bomber parked on the tarmac at an American Air Force base in Louisiana. In 2008, for the first time in sixteen years, America resumed small-scale production of nuclear weapons at Los Alamos.

The fear remains. If the time for protest has ended, the time for vigilance has not.

TIMELINE

1939	Germany	UK	USA	Italy	USSR
		Christmas 1938: Discovery of nuclear fission in uranium			
January			Bohr arrives at Princeton Bohr publicly announces discovery of fission		
February			Fission verified		
March				Fission verified	
April	Uranverein formed		Role of U-235 identified		Fission verified
July		Frisch moves to Birmingham	Heisenberg visit to America		
August		Lindemann downplays potential for atomic weapons	Einstein letter to Roosevelt	Joliot-Curie, Halban and Kowarski observe increased fission in uranium oxide Halban identifies potential for heavy water as a moderator	Molotov–Ribbentrop non-aggression pact

1939	🏴 (Germany)	🏴 (Britain)	🏳 (USA)	🏳 (France)	☭ (Soviet Union)
September	Germany attacks Poland – war begins Army Ordnance consolidates nuclear research Heisenberg joins Uranverein				Soviet Union invades Eastern Poland
October			Sachs presents Einstein's letter to Roosevelt Advisory Committee on Uranium established		
November		Rotblat identifies potential for fast-neutron fission			Soviet Union invades Finland
December	Heisenberg's first report to the War Office		Lawrence bids for funding for 184-inch cyclotron		

1940	Germany	Britain	USA	France	USSR
January		Frisch identifies potential for fast-neutron fission in U-235			
February	Heisenberg's second report to the War Office Heavy water identified as potential moderator				
March		Frisch–Peierls memorandum	Einstein letter to Sachs Investigation of graphite as potential moderator begins	Allier retrieves stocks of heavy water from Vemork	
April	German forces invade Denmark and Norway	Advisory Committee on uranium formed – later renamed MAUD Committee			
May	German forces invade France and the Low Countries	British and French forces evacuate from Dunkirk			

1940	Germany	Britain	USA	France	USSR
June	German forces enter Paris		McMillan–Abelson paper on element 93 (neptunium)	Uranverein physicists at Joliot-Curie's laboratory	
			Turner identifies potential for element 94	Halban and Kowarski escape to Britain with heavy water	
			Bush forms NDRC		
July	Weizsäcker identifies element 93 as potentially fissionable	Frisch joins Chadwick in Liverpool	Briggs report on uranium to NDRC		
		Battle of Britain begins			
September	Germany, Italy and Japan sign Tripartite Pact	London Blitz begins			
October	The Virus House is completed				
November			Roosevelt re-elected		
			Work begins on uranium–graphite pile at Columbia University		
December	First nuclear reactor experiments (B1 and L1) are conducted	Simon report on gaseous diffusion			

1941	🇩🇪	🇬🇧	🇺🇸	☭
January	Bothe concludes that graphite will not function as a moderator; Hanle disagrees			
March	Uranverein meeting concludes Clusius–Dickel technique will not work for uranium Houtermans alerts the Allies via Reiche	Conant visits MAUD physicists	Lawrence agitates for greater effort on atomic energy Seaborg and Wahl isolate element 93 (neptunium)	
April	Reiche delivers Houtermans' warning Compton heads National Academy review group	Experiments with scale model of a gaseous diffusion plant		Stalin signs neutrality pact with Japan
May	First National Academy report OSRD established Seaborg and Segrè measure fission rate of element 94 (plutonium)	Fuchs joins Peierls in Birmingham		

1941	🇩🇪	🇬🇧	🇺🇸	☭
June	German forces invade the Soviet Union			
July		MAUD Committee Report Anglo–Soviet Treaty of Mutual Assistance	Second National Academy report	
August			Work commences at Columbia to build sub-critical pile	German Army Group North advances on Leningrad Army Group South captures Kiev
September	Heisenberg visits Bohr in Copenhagen	MAUD Report approved by Defence Services Panel and Chiefs of Staff Tube Alloys project initiated Cairncross informs Moscow of decision	Oliphant visits America and discusses MAUD Report	
October		Tronstad escapes to Britain	Bush receives official copy of MAUD Report	German forces advance on Moscow

1941	🇩🇪	🇬🇧	🇺🇸	☭
November	German advance on Moscow grinds to a halt		Third National Academy report S-1 Committee established U-235 enrichment using gaseous diffusion is demonstrated	Soviet counter-attack on German positions outside Moscow
December	Conference concludes fission unlikely to support war effort Control passed back to Reich Research Council	Fuchs becomes a Soviet spy	Japanese attack Pearl Harbor America enters the war First meeting of the S-1 project	

1942	🇩🇪	🇬🇧	🇺🇸	☭
January			The Met Lab is established in Chicago UN Declaration	
February	Uranverein reports that a bomb is feasible Army Ordnance conferences		Tube Alloys physicists visit S-1 project laboratories	Flerov writes to Kurchatov about atomic energy
March				Beria reviews espionage materials provided by Cairncross
April			Fermi moves to Chicago Work commences on uranium–graphite pile CP-1	Flerov writes letter to Stalin Beria decides to re-start Soviet atomic research
May			Oppenheimer leads S-1 work on bomb physics	
June	Meeting at Harnack House Experimental pile L-IV explodes	Proposal for full collaboration between Tube Alloys and S-1		

1942	🕂 (Nazi)	🇬🇧 (Britain)	🇺🇸 (USA)	☭ (USSR)
August			Report on Berkeley summer school Seaborg separates plutonium	
September			Groves appointed to lead atomic project S-1 becomes the Manhattan Project Preliminary construction work at Oak Ridge	Battle for Stalingrad begins
October		Grouse party parachutes on to Hardanger Plateau	Oppenheimer appointed as scientific head of weapons laboratory 'Site Y'	
November		Operation Freshman ends in disaster	Construction of CP-1	Red Army counter-offensive at Stalingrad: 'Uran'
December			First successful nuclear reactor (CP-1) demonstrated in Chicago	

1943	🇩🇪	🇬🇧	🇺🇸	☭
January		Tube Alloys physicists relocate to Montreal Laboratory	Kvasnikov arrives in New York Chevalier 'incident' (winter 1942–43)	Fitin initiates intelligence project ENORMOZ
February		Swallow/Gunnerside teams sabotage heavy water plant	Clarke initiates project to decrypt Soviet messages Construction of electromagnetic separation plant at Oak Ridge commences	German surrender at Stalingrad Special resolution on atomic energy Kurchatov appointed scientific director
March			Construction of large-scale reactors commences at Hanford Physicists begin to arrive at Los Alamos	Kurchatov reviews espionage materials
April			Inaugural lectures at Los Alamos Neddermeyer suggests implosion	
May	Heisenberg acknowledges results of G-II experiment			

1943				
June			Oppenheimer meets Tatlock in Berkeley	
July			Small-scale work on implosion begins	
August		Quebec Agreement on Anglo–American collaboration	Oppenheimer interview with Pash and Johnson	
September			Oppenheimer interview with Lansdale Alsos mission launched	
October		Bohr escapes to Britain	Construction of gaseous diffusion plant at Oak Ridge commences	
November	Heaviest Allied bombing raid on Berlin	Bombing raid on the Vemork plant Tube Alloys delegation departs for America	Electromagnetic isotope separation commences at Oak Ridge AZUSA project initiated	
December	Heisenberg visits Frank in Poland		Segrè measurements on spontaneous fission in U-235 Alsos mission arrives in Italy	

1944	☭ (Germany)	✠ (UK)	★ (USA)	☭ (USSR)
January	Gerlach appointed plenipotentiary for nuclear science Heisenberg returns to Copenhagen		Tatlock commits suicide Kistiakowsky joins Los Alamos Teller is assigned to lead implosion theory group	
February	Allied air raid severely damages Kaiser Wilhelm Institute (KWI) for Chemistry	*Hydro* sabotaged	Priority of fusion research downgraded Teller resigns from Theoretical Division Fuchs makes contact with Gold	
March	RAF bombs Hamburg			
April		Bohr meets Churchill	Tuck identifies potential for explosive lenses	Laboratory No. 2 expands and relocates
May				
June		D-Day: Allied invasion of Normandy	Pash confronts Berg in Rome Goudsmit appointed scientific leader of Alsos II Groves contracts construction of liquid thermal diffusion plant at Oak Ridge	

1944	🇩🇪	🇬🇧	🇺🇸	☭
July	KWI for Physics relocates to Hechingen		Segrè measurements on spontaneous fission in Pu-240 triggers crisis	
	KWI for Chemistry relocates to Tailfingen			
	Reactor research moved to Haigerloch			
	Failed attempt to assassinate Hitler			
August			Fuchs relocates to Los Alamos with Peierls	
			Oppenheimer reorganises work on implosion	
			Bohr meets with Roosevelt	
			Modification of B-29s for delivery of atomic bombs commences	
September		Hyde Park aide-mémoire	Hanford B reactor operational	Kurchatov letter to Beria
			First Ra-La implosion tests	Soviet cyclotron operational

1944	☭	🇬🇧	🇺🇸	☭
October			Hall becomes a Soviet spy	
November			Greenglass recruited as a spy by Rosenberg	
			Documents recovered in Strasbourg convince Goudsmit there is no Nazi A-bomb	
December	Battle of the Bulge		Berg attends Heisenberg lecture in Switzerland	
			Hall passes details of implosion to Sax	
			Hanford D-pile goes critical	

1945	卐	✠	▦	☭
January			Dragon experiments confirm critical mass for U-235	
			Gaseous diffusion plant at Oak Ridge commences operations	
February	Experimental reactor B-VIII reassembled in Haigerloch		Hanford reactors achieve designed output	
	Allied forces cross the Rhine		Uranium gun design completed	
	Bombing of Dresden		Composition of explosive lenses finalised	
			Fuchs re-establishes contact with Gold	
March	Alsos captures Bothe		Cowpuncher Committee established	Beria establishes 'Soviet Alsos' mission
			Experiments confirm feasibility of solid-core compression for plutonium	

1945	🇩🇪 (Germany)	🇬🇧 (Britain)	🇺🇸 (USA)	☭ (USSR)
April	Bagge, Wirtz, Weizsäcker, Korsching, Laue and Hahn captured		Roosevelt dies	
			Truman becomes president	
	British forces enter Belsen and Buchenwald concentrations camps		Preparations begin on Tinian island	
	Soviet offensive against Berlin begins		Target Committee identifies potential targets for A-bomb	
	Hitler commits suicide			
May	Red Army captures Berlin	May is reactivated as GRU spy in Montreal	Cohen hands Hall materials to Yatskov	
	Pash arrests Heisenberg	V-E Day	Szilard meets Byrnes	
	Gerlach, Diebner and Harteck captured		Meeting of the Interim Committee	
	Ardenne leaves for Moscow			
	Germany surrenders		Target Committee produces shortlist	

1945	Germany	United Kingdom	United States	USSR
June	Riehl leaves for Moscow		Gold meets Fuchs and Greenglass	
	Soviets find 300 tons of uranium		Franck Report	
			Interim Committee supports earlier decisions	
			Modified B-29s arrive on Tinian	
July	Potsdam Declaration	German physicists arrive at Farm Hall	First Szilard petition	
		Attlee becomes prime minister	Trinity test	
			Truman tells Stalin about the bomb at Potsdam Conference	
			Second Szilard petition	
August		German physicists at Farm Hall learn of Allied success	Bombing of Hiroshima and Nagasaki	Special State Committee established to fast-track Soviet A-bomb
			Smyth Report published	
			Japan surrenders	
			Daghlian killed in accident at Los Alamos	
			Cohen meets Hall	

1945	🇳🇿 (Nazi)	🇬🇧 (UK)	🇺🇸 (US)	☭ (Soviet)
September		Council of Foreign Ministers meeting Gouzenko defects in Ottawa, unmasking May	Greenglass meets Rosenberg Fuchs meets Gold for the last time	
October			Oppenheimer resigns from Los Alamos Bradbury is appointed as successor	Beria receives comprehensive report based on Smyth Report and espionage materials
November		Hahn learns he has won Nobel prize at Farm Hall	Bentley confesses to being an NKVD courier Truman–Attlee–King Declaration Truman–Attlee Memorandum	Terletsky visits Bohr in Copenhagen
December		Attlee establishes UK atomic project		Council of Foreign Ministers meeting

510

1946	🇬🇧	🇺🇸	☭
January	Atomic Energy Research Establishment (AERE) founded Farm Hall detainees released	UN establishes Atomic Energy Commission (AEC)	Kurchatov meeting with Stalin
February		Greenglass receives honourable discharge	Stalin speech at the Bolshoi Theatre Kennan 'long telegram'
March	May arrested at King's College, London	Acheson–Lilienthal Report Churchill 'iron curtain' speech	
April			Arzamas-16 established in Sarov
May		Teller reports 'prima facie' proof of the feasibility of the Super Slotin killed in second accident at Los Alamos	
June	May tried and sentenced at the Old Bailey Fuchs arrives at AERE	Baruch delivers his plan to UN AEC Hall receives honourable discharge	
July		Crossroads A-bomb tests	
		Oppenheimer moves to Princeton	
August	Britain resolves to build its own nuclear deterrent	Truman signs US Atomic Energy Act	
September		Oppenheimer interviewed by FBI	
December		First Soviet message relating to atomic espionage decrypted	Soviet Union's first experimental nuclear reactor goes critical

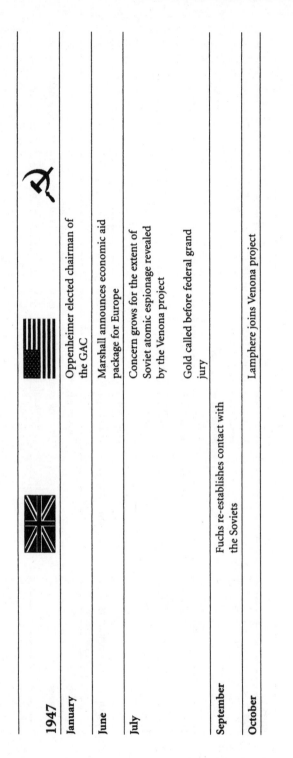

1947	 🇬🇧	🇺🇸	☭
January		Oppenheimer elected chairman of the GAC	
June		Marshall announces economic aid package for Europe	
July		Concern grows for the extent of Soviet atomic espionage revealed by the Venona project	
		Gold called before federal grand jury	
September	Fuchs re-establishes contact with the Soviets		
October		Lamphere joins Venona project	

1948	🇬🇧	🇺🇸	☭
January		Modus vivendi signed with Britain	
March	Fuchs passes information about the Super to Feklisov		Work on plutonium production reactor (Installation-A) commences
April		A-bomb tests at Eniwetok	Work begins on a Soviet 'Super'
May		UN AEC is wound up	
June			Installation-A achieves designed output
October		LeMay takes command of SAC	
December			Installation-B completed

	Britain	USA	USSR
1949			
March		SAC Emergency War Plan 1-49	
April	Fuchs holds last meeting with Feklisov		
May			Berlin blockade lifted
August			First Lightning test
September		Fallout from First Lightning detected	
		Lamphere identifies Fuchs as prime suspect	
		Truman announces that Soviets have the A-bomb	
October		GAC rejects the Super	
December	Skardon interrogates Fuchs		
1950			
January	Fuchs confesses at the War Office	Truman announces programme to build the Super	
February	Fuchs is arrested		

LIST OF KEY CHARACTERS

Abelson, Philip Hauge. American physicist. Verified nuclear fission in 1939 and co-discovered neptunium in 1940.

Acheson, Dean Gooderham. American politician. Secretary of State in the Truman administration from 1949 to 1953. Co-author of the Acheson–Lilienthal report.

Akers, Wallace Alan. British chemist and industrialist. Appointed head of the British Tube Alloys project in 1941.

Alikhanov, Abram Isaakovich. Soviet physicist. Organised the Soviet atomic project's Laboratory No. 3 which was founded in 1945.

Allier, Jacques. French banker and agent of the Deuxième Bureau, French military intelligence. Salvaged quantities of heavy water from Norway and Nazi-occupied France in 1940.

Allison, Samuel King. American physicist. Headed the 'Cowpuncher' Committee in the run up to the Trinity test.

Altshuler, Lev Vladimirovich. Soviet physicist. Developed pulsed X-ray radiography with Veniamin Tsukerman and worked on implosion at Arzamas-16.

Alvarez, Luis Walter. American physicist. Joined the Met Lab in 1943 before moving to Los Alamos in 1944. Won the 1968 Nobel prize for physics.

Anami, Korechika. Japanese Imperial Army general and Army Minister.

Anderson, Herbert. American physicist. Worked on the first experimental nuclear reactor in Chicago.

Anderson, John. British politician. Minister in Winston Churchill's wartime cabinet responsible for the UK atomic energy programme.

Angelov, Pavel. Soviet diplomat and GRU spy. Worked at the Soviet Embassy in Ottawa. Reactivated Alan Nunn May in 1945.

Apresyan, Stepan. Soviet diplomat and NKGB station chief in New York.

Ardenne, Manfred von. German physicist and inventor. Secured funding for nuclear research from the Reich Postal Ministry. Relocated to Moscow in May 1945.

Arnold, Henry. Security officer at the UK's Atomic Energy Research Establishment at Harwell.

Artsimovich, Lev Andreyevich. Soviet physicist. Worked on electromagnetic isotope separation.

Bacher, Robert. American physicist. Headed the Los Alamos G Division in 1944.

Bagge, Erich Rudolf. German physicist. Former student of Werner Heisenberg. Founding member of the Uranverein. Worked on isotope separation. Captured by the Alsos mission and detained at Farm Hall.

Baruch, Bernard Mannes. American financier. Led the US delegation to the UN Atomic Energy Commission in 1946.

Bainbridge, Kenneth Tompkins. American physicist. Directed the Trinity test.

Bentley, Elizabeth Terrill. Vassar graduate and NKVD spy. Defected in 1945, uncovering two major spy networks.

Berg, Morris 'Moe'. American Major League Baseball player and later an OSS spy. Part of the AZUSA mission. Visited Zurich in 1944 with instructions to assassinate Werner Heisenberg.

Beria, Lavrenty Pavlovich. Soviet politician and head of the NKVD. Oversaw Soviet atomic espionage and chaired the Committee responsible for the Soviet atomic project. Executed in 1953.

Bethe, Hans Albrecht. German émigré physicist. Headed the Los Alamos Theoretical Division. Won the Nobel prize for physics in 1967.

Beurton, Ruth (née Kuczynski). GRU spy. Courier for Klaus Fuchs from 1942 until his departure for America with the British mission in late 1943.

Bevin, Ernest. British Foreign Secretary in the Attlee government.

Blackett, Patrick Maynard Stuart. British physicist. Served on the MAUD Committee 1940–41.

Bloch, Felix. Swiss physicist. Contributed to Oppenheimer's summer study group at Berkeley. Joined Los Alamos but left shortly afterwards.

Blunt, Anthony Frederick. British art historian and Soviet spy. One of the Cambridge spy 'ring'.

Bohm, David Joseph. American physicist. Worked on electromagnetic separation at the Rad Lab. Left America following his trial in 1951.

Bohr, Aage Niels. Danish physicist. Son of Niels Bohr. Won the Nobel prize for physics in 1975.

Bohr, Erik. Danish chemical engineer. Son of Niels Bohr

Bohr, Harald August. Danish mathematician and footballer. Brother of Niels Bohr.

Bohr, Niels Henrik David. Danish physicist and Nobel laureate. Elucidated the origin of nuclear fission in uranium. Joined the British Tube Alloys delegation following his escape from Nazi-occupied Copenhagen. Worked to pre-empt nuclear weapons proliferation.

Borden, William Liscum. American lawyer and executive director of the Joint Committee on Atomic Energy. Summarised the evidence gathered by the FBI and concluded that J. Robert Oppenheimer was 'more probably than not' a Soviet spy.

Born, Max. German émigré physicist. Worked with Klaus Fuchs in Edinburgh.

Bothe, Walther Wilhelm Georg. German physicist. Founder member of the Uranverein. Studied the properties of nuclear materials and built Germany's first cyclotron.

Bradbury, Norris Edwin. American physicist. Succeeded J. Robert Oppenheimer as director of the Los Alamos laboratory in 1945.

Briggs, Lyman James. American engineer and administrator. Director of the US National Bureau of Standards. Chaired the Uranium Committee from its inception in 1939 to 1941.

Brothman, Abraham. American industrialist and Soviet spy.

Brun, Jomar. Norwegian chemist. Directed operations at the Norsk Hydro heavy water plant at Vemork. Carried out acts of sabotage before fleeing to Sweden in 1940. Planned subsequent commando raids on the plant with Lief Tronstad.

Burgess, Guy Francis De Moncy. British civil servant and Soviet spy. Part of the Cambridge spy 'ring'.

Burt, Leonard. British police officer. Involved in the cases of both Alan Nunn May and Klaus Fuchs.

Bush, Vannevar. American engineer and administrator. Directed the NDRC and OSRD, which controlled the Manhattan Project.

Byrnes, James Francis. American politician. Secretary of State in the Truman administration from 1945–47.

Cairncross, John. British civil servant and Soviet spy. Secretary to Lord Hankey in 1941. Passed notes of the meeting signalling Britain's decision to build an atomic bomb to Moscow in September 1941.

Chadwick, James. British physicist and Nobel laureate. Joined the MAUD Committee in 1940 and the Manhattan Project as part of the Tube Alloys delegation in late 1943. Knighted in 1945.

Chambers, Whittaker. American journalist and *Time* editor. Acted as a Soviet courier. Testified against Alger Hiss in 1948.

Chevalier, Haakon Maurice. Translator, author and professor of French literature at Berkeley. Close friend of J. Robert Oppenheimer. Sponsored the notorious 'Chevalier incident' in 1942–43.

Clarke, Carter. Chief of the Special Branch of the US Army's Military Intelligence Division. Initiated project to decrypt Soviet messages in 1943.

Clusius, Klaus. German chemist. Co-inventor (with Gerhard Dickel) of the Clusius–Dickel thermal diffusion technique used to separate isotopes. Worked on isotope separation as part of the Uranverein.

Cockcroft, John Douglas. British physicist and Nobel laureate. Director of the Anglo–Canadian atomic energy project in Montreal from 1944 until 1946, when he returned to set up Britain's Atomic Energy Research Establishment at Harwell. Knighted in 1948.

Cohen, Lona Theresa. Soviet spy, recruited by her husband, Morris Cohen. Acted as courier for Theodore Hall from 1945.

Compton, Arthur Holly. American physicist and Nobel laureate. Directed the OSRD's S-1 programme, which became the Manhattan Project in 1942. Thereafter Compton was responsible for work at the Met Lab in Chicago.

Compton, Karl Taylor. American physicist. Older brother of Arthur Compton. President of MIT. Karl contributed to the Interim Committee deliberations on the first use of the atomic bomb.

Conant, James Bryant. American chemist and administrator. President of Harvard University. Became chairman of the NDRC when Vannevar Bush established the OSRD. Member of the US Atomic Energy Authority's General Advisory Committee.

Condon, Edward Uhler. American physicist. Served as deputy director of Los Alamos for a short time until conflicts with Leslie Groves led to his resignation.

Daghlian, Harry Krikor, Jr. American physicist. Suffered a fatal dose of radiation in an accident at Los Alamos in 1945.

Debye, Pieter Joseph William. Dutch physicist and Nobel laureate. Director of the Kaiser Wilhelm Institute for Physics in Berlin at the outbreak of war. Left Germany for America in 1940.

Diebner, Kurt. German physicist. Administrative director of the German atomic project. Pursued experimental work on nuclear reactors at the German Army Ordnance laboratory in Gottow. Captured by the Alsos mission and detained at Farm Hall.

Dirac, Paul Adrien Maurice. British physicist and Nobel laureate. Advised the MAUD Committee on aspects of isotope separation and bomb physics.

Döpel, Georg Robert. German physicist. Worked in Leipzig with Werner Heisenberg on experimental nuclear reactors.

Donovan, William Joseph 'Wild Bill'. American wartime head of the Office of Strategic Services (OSS).

Dunning, John Ray. American physicist. Worked on the fission properties of U-235 and gaseous diffusion.

Eifler, Carl F. American Army officer. Commanded the OSS Detachment 101 in Burma from 1942 until 1943. Assigned to capture or assassinate Werner Heisenberg in 1944 but the mission was scrubbed.

Einstein, Albert. German-born physicist and Nobel laureate. His special theory of relativity identified the equivalence of mass and energy. With prompting from Leo Szilard, Edward Teller and Eugene Wigner, Einstein wrote to US President Franklin Roosevelt warning of the dangers of atomic weapons.

Eisenhower, Dwight David. US Army Chief of Staff 1945–48. President 1953–61.

Eltenton, George C. British chemical engineer and Soviet spy. Worked at the Shell Development Company's Emeryville laboratory in San Francisco. Approached Haakon Chevalier with a view to recruiting J. Robert Oppenheimer.

Esau, Robert Abraham. German physicist. Head of the physics section of the Reich Research Council in the months before the outbreak of war. Convened the first meeting of the Uranverein in 1939. Plenipotentiary for nuclear physics 1942–43.

Feklisov, Alexander Semyonovich. Soviet diplomat and NKVD spy. Managed the Rosenberg network in New York before becoming Klaus Fuchs' controller in London. Returned to the US as station chief in Washington in 1960.

Ferebee, Thomas. Bombardier aboard the *Enola Gay* B-29 Superfortress.

Fermi, Enrico. Italian physicist and Nobel laureate. Supervised the first successful nuclear reactor in Chicago in December 1942.

Feynman, Richard Phillips. American physicist. Joined Hans Bethe's Theoretical Division at Los Alamos in 1943. Won the Nobel prize for physics in 1965.

Fitin, Pavel Mikhailovich. Head of the NKVD's First Chief Directorate, responsible for foreign operations and intelligence, from 1939 to 1951.

Flerov, Georgei Nikolayevich. Soviet physicist. Verified nuclear fission and co-discovered spontaneous fission in uranium. Went on to become a prominent member of the Soviet atomic bomb project.

Franck, James. German émigré physicist and Nobel laureate. Directed the Met Lab's chemistry division. Chaired the committee on political and social problems and together with other Met Lab physicists drafted the Franck report in 1945.

Friedman, Max. American physicist. Worked on electromagnetic isotope separation at the Rad Lab until 1943.

Frisch, Otto Robert. Austrian émigré physicist. Co-discovered the origin of nuclear fission in uranium with his aunt, Lise Meitner. Independently discovered spontaneous fission. With Rudolf Peierls, co-authored the Frisch–Peierls memorandum on critical mass. Joined the British Tube Alloys delegation to the Manhattan Project in late 1943.

Fromm, Friedrich. German chief of the reserve army.

Fuchs, Klaus Emil Julius. German émigré physicist and Soviet spy. Provided the Soviet Union with information on the design of the Fat Man plutonium bomb and the 'classical' Super thermonuclear bomb. Unmasked by the Venona project decryptions in 1949. Arrested and tried in 1950. Sentenced to fourteen years in prison.

Furman, Robert Ralph. American civil engineer and chief of foreign intelligence for the Manhattan Project. Co-ordinated the AZUSA project aimed at gathering intelligence on German physicists and, ultimately, kidnapping or assassinating Werner Heisenberg.

Gamow, George. Russian-born physicist. Proposed the 'liquid-drop' model of the nucleus. Contributed to work on the design of the Super thermonuclear bomb. Known for his popular 'Mr. Thompkins' science series, work on the cosmic microwave background radiation and the big bang theory of the origin of the universe.

Gardner, Meredith Knox. American linguist and code-breaker. Joined the US Army's Signals Intelligence Service in 1942. His decryption of Soviet messages led to the identification of several Soviet spies that had worked on the Manhattan Project.

Garrison, Lloyd. American lawyer. Chief defence council in J. Robert Oppenheimer's security clearance hearing.

Gerlach, Walther. German physicist. Became the plenipotentiary for nuclear physics in late 1943 following the resignation of Abraham Esau. Captured by the Alsos mission and detained at Farm Hall.

Ginzburg, Vitaly Lazarevich. Soviet physicist. Worked on the early design of the Soviet Super in Igor Tamm's group. Won the Nobel prize in 2003.

Gold, Harry. Swiss-born American chemist and Soviet spy. Acted as courier for Klaus Fuchs.

Golos, Jacob. Soviet NKVD operative. Senior official in the American Communist Party. Managed Elizabeth Bentley and other intelligence networks.

Göring, Hermann Wilhelm. German politician and military leader. Commander of the Luftwaffe. Placed in charge of the Reich Research Council in 1942.

Gorsky, Anatoly Veniaminovitch. Attaché at the Soviet Embassy in London, and NKVD station chief. Controlled the Cambridge spy 'ring' and was involved with early Soviet penetration of the British atomic project. Became the NKVD station chief in Washington in 1944.

Goudsmit, Samuel Abraham. Dutch physicist. Emigrated to America in 1927. Scientific head of the second Alsos mission.

Gouzenko, Igor Sergeyevich. Cipher clerk at the Soviet Embassy in Ottawa. Defected in 1945. Gouzenko's defection led to the arrest of Alan Nunn May.

Greenglass, David. Machinist assigned to Los Alamos as part of the Special Engineering Detachment. Was also a Soviet spy recruited by Julius Rosenberg.

Gromyko, Andrei Andreyevich. Soviet politician and diplomat. Soviet representative on the UN Security Council, 1946 . Became Minister for Foreign Affairs in 1957.

Groves, Leslie Richard. US Army engineer. Became the military head of the Manhattan Project in 1942.

Gyth, Volmer. Captain in the Danish General Staff intelligence section. Helped Niels Bohr escape from occupied Copenhagen.

Hahn, Otto. German chemist. Discovered (with Fritz Strassmann) evidence for nuclear fission in uranium. Joined the Uranverein in 1939. Captured by the

Alsos mission and detained at Farm Hall, where he learned he had won the 1944 Nobel prize for chemistry.

Halban, Hans von. French physicist, of Austrian descent. Verified nuclear fission in Paris in 1939. Escaped from France with a quantity of heavy water in 1940 and joined the British Tube Alloys project. Helped to establish the nuclear research laboratory in Montreal in 1942.

Hall, Theodore Alvin. American physicist. Became a Soviet spy in 1944, just eight months after joining the Manhattan Project at Los Alamos. Hall continued spying for the Soviet Union until 1953, and was only revealed when the Venona decrypts were published by the NSA in 1995.

Hallock, Richard. Made the first breakthrough on the Venona project when he discovered that Soviet messages had been enciphered using one-time pads that had been used more than once.

Halperin, Israel. Canadian mathematician. Arrested in 1946 and accused of espionage as part of the Royal Commission investigation following the defection of Igor Gouzenko. Cleared of charges following his trial in 1947.

Hanle, Wilhelm. German physicist. Contributed to the work of the Uranverein, primarily through the measurements of the properties of potential reactor moderators.

Harteck, Paul Karl Maria. German chemist. Alerted the German War Office to the possibilities of nuclear weapons and joined the Uranverein in 1939. Captured by the Alsos mission and detained at Farm Hall.

Haugland, Knut. Norwegian commando. Member of the advance Grouse/Swallow party parachuted onto the Hardanger Plateau in 1942. Supported the successful Gunnerside raid on the Vemork heavy water plant. In 1947 Haugland joined Thor Heyerdahl on the *Kon-Tiki* expedition.

Haukelid, Knut. Norwegian commando. Member of the successful Gunnerside team. Subsequently sabotaged the *Hydro*.

Heineman, Kristel (née Fuchs). Klaus Fuchs' sister. Living in Cambridge, Massachusetts during the war.

Heisenberg, Werner. German physicist and Nobel laureate. Joined the Uranverein in 1939. Worked on experimental nuclear reactors in Leipzig, Berlin and Haigerloch. Captured by the Alsos mission and detained at Farm Hall.

Helberg, Claus. Norwegian commando. Member of the advance Grouse/Swallow party parachuted onto the Hardanger Plateau in 1942. Supported the successful Gunnerside raid on the Vemork heavy water plant.

Hida, Shuntaro. Japanese Army medical officer at Hiroshima Military Hospital.

Himmler, Heinrich Luitpold. German politician and head of the SS.

Hirohito. Emperor of Japan 1926–89.

Hollis, Roger Henry. MI5 chief of counter-intelligence. Became Director General in 1956.

Houtermans, Friedrich Georg 'Fritz'. German physicist. Worked on nuclear problems in Manfred von Ardenne's private laboratory. Alerted physicists in America to German progress.

Howard, Charles Henry George. Earl of Suffolk and Berkshire. Worked for the DSIR as liaison to the French Ministry of Armaments. Helped French physicists Hans von Halban and Lew Kowarski to escape to Britain.

Idland, Kasper. Norwegian commando. Member of the successful Gunnerside team.

Ioffe, Abram Fedorovich. Soviet physicist. Founder of the Leningrad Institute of Physics and Technology.

Ivanov, Pyotr. Soviet diplomat and NKVD spy. Third secretary at the Soviet Consulate in San Francisco.

Jensen, Johannes Hans Daniel. German physicist. Worked on isotope separation as part of the Uranverein.

Jensen, Peter Herbert. German physicist. Worked on the German atomic project with Walther Bothe.

Johnson, Lyall. American counter-intelligence officer. Former FBI agent working for Army G-2 on the Berkeley campus.

Joliot-Curie, Frédéric. French physicist and Nobel laureate. Verified nuclear fission in uranium in 1939. Active in the French resistance during the Nazi occupation of Paris.

Kapitza, Peter Leonidovich. Soviet physicist and Nobel laureate. Got involved with the Soviet atomic project in 1945 but disagreed with Lavrenty Beria about how the project was being run. Resigned from the project later that year.

Kayser, Fredrik. Norwegian commando. Member of the successful Gunnerside team.

Kennan, George Frost. American diplomat. Author of the 'long telegram' and a 1947 article on Soviet politics under the pseudonym 'X'.

Khariton, Yuli Borisovich. Soviet physicist. Scientific head of the Soviet weapons laboratory Arzamas-16. Worked on weapons design.

Kheifets, Gregori. Soviet diplomat and NKVD spy. Vice Consul at the Soviet Consulate in San Francisco.

Kikoin, Issak Konstantinovich. Soviet physicist. Worked on isotope separation.

Kistiakowsky, George Bogdan. Ukranian-born chemist and explosives expert. Worked at Los Alamos on implosion. Headed X Division.

Kjelstrup, Arne. Norwegian commando. Member of the advance Grouse/Swallow party parachuted onto the Hardanger Plateau in 1942. Supported the successful Gunnerside raid on the Vemork heavy water plant.

Konopinski, Emil John. American physicist of Polish descent. Contributed to Oppenheimer's summer study group at Berkeley and worked with Teller on the early theory of the Super thermonuclear bomb.

Korsching, Horst. German physicist. Worked on isotope separation as part of the Uranverein. Captured by the Alsos mission and detained at Farm Hall.

Kowarski, Lew. French physicist of Russian-Polish descent. Worked in Frédéric Joliot-Curie's research team in Paris. Escaped to Britain with Hans von Halban following the German occupation. Subsequently worked at the Montreal laboratory and supervised construction of Canada's first nuclear reactor at Chalk River.

Kremer, Simon. Soviet diplomat and GRU spy. Secretary to the military attaché at the Soviet embassy in London. Acted as Klaus Fuchs' Soviet controller in 1941–42.

Kuczynski, Jurgen. German economist, historian and Soviet spy. Brother of Ruth Kuczynski, Klaus Fuchs' Soviet controller in 1942–43.

Kurnakov, Sergei Nikolaevich. Soviet journalist who wrote on military affairs for the *Daily Worker*. Former Tsarist cavalry officer.

Kurchatov, Boris Vasilyevich. Soviet chemist. The first Soviet scientist to separate plutonium. Igor Kurchatov's brother.

Kurchatov, Igor Vasilyevich. Soviet physicist. Scientific head of the Soviet atomic project. Oversaw the first successful Soviet atomic bomb test in 1949.

Kvasnikov, Leonid Romanovich. Soviet diplomat and NKVD spy. 'XY' station chief at the Soviet Embassy in New York charged with gathering intelligence on atomic energy.

Lamphere, Robert Joseph. American FBI counter-intelligence agent. Worked as FBI liaison to the Venona project and pursued the leads revealed in the decrypts.

Lansdale, John, Jr. American lawyer and Army G-2 counter-intelligence expert. Head of security on the Manhattan Project.

Laue, Max Theodor Felix von. German physicist and Nobel laureate. Although a strong opponent to Nazism who had never contributed to the work of the Uranverein, he was nevertheless captured by the Alsos mission and detained at Farm Hall.

Lawrence, Ernest Orlando. American physicist and Nobel laureate. Inventor of the cyclotron. Actively supported the establishment of the American bomb project. Worked on electromagnetic isotope separation.

LeMay, Curtis Emerson. American military leader and politician. Commander of strategic air operations against Japan towards the end of the war. Took charge of the US Strategic Air Command in 1948.

Lewis, Robert Alvin. Co-pilot aboard the *Enola Gay* B-29 Superfortress.

Lilienthal, David Eli. American lawyer and public servant. Director of the Tennessee Valley Authority. Appointed to an advisory board on atomic energy in 1946. Co-author of the Acheson–Lilienthal report. Went on to become the first chairman of the US Atomic Energy Authority.

Lindemann, Frederick Alexander (Lord Cherwell). German-born British physicist and scientific adviser to Winston Churchill.

Lomanitz, Giovanni Rossi. American physicist. Worked on electromagnetic isotope separation at the Rad Lab before he was drafted in 1943.

Maclean, Donald Duart. British diplomat and Soviet spy. One of the Cambridge spy 'ring'. Appointed co-secretary of the Combined Policy Committee in Washington in 1947.

Makhnev, Vitaly A. Soviet NKVD general. Headed the secretariat of the Special State Committee formed in 1945 to develop the Soviet atomic bomb.

Malenkov, Georgei Maximilianovich. Soviet politician. Appointed to the Special State Committee formed in 1945 to develop the Soviet atomic bomb.

Marshall, George Catlett, Jr. American military leader and politician. Served as Army Chief of Staff during the war. Replaced James Byrnes as Secretary of State in the Truman administration and set out an economic recovery programme for war-devastated Europe that came to be known as the Marshall Plan.

May, Alan Nunn. British physicist and GRU spy. Worked as part of the Tube Alloys project before relocating to the Montreal laboratory in early 1943. Unmasked by the defection of Igor Gouzenko.

McCarthy, Joseph Raymond. American politician. Republican Senator. Noted for his 1950 speech denouncing Communists in the State Department and his sponsoring of fervent anti-Communism ('McCarthyism').

McMahon, Brien. American lawyer and politician. Author of the McMahon Bill and the US Atomic Energy Act, 1946.

McMillan, Edwin Mattison. American physicist. Co-discovered neptunium in 1940, for which he shared the Nobel prize in chemistry in 1951 with Glenn Seaborg.

Meitner, Lise. Austrian émigré physicist. Co-discovered the origin of nuclear fission in uranium with her nephew, Otto Frisch.

Menzies, Stewart Graham. British head of the Secret Intelligence Service (also known as MI6).

Milch, Erhard. German field marshal responsible for the development of the Luftwaffe and mass production of German 'vengeance weapons'.

Møller, Christian. Danish physicist. Worked at Niels Bohr's Institute for Theoretical Physics.

Molotov, Viacheslav Mikhailovich. Soviet politician. First Deputy Chairman of the Soviet Council of Ministers. Co-signatory of the Molotov–Ribbentrop pact. Supervised the early Soviet atomic programme.

Morrison, Philip. American physicist. Joined the Manhattan Project in 1942.

Neddermeyer, Seth Henry. American physicist. Suggested implosion as a means of creating a super-critical mass. Worked on implosion at Los Alamos until 1944.

Nelson, Stephen (aka Mesarosh, Stephen). Croatian-born political activist and Soviet spy. Key figure in the American Communist Party, based in Oakland, California.

Neumann, John von. Hungarian émigré mathematician. Worked on the design of explosive lenses required to produce symmetrical implosion in the Fat Man bomb. Went on to become a Cold War 'hawk' and one of the models for Peter Sellers' portrayal of Dr Stangelove.

Nichols, Kenneth David. American engineer and aide to Manhattan Project head Leslie Groves. Subsequently became General Manager of the US Atomic Energy Commission.

Nier, Alfred Otto Carl. American physicist. Worked on the fission properties of U-235.

Oliphant, Marcus 'Mark' Lawrence Elwin. Australian physicist. Instrumental in drawing attention to the Frisch–Peierls proposals concerning critical mass for a uranium bomb and active participant on the MAUD Committee and Tube Alloys project. Joined the work on electromagnetic isotope separation at the Rab Lab in 1943.

Oppenheimer, Frank Friedman. American physicist. Worked on preparations for the Trinity test. Brother of J. Robert Oppenheimer.

Oppenheimer, Julius Robert. American physicist. Scientific head of the Los Alamos laboratory and widely acknowledged as the 'father of the atom bomb'. Chairman of the US Atomic Energy Authority's General Advisory Committee. Campaigned for international control of atomic weapons. Lost his security clearance in 1954.

Ovakimyan, Gaik Badalovich. Soviet NKVD station chief in New York, responsible for managing several spy rings. Exposed as a spy in 1941, he returned to Moscow to head the NKVD's American desk.

Parsons, William Sterling 'Deke' or 'Deak'. American engineer, Navy captain and head of the Manhattan Project's ordnance division. Oversaw delivery of the first atomic bombs to Tinian Island. Armed the Little Boy bomb dropped on Hiroshima.

Pash, Boris T. American Army intelligence officer. Responsible for counter-intelligence for the Manhattan Project. Military head of the first and second Alsos missions.

Pauli, Wolfgang Ernst. Austrian émigré physicist and Nobel laureate. Emigrated to America in 1940 and became a naturalised US citizen in 1946.

Pavlov, Vasily. Soviet diplomat and NKVD spy. Operated the NKVD spy rings from the Soviet Embassy in Ottawa.

Peierls, Rudolf Ernst. German émigré physicist. With Otto Frisch, co-authored the Frisch–Peierls memorandum on critical mass. Joined the British Tube Alloys delegation to the Manhattan Project in late 1943. Knighted in 1968.

Penney, William George. British physicist. Joined the British Tube Alloys delegation to the Manhattan Project in late 1943. Returned to Britain in 1946 to become Chief Superintendent Armament Research at Fort Halstead in Kent and went on to build Britain's first atomic bomb.

Perrin, Michael Willcox. British chemist and industrialist. Patented the first industrial method for producing polyethylene in 1935. Joined Tube Alloys in 1940 to assist Wallace Akers.

Pervukhin, Mikhail Georgievich. Soviet politician. People's Commissar for the Chemical Industry.

Philby, Harold Adrian Russell 'Kim'. British intelligence officer and NKVD spy. One of the Cambridge spy 'ring'. Held various positions in the SOE and SIS before becoming head of Section IX, responsible for Soviet counter-intelligence.

Phillips, Cecil J. American cryptanalyst at the Army Signals Intelligence Service. Identified a pattern in coded Soviet messages which provided a crucial breakthrough.

Placzek, George. Czech physicist. Joined the British Tube Alloys delegation and headed the Theoretical Division at the Montreal laboratory from 1943–45. Replaced Hans Bethe as head of the Theoretical Division at Los Alamos in 1945.

Planck, Erwin. German politician. Son of Max Planck. Involved in the 20 July 1944 plot to assassinate Adolf Hitler. Subsequently executed.

Planck, Karl Ernst Ludwig Marx 'Max'. German physicist. Discovered the quantisation of energy in 1900 and remained an influential figure in German science through two world wars.

Pontecorvo, Bruno. Italian physicist and Soviet spy. Worked in Enrico Fermi's team in Rome. Joined the Montreal laboratory in 1943, where he worked on reactor design. Defected to the Soviet Union with his family in 1950.

Poulsson, Jens Anton. Norwegian commando. Leader of the advance Grouse/Swallow party which parachuted onto the Hardanger Plateau in 1942. Supported the successful Gunnerside raid on the Vemork heavy water plant.

Rabi, Isidor Isaac. Galician-born émigré physicist and Nobel laureate. Associate director of the MIT Radiation Laboratory, which worked on radar during the war. Served as a visiting consultant to Los Alamos and as a member of the US Atomic Energy Authority's General Advisory Committee.

Reiche, Fritz. German physicist. Carried Fritz Houtermans' warning message to America.

Ribbentrop, Ulrich Friedrich Wilhelm Joachim von. German Foreign Minister, 1938–45.

Riehl, Nikolaus. Russian-born German industrial chemist. Worked on the production of uranium at the Auer company's Oranienburg plant. Captured by the Soviets in 1945. Riehl contributed to the Soviet atomic programme for the next ten years.

Robb, Roger. American lawyer. Prosecuting attorney in Oppenheimer's security clearance hearing.

Rønneberg, Joachim Holmboe. Norwegian commando. Led the successful Gunnerside raid on the Vemork heavy water plant.

Rosbaud, Paul. Austrian chemist, editor of the scientific journal *Die Naturwissenschaften*, adviser to German publisher Springer Verlag and an agent for the British SIS. Helped Lise Meitner escape from Nazi Germany.

Rosenberg, Julius. American engineer and Soviet spy. Acted as a courier and recruited a network of industrial spies, including his brother-in-law, Los Alamos machinist David Greenglass. Executed with his wife Ethel in 1953.

Rosenfeld, Léon. Belgian physicist. Collaborated with Niels Bohr and worked at Bohr's Institute for Theoretical Physics in Copenhagen.

Rotblat, Joseph. Polish physicist. Worked with James Chadwick in Liverpool and joined the British delegation to the Manhattan Project in early 1944. Resigned from the project in 1945 when it became obvious that there was no threat from a Nazi weapon. A noted campaigner for nuclear disarmament, he became secretary general of the Pugwash Conferences on Science and World Affairs and in 1995 won the Nobel peace prize.

Rozental, Stefan. Polish physicist. Emigrated to Denmark in 1938 and became Niels Bohr's personal assistant.

Sachs, Alexander. American economist and banker. Delivered Einstein's letter to US President Franklin Roosevelt in 1939.

Sakharov, Andrei Dmitrievich. Soviet physicist. Led the development of the Soviet Union's first thermonuclear weapons. Joined Arzamas-16 in 1950. Subsequently became a noted campaigner against nuclear weapons proliferation and civil rights activist. Won the 1975 Nobel peace prize.

Sato, Naotake. Japanese diplomat. Ambassador to the Soviet Union.

Sax, Saville. American teacher and Soviet spy. Friend of Theodore Hall. Acted as courier for Hall.

Scherrer, Paul. Swiss physicist. Acted as an informant for both the British SIS and American OSS.

Schumann, Erich. German physicist and administrator. Grandson of composer Robert Schumann. Worked for German Army Ordnance and supervised the German atomic programme from 1939 to 1942.

Seaborg, Glenn Theodore. American chemist. Pioneer of nuclear chemistry. Developed chemical methods for separating plutonium and went on to discover and co-discover many new elements. Shared the 1951 Nobel prize in chemistry with Ed McMillan. Became chairman of the US Atomic Energy Commission in 1961.

Segrè, Emilio Gino. Italian émigré physicist. Worked in Enrico Fermi's research team in Rome. Emigrated to America in 1938 and joined Ernest Lawrence's team at the Rad Lab. Worked at Los Alamos on problems relating to spontaneous fission in U-235 and plutonium. Won the 1959 Nobel prize for physics.

Serber, Robert. American physicist. Former student of J. Robert Oppenheimer. Worked on aspects of atomic weapons design at Los Alamos and was part of the scientific team that assembled the bombs on Tinian Island and prepared them for delivery. Author of the 'Los Alamos Primer'.

Siegbahn, Karl Manne Georg. Swedish physicist and Nobel laureate. Provided a research post and laboratory facilities to Lise Meitner following her escape from Germany.

Silva, Peer de. American counter-intelligence operative with Army G-2.

Simon, Franz Eugen. German émigré chemist. Worked on the gaseous diffusion technique for separating U-235 as part of the MAUD Committee and Tube Alloys project. Knighted in 1954.

Skinnarland, Einar. Norwegian SOE operative. Wireless operator for the commando raids on the heavy water plant at Vemork.

Slater, John Clarke. American physicist. Contributed to the National Academy review group.

Slotin, Louis Alexander. Canadian physicist. Suffered a fatal dose of radiation in an accident at Los Alamos in 1946.

Smyth, Henry DeWolf. American physicist. Author of the Smyth report, the first official history of the Manhattan Project.

Snow, Charles Percy. British physicist and novelist. Held several technical positions in the British government from 1940–60. Knighted in 1957 and made a life peer in 1964.

Sørlie, Rolf. Norwegian engineer. Assisted the Gunnerside team in their successful raid on the heavy water plant at Vemork. Involved in sabotage of the *Hydro*.

Speer, Albert. German architect and Minister for Armaments and War Production.

Stimson, Henry Lewis. American politician. Secretary of War in both the Roosevelt and Truman administrations, 1940–45.

Storhaug, Hans. Norwegian commando. Member of the successful Gunnerside team.

Strassman, Friedrich Wilhelm 'Fritz'. German chemist. Discovered (with Otto Hahn) evidence for nuclear fission in uranium.

Strauss, Lewis Lichtenstein. American businessman, Naval officer and administrator. Became chairman of the US Atomic Energy Commission in 1953.

Strømsheim, Birger. Norwegian commando. Member of the successful Gunnerside team.

Sudoplatov, Pavel Anatolyevich. Soviet NKVD lieutenant general. Responsible for 'special tasks', including sabotage and assassinations, 1941–44. Appointed to lead Department S, a joint NKVD–GRU atomic intelligence-gathering operation in 1945.

Suzuki, Kantaro. Japanese Prime Minister April–August 1945.

Szilard, Leo. Hungarian émigré physicist. Anticipated the development of atomic weapons and worked to alert US President Franklin Roosevelt to the danger in 1939. Involved in the fledgling American bomb project, he joined the Met Lab in early 1942. In 1945 he vigorously campaigned against the first use of atomic bombs by America.

Tamm, Igor Yevgenyevich. Soviet physicist. Studied the theory of the Super thermonuclear bomb. Won the Nobel prize for physics in 1958.

Tatlock, Jean. American psychologist and physician. Former fiancée and lover of J. Robert Oppenheimer. She committed suicide in 1944.

Teller, Edward. Hungarian émigré physicist. Involved in the American atomic bomb project from its inception. Worked at Los Alamos on the theory of the Super thermonuclear bomb. Co-founded the Lawrence Radiation Laboratory at Livermore (subsequently called the Lawrence Livermore National Laboratory) in 1952. He remained an advocate of strong American military security, inspiring Ronald Reagan's Strategic Defense Initiative in 1983. Teller is another possible model for Peter Sellers' portrayal of Dr Stangelove.

Terletsky, Yakov Petrovich. Soviet physicist. Scientific adviser to Department S, a joint NKVD–GRU atomic intelligence-gathering operation. Visited Niels Bohr in 1945.

Tibbets, Paul Warfield. American pilot. Commander of the 509th Composite Group. Piloted the B-29 Superfortress bomber *Enola Gay* which dropped the first atom bomb on Hiroshima.

Tizard, Henry Thomas. British chemist and administrator. Chairman of the Aeronautical Research Committee, responsible for the development of radar during the war.

Thomson, George Paget. British physicist and Nobel laureate. Son of J.J. Thomson. Chairman of the MAUD Committee. Knighted in 1943.

Togo, Shigenori. Japanese Minister of Foreign Affairs.

Toyoda, Soemu. Japanese Imperial Navy admiral and Navy Chief of Staff.

Tronstad, Leif. Norwegian chemist. Developed the process for heavy water production at the Vemork plant. Joined the Norwegian resistance in 1941 and, following his escape to England, advised the SOE on sabotage operations. Killed in Norway in 1945 while on a mission to protect local power plants and industrial facilities from the retreating Germans.

Tsukerman, Veniamin Aronovich. Soviet physicist. Developed techniques for the X-ray radiography of explosions used to investigate implosion at Arzamas-16.

Tuck, James Leslie. British physicist. Expert on armour-piercing shaped charges. Joined the British delegation to the Manhattan Project in late 1943. Suggested implosion using explosive lenses as a method for assembling a super-critical mass of plutonium in the Fat Man bomb design.

Turner, Louis A. American physicist. Independently concluded that neutron capture by U-238 could lead to the production of a fissionable isotope of plutonium.

Ulam, Stanislaw Marcin. Polish mathematician. Worked on the hydrodynamics of implosion and the theory of the Super thermonuclear bomb at Los Alamos. Proposed the Teller–Ulam design in 1951.

Umezu, Yoshijiro. Japanese Imperial Army general and Army Chief of Staff.

Urey, Harold Clayton. American chemist and Nobel laureate. Worked on gaseous diffusion.

Van Vleck, John Hasbrouk. American physicist. Contributed to the National Academy review group and Oppenheimer's summer study group at Berkeley. Won the Nobel prize in 1977.

Vannikov, Boris Lvovich. Soviet politician. People's Commissar of Munitions. Joined the Special State Committee formed in 1945 to develop the Soviet atomic bomb.

Voznesensky, Nikolai Alexeyevich. Soviet politician. Head of the Soviet State Planning Committee (Gosplan). Joined the Special State Committee formed in 1945 to develop the atomic bomb.

Wahl, Arthur. American chemist. Worked with Glenn Seaborg to isolate plutonium.

Weinberg, Joseph. American physicist. Worked on electromagnetic isotope separation at the Rad Lab until 1943. Was caught betraying atomic secrets to the Soviet Union by an FBI bug. Acquitted in 1953 when HUAC refused to make evidence available in court.

Weizsäcker, Carl Friedrich Freiherr von. German physicist and philosopher. Son of Ernst von Weizsäcker and brother of Richard von Weizsäcker, German President 1984–94. Joined the Uranverein in 1939 and independently concluded that neutron capture by U-238 could lead to the production of a fissionable isotope of neptunium (and, subsequently, plutonium). Close friend of Werner Heisenberg. Captured by the Alsos mission and detained at Farm Hall.

Weizsäcker, Ernst Freiherr von. German diplomat. Secretary of State under Foreign Minister Joachim von Ribbentrop.

Welsh, Eric. British SIS agent. Supervised atomic intelligence-gathering and sabotage of the heavy water plant at Vemork. Followed the Alsos mission into Germany.

Wheeler, John Archibald. American physicist. Elaborated the origin of nuclear fission in uranium with Niels Bohr in 1939. Joined the Met Lab and worked on the first full-scale nuclear reactors at Hanford. Subsequently made many contributions to quantum theory, relativity and cosmology.

Wigner, Eugene Paul. Hungarian émigré physicist. Part of the 'Hungarian conspiracy' that helped to raise the profile of atomic energy in America in 1939. Worked on reactor design at the Met Lab in Chicago. Subsequently directed research and development at the Clinton laboratory (now Oak Ridge National Laboratory) in Tennessee. Won the Nobel prize for physics in 1963.

Wirtz, Karl Eugen Julius. German physicist. Worked on reactor design as part of the Uranverein. Captured by the Alsos mission and detained at Farm Hall.

Yatskov, Anatoly Antonovich. Soviet diplomat and NKVD spy. Senior case officer for a number of Soviet spy rings in New York. Supervised the handling of key atomic spies, including Klaus Fuchs, Theodore Hall and David Greenglass.

Yonai, Mitsumasa. Japanese Imperial Navy admiral and Navy Minister.

Zabotin, Nikolai. Military attaché at the Soviet Embassy in Ottawa and GRU spy. The boss of cipher clerk Igor Gouzenko, who defected in 1945.

Zarubin, Vasily Mikhailovich. Soviet diplomat and NKVD spy. Third secretary at the Soviet Embassy in Washington, 1941–44. Recalled to Moscow following allegations made in a letter to FBI director J. Edgar Hoover.

Zavenyagin, Avram Pavlovich. Soviet politician. Deputy People's Commissar of Internal Affairs 1941–50. Led the Soviet search mission into Germany in 1945. Joined the Special State Committee formed later that year to develop the Soviet atomic bomb.

Zeldovich, Yakov Borisovich. Soviet physicist. Worked on the theory of nuclear chain reactions with Yuli Khariton. Joined the Soviet atomic weapons laboratory Arzamas-16 in 1946. Worked on the design of the 'classical' Super thermonuclear bomb.

Zernov, Pavel Mikhailovich. Deputy People's Commissar of the Tank Industry. First head of the Soviet atomic weapons laboratory Arzamas-16.

NOTES AND SOURCES

PROLOGUE: LETTER FROM BERLIN
4   'The most momentous visit of my whole life': Frisch, p. 114.
7   'I don't believe it ...': Rhodes, *The Making of the Atomic Bomb*, p. 257.
7   'But in nuclear physics ...': Sime, p. 235.
8   'A very wobbly, unstable drop ...': Frisch, p. 116.
9   'Oh what idiots we have all been ...': Frisch, p. 116.
13   'Instantly pronounced the reaction impossible ...': Alvarez, p. 75.
14   'What kind of crazy thing is this ...': Wheeler, p. 27.
15   'Yes it would be possible to make a bomb ...': Wheeler, p. 44.
16   'We take the liberty of calling your attention to ...': Irving, p. 32.
18   'When things got too bad.': Lanouette, p. 112.
19   'Entered history as Szilard's chauffeur.': Goodchild, *Edward Teller*, p. 52.
19   'Extremely powerful bombs of a new type.': Einstein letter to Roosevelt, 2 August 1939. The letter is reproduced in Snow, p. 178. See also www.atomicarchive.com

CHAPTER 1: THE URANVEREIN
24   'How secure the "White Jews" feel ...': Heisenberg, Elisabeth, p. 47.
25   'Now I actually see no other possibility ...': Cassidy, p. 384.
25   'We mothers know nothing about politics ...': Cassidy, p. 386.
26   'Heisenberg is decent ...': Cassidy, p. 393.
27   'I believe that the war will be over ...': Heisenberg, Werner, p. 170.
27   'People must learn to prevent catastrophes ...': Heisenberg, Werner, p. 171. Heisenberg attributes his comment to a conversation he had with Enrico Fermi.
28   'We must make use of warfare for physics ...': Cassidy, p. 427.
32   'The only method of producing explosives ...': Irving, p. 53.
34   'Say that our company will accept ...': Irving, p. 65.
38   'Had Nazis in the Institute.': Irving, p. 57.

CHAPTER 2: ELEMENT 94
41   'Bombs of hitherto unenvisaged potency and scope ...', and 'can only hope ...': Rhodes, *The Making of the Atomic Bomb*, p. 314.
42   'Alex, what you are after ...': Rhodes, *The Making of the Atomic Bomb*, p. 314.
42   'My friends blamed me ...': Rhodes, *The Making of the Atomic Bomb*, p. 316.
43   'Gentlemen, armaments are not what decides war ...': Szanton, p. 203.
43   'If that is true ...': Szanton, p. 203.
46   'You wouldn't put boron into your graphite, or would you?': Lanouette, p. 222.
46   'Fermi and I had disagreed ...': Lanouette, p. 218.
51   'It seems as if it was wild enough ...': Rhodes, *The Making of the Atomic Bomb*, p. 346.

51 'With this remark of Turner ...': Lanouette, p. 220.
53 'Undoubtedly a fascist ...' and further quotations: Lanouette, p. 223.
54 'At that time ...': Rhodes, *The Making of the Atomic Bomb*, p. 335.
55 'You who are scientists ...': Rhodes, *The Making of the Atomic Bomb*, p. 336.
55 'I had the obligation ...': Goodchild, *Edward Teller*, p. 56.
56 'Before this war is over ...': Lanouette, p. 224.

CHAPTER 3: CRITICAL MASS
58 'Just like any tourist.': Frisch, p. 120.
59 'The most valuable English scientific innovation ...': Snow, p. 105.
59 'If you were faced with the problem ...', and 'Oliphant knew that Peierls knew ...': Frisch, p. 123.
59 'The result would be no worse ...': Frisch, p. 125.
61 'The work of Bohr and Wheeler ...': Peierls, p. 154.
61 'At that point we stared at each other ...': Frisch, p. 126.
61 'I had no doubt that the Nazis would not hesitate ...': Preston, p. 213.
62 'It was a terrible time for me ...': quotation taken from the profile of Joseph Rotblat on the Peace Pledge Union website: www.ppu.org.uk
63 'Practically irresistible ...' and other quotations from the Frisch–Peierls memorandum, see: www.atomicarchive.com
65 'Why start on a project ...': Frisch, p. 126.
66 'With cheerful intelligence ...': Frisch, p. 130.
68 'Met Niels and Margrethe recently ...': Sime, p. 284.
71 'His infectious good humour ...': Howard, p. 142.
72 'Not only had our report started the whole thing ...': Frisch, p. 131.

CHAPTER 4: A VISIT TO COPENHAGEN
78 'Only for special applications ...': Irving, p. 97.
81 'About ten million times': Karlsch and Walker, p. 17.
82 'Please say all this ...'; Powers, p. 107.
83 'For heaven's sake keep this under your hat ...': Kramish, p. 158.
84 'If you can assure us that it is of immediate importance ...': Kramish, p. 159.
85 'It was from September 1941 that we saw an open road ...': Irving, p. 114.
87 'It might be a good thing ...': Heisenberg, Werner, p. 181.
87 'Grave consequences in the technique of war': Rhodes, *The Making of the Atomic Bomb*, p. 384.
87 '[Heisenberg] had agreed to sup with the devil ...': Rhodes, *The Making of the Atomic Bomb*, p. 386.
88 'In Tisvilde, the beautiful vacation home of the Bohrs ...': Heisenberg, Elisabeth, p. 77.
88 'Here I am once again in the city which is so familiar to me ...' and subsequent quotations: Heisenberg, letter to Elisabeth Heisenberg, September 1941. A facsimile, transcription and English translation of this letter can be viewed online at www.werner-heisenberg.unh.edu
90 '[Heisenberg] stressed how important it was ...': Pais, *Niels Bohr's Times*, p. 483.

91 'In vague terms you spoke in a manner ...': reproduced in Dörries, p. 109. The Bohr drafts can also be viewed online at www.nba.nbi.dk

91 'It had to make a very strong impression on me ...': reproduced in Dörries, p. 163. See also www.nba.nbi.dk

91 'And that it was, so to speak, the natural course in this world ...': Helmut Rechenberg, 'Documents and Recollections of the Bohr–Heisenberg Meeting in 1941', in Dörries, p. 69.

92 'You know, I'm afraid it went badly wrong.': Rechenberg in Dörries, p. 70.

92 'Our relations with scientific circles in Scandinavia ...': Rechenberg in Dörries, p. 69.

CHAPTER 5: TUBE ALLOYS

93 'The first test of theory ...': Gowing, *Britain and Atomic Energy*, p. 67.

99 'Put a question in, and you get an answer out ...': Moss, p. 48.

100 'We have now reached the conclusion ...' and subsequent quotations from the MAUD Committee Report: Gowing, *Britain and Atomic Energy*, p. 394. See also www. atomicarchive.com

101 'I would not bet more than two-to-one against ...': Gowing, *Britain and Atomic Energy*, p. 96.

101 'Although personally I am quite content ...': Gowing, *Britain and Atomic Energy*, p. 106.

102 'We have been impressed by the unanimity ...': Gowing, *Britain and Atomic Energy*, p. 100.

103 'VADIM has relayed a report ...': from Report No. 6881/1065 of 25 September 1941, from the archives of the Foreign Intelligence Service of Russia. This report is available as Document No. 1 in Appendix Two of Sudoplatov, p. 437. In this memoir, it was claimed that Gorsky's source was LEAF and incorrectly attributed to Donald Maclean. This was either a possible mistranslation of LIST or deliberate misinformation.

103 'The Chiefs of Staff Committee ...': from Report No. 6881/1065 of 25 September 1941, from the archives of the Foreign Intelligence Service of Russia. See Sudoplatov, p. 437.

104 'If Congress knew the true history of the atomic energy project ...': quoted in Rhodes, *The Making of the Atomic Bomb*, p. 372.

105 'Light a fire under the Briggs Committee': Rhodes, *The Making of the Atomic Bomb*, p. 360.

106 'Energetic but dispassionate review ...': Rhodes, *The Making of the Atomic Bomb*, p. 362.

108 'If large amounts of element 94 were available ...': quoted in Compton, p. 50.

108 'This inarticulate and unimpressive man ...': Rhodes, *The Making of the Atomic Bomb*, p. 372.

109 'Said "bomb" in no uncertain terms ...': Rhodes, *The Making of the Atomic Bomb*, p. 373.

109 'Ernest, you say you are convinced ...': Compton, p. 8.

110 'A fission bomb of superlatively destructive power ...': Rhodes, *The Making of the Atomic Bomb*, p. 386.

111 'V.B. OK – returned – I think you had best keep this in your own safe. FDR.': Rhodes, *The Making of the Atomic Bomb*, p. 388.

112 'Assist one another with all political, economic and military means ...': the text of the Tripartite Pact can be viewed on the Yale Law School Avalon Project website – www.yale.edu/lawweb/avalon

113 'When this war is over ...': quoted by Dennis Showalter, 'Storm over the Pacific: Japan's Road to Empire and War' in Marston, p. 29.

CHAPTER 6: A MODEST REQUEST

119 'Only if there is a certainty ...': Irving, p. 117.

120 'A million times greater than the same weight of dynamite' and '10 to 100 kilograms of fissionable material': Powers, p. 136.

122 'As I will not be in Berlin at the time in question ...': Irving, p. 119.

122 'The behaviour of neutrons in uranium ...': Irving, p. 120.

122 'Pure uranium-235 is thus seen to be an explosive of quite unimaginable force.': Irving, p. 121.

123 'Realm of atomic destruction': Mark Walker, *German National Socialism*, p. 58.

124 'With the aid of Norwegian compatriots ...': Gallagher, p. 28.

124 'So we have at last succeeded in building a pile configuration ...': Irving, p. 131.

125 'I have confidence in you ...': Speer, p. 276.

125 'Annihilate whole cities': Speer, p. 314.

126 'Given the positive results achieved up until now ...': Karlsch and Walker, p. 16.

126 'About the size of a pineapple': Irving, p. 134.

127 'It was such a ridiculously low figure ...': Irving, p. 349.

128 'Rather put out by these modest requests ...': Speer, p. 316.

131 'No orders were given to build atom bombs ...': Heisenberg, Werner, p. 183.

131 'Hurry up. We are on the track.': Szanton, p. 241.

131 'We have become convinced that there is a real danger ...': Powers, p. 162.

133 'The valley is so deep ...': Gallagher, p. 31.

134 'We not only learned to force locks ...': Haukelid, p. 43.

135 'It was always curious to the Norwegians ...': Haukelid, p. 49.

135 'In good weather, it would have taken us a couple of days ...': Mears, p. 61.

135 'Lake covered with ice and partly covered with snow ...': Gallagher, p. 53.

CHAPTER 7: THE ITALIAN NAVIGATOR

137 'We'll have the chain reaction going here [in Chicago] by the end of the year' and subsequent quotations: Compton, p. 81.

137 'Except for this afterthought ....': Compton, p. 71.

138 'Even this piece of information was not to be divulged ...': Fermi, p. 176.

139 'We considered names like extremium and ultimium ...': Seaborg, p. 72.

139 'I like to say that she was so efficient as a secretary that I began to date her ...': Seaborg, 'An Early History of LBNL', http://acs.lbl.gov

140 'At worst in Canada.': Brown, p. 218.

140 'One thing is clear ...': Gowing, *Britain and Atomic Energy*, p. 131.

141 'To embark on this Napoleonic approach ...': Rhodes, *The Making of the Atomic Bomb*, p. 407.

142 'We must, however, face the fact ...': minute from Sir John Anderson to the Prime Minister, 30 July 1942, reproduced in Gowing, *Britain and Atomic Energy*, p. 437.

143 'A specialist in the problems of nuclear physics ...': Compton, p. 125.

144 'I had had a continuing, smouldering fury ...': Pais, *J. Robert Oppenheimer: A Life*, p. 36.

146 'Although Mici and I were both [American] citizens ...': Goodchild, *Edward Teller*, p. 61.

148 'At this point something remarkable happened ...': Serber, p. 71.

149 'Should have a destructive effect equivalent ...': Rhodes, *The Making of the Atomic Bomb*, p. 421.

151 'Alternatively, we may take the stand ...': Lanouette, p. 236.

151 'Probably the angriest officer in the United States Army.': Rhodes, *The Making of the Atomic Bomb*, p. 425.

151 'If you do the job right ...': Groves, p. 4.

152 'The biggest sonovabitch I've ever met in my life ...': Rhodes, *The Making of the Atomic Bomb*, p. 426.

152 'I fear we are in the soup.': Rhodes, *The Making of the Atomic Bomb*, p. 427.

152 'Before German bombs wipe out American cities.': Lanouette, p. 232.

153 'On the night of November 19–20th ...': Haukelid, p. 59.

156 'Congratulations? What for?' and subsequent quotations: Fermi, p. 177.

157 'We were in the high intensity regime and the counters were unable to cope with the situation anymore ...': Rhodes, *The Making of the Atomic Bomb*, p. 440.

157 'Jim, you'll be interested to know ...': Compton, p. 144.

158 'I shook hands with Fermi ...': Lanouette, p. 245.

CHAPTER 8: LOS ALAMOS RANCH SCHOOL

161 'Importance and urgency of the practical utilization ...': Sudoplatov. Beria's memorandum is reproduced in Appendix 2, pp. 439–41.

163 'How can you work with people like that?': Lanouette, p. 238.

163 'There are no experts ...': Pais, *J. Robert Oppenheimer: A Life*, p. 40.

165 'He couldn't run a hamburger stand.': Alvarez, p. 78.

165 'An important weapon that was being developed.': Goodchild, *J. Robert Oppenheimer*, p. 66.

167 'Nobody could think straight in a place like that ...': Rhodes, *The Making of the Atomic Bomb*, p. 451.

169 'Had thought it best to explain the whole situation to us ...': Haukelid, p. 71.

169 'None of us had been to the plant in our lives ...': Mears, p. 118.

172 'I almost killed him ...': Gallagher, p. 148.

172 'The explosion itself was not very loud ...': Mears, p. 163.

173 'The explosion was tremendous ...': Gallagher, p. 160.

174 'The finest coup I have seen in this war.': Gallagher, p. 169.

174 'Gives sufficient basis in itself for insistence ...': Kramish, p. 170.

175 'Then I decided to give him our intelligence materials.': Holloway, p. 90.

176 'The prospects of this direction are unusually captivating.': Sudoplatov. Kurchatov's memoranda are reproduced in Appendix 2, pp. 446–53.

176 'They fill in just what we are lacking.': Holloway, p. 95.

176 'In this connection I am asking you to instruct Intelligence Bodies ...': Sudoplatov, p. 453.

CHAPTER 9: ЭНОРМОЗ

180 'Derives from the basic principle that interchange on design and construction...': Gowing, *Britain and Atomic Energy*, p. 156. Emphasis added.

183 'When you know you are being taken advantage of...': Feklisov, p. 30.

184 'Do you know any of the guys or any others connected with it?': Herken, p. 92.

185 'Eltenton's manner was somewhat embarrassed...': Chevalier, p. 53.

185 'Means of getting technical information to Soviet scientists.': Pais, *J. Robert Oppenheimer: A Life*, p. 237.

185 'I was not, of course, in the kitchen...': Barbara Chevalier, unpublished diaries. Extracts have been reproduced with notes by Gregg Herken and are available to view at www.brotherhoodofthebomb.com. See also Bird and Sherwin, p. 197.

186 'To my sorrow, his wife is influencing him in the wrong direction.': Herken, p. 96.

189 'The object of the project...': *Los Alamos Primer*. A scanned version of the original document is available on the US Department of Energy website, www.cfo.doe.gov.

191 'For example, it has been suggested that the pieces might be mounted...': *Los Alamos Primer*, p. 22.

192 'From the preceding outline we see that the immediate program...': *Los Alamos Primer*, p. 24.

193 'I believe your people actually *want* to make a bomb.': Rhodes, *The Making of the Atomic Bomb*, p. 468.

CHAPTER 10: ESCAPE FROM COPENHAGEN

200 'Indeed I have in mind a particular problem ...': Pais, *Niels Bohr's Times*, p. 486.

200 'Professor Bohr should gently file the keys ...': Kramish, p. 192. A photograph of the keys and a facsimile of the instructions are reproduced in R.V. Jones, 'Meetings in Wartime and After', in French and Kennedy (eds), p. 279.

200 'However, there may, and perhaps in a near future, come a moment ...': Brown, p. 243.

203 'Somewhat improved apparatus ...': Irving, p. 202.

204 'Oh, I think that is true ...': Bird and Sherwin, p. 238.

205 'Could you give me a little more specific information ...' and 'Well, I might say, that the approaches were always to other people ...': Goodchild, *J. Robert Oppenheimer*, p. 93.

205 'Who had a lot of experience in microfilm work ...', 'I think it would be a mistake ...' and 'I want to again sort of explore the possibility ...': Bird and Sherwin, p. 240.

207 'The writer wishes to go on record ...'; Conant, p. 177.

208 'The case of Dr J.R. Oppenheimer ...', 'I've made up my mind that you, yourself, are OK', and 'I'd better be – that's all I've got to say.': Goodchild, *J. Robert Oppenheimer*, p. 95.

216 'Scores of American bombers were flying across Norway ...': Haukelid, p. 177.

217 'Seems out of all proportion ...': Gallagher, p. 211.

217 'I would like that very much' and subsequent quotations: Frisch, p. 145.

217 'I will have nothing to do with a bomb.': Sime, p. 305.

218 'Oh, about once a year.': Kragh, p. 158

219 'Some of us got seasick ...': Frisch, p. 147.

CHAPTER 11: UNCLE NICK

224 'Someone obviously has it in for you.': Chevalier, p. 58.

225 'Got paralyzed somehow.': Bird and Sherwin, p. 250.

228 'Good evening, Mrs von Halban' and subsequent quotations: Serber, p. 85. Note, however, that Peierls thought this encounter rather apocryphal. See Peierls, p. 188.

228 'To throw a reactor down on London.': Rhodes, *The Making of the Atomic Bomb*, p. 524.

228 'They did not need my help ...': Pais, *Niels Bohr's Times*, p. 496.

229 'He made the enterprise seem hopeful ...': Rhodes, *The Making of the Atomic Bomb*, p. 524. This is a quotation from an edited version of a post-war Oppenheimer speech.

229 'We must hear all the rumours ...': Conant, p. 207.

229 'The sense of betrayal of an ally.': Brown, p. 262.

232 'I think that he really realised that the other person knew ...': Bird and Sherwin, p. 219.

234 'So finally they sent me a note ...': Feynman, p. 117.

236 'Although Hans did not criticise me directly ...': Goodchild, *Edward Teller*, p. 88.

237 'One of the blackest comedies of the war.': Snow, p. 112.

237 'I cannot see what you are talking about ...': Rhodes, *The Making of the Atomic Bomb*, p. 530.

237 'We are in a completely new situation ...': Rhodes, *The Making of the Atomic Bomb*, p. 532.

238 'He was only too conscious ...': Snow, p. 116.

CHAPTER 12: MORTAL CRIMES

240 'Can you tell me the way to Grand Central Station?' and subsequent quotations: Moss, p. 64.

241 'Somewhere in Mexico.': Albright and Kunstel, p. 79.

243 'It appears reasonable ...': Rhodes, *The Making of the Atomic Bomb*, p. 548.

244 'We finally arrived at the conclusion ...': Conant, p. 228.

244 'I think [Oppenheimer] felt very badly ...': Bird and Sherwin, p. 280.

244 'I am old, I am tired, and I am disgusted.': Conant, p. 210.

244 'The two never agreed about anything ...': Rhodes, *The Making of the Atomic Bomb*, p. 543.

246 'Oppenheimer lit into me ...': Goodchild, *J. Robert Oppenheimer*, p. 116.

246 'Parsons was furious ...': Rhodes, *The Making of the Atomic Bomb*, p. 549.

249 'But in our country, in spite of great progress ...': Holloway, p. 102

249 'Forestalling a fateful competition ...': Pais, *Niels Bohr's Times*, p. 501.

250 'Roosevelt agreed that an approach to the Soviet Union ...': Rhodes, *The Making of the Atomic Bomb*, p. 536.

250 'Enquiries should be made regarding the activities of Professor Bohr ...': Rhodes, *The Making of the Atomic Bomb*, p. 537.

251 'It seems to me Bohr ought to be confined …': Pais, *Niels Bohr's Times*, p. 502.
253 'Thinking back to the rather arrogant 19-year-old I then was …': Albright and Kunstel, p. 89.
253 'Do you understand what you are doing?' and subsequent quotations: Haynes and Klehr, p. 315.
255 'Victory shall be ours …': Rhodes, *Dark Sun*, p. 136.
255 'I have been very reticent in my writing …': Rhodes, *Dark Sun*, p. 137.
256 'Most certainly will be glad to be part of the community …': Rhodes, *Dark Sun*, p. 138.
257 'On the basis of theoretical data …': Albright and Kunstel, p. 109.

CHAPTER 13: ALSOS AND AZUSA
260 'No practical utilisation of fission chain reactions …': Powers, p. 305.
262 'We told people generally what to look for …': Dawidoff, p. 162.
263 'Deny the enemy his brain.': Powers, p. 266.
264 'Schwindel.': Irving, p. 234.
266 'Out of their comfortable quarters.': Powers, p. 339.
269 'Colonel, looks like you and I are going to have to reach an understanding' and subsequent quotations: Pash, pp. 31–2. Pash does not name Berg as the captain on the receiving end of his abuse.
271 'For the sake of method.': Irving, p. 275.
273 'Some valuable assets, some liabilities.': Goudsmit, p. 15.
274 'Thank God I didn't know them personally …': Irving, p. 305.
275 'We found references to "special metal" …': Goudsmit, p. 69.
275 'We've got it' and subsequent quotations: Pash, p. 157.
276 'Isn't it wonderful that the Germans have no atom bomb?' and subsequent quotations: Goudsmit, p. 76.
277 'Gun in my pocket' and subsequent quotations: Powers, p. 392.
277 'If anything Heisenberg said convinced him …': Powers, p. 393.
277 'As I listen, I am uncertain …': Powers, p. 399.
278 'They're coming on!': Powers, p. 400.
278 'Yes, but it would have been so good …': Powers, p. 402.

CHAPTER 14: THE FINAL PUSH
279 'Welcome to Los Alamos …': Frisch, p. 150
280 'A terrible scientific blunder.': Rhodes, *The Making of the Atomic Bomb*, p. 552.
281 'Like tickling the tail of a sleeping dragon' and subsequent quotations: Frisch, p. 159.
283 'The reactor was not exactly following the script,': Wheeler, p. 54.
283 'It was as if the engine of your car got sick …': Wheeler, p. 54.
286 'On one occasion I was forced to say to Oppie …': Goodchild, *J. Robert Oppenheimer*, p. 134.
287 'Man is a creature whose substance is faith …': Rhodes, *The Making of the Atomic Bomb*, p. 614.
287 'Is there anything I can do for you?': www.trumanlibrary.org/eleanor

288 'Again and again the officials overseeing research and development …': Walker, 'Nuclear Weapons and Reactor Research at the Kaiser Wilhelm Institute for Physics'.

288 'With the heavy stuff': Irving, p. 315.

290 'It is difficult to give such a conclusion a final assessment …': Sudoplatov, pp. 458–9.

290 'Oppenheimer was agitating to get sidewalks …': Albright and Kunstel, p. 116.

291 'All these are very valuable data …': Sudoplatov, p. 461.

292 'Grope in Germany and search there …': Oleynikov, p. 4.

292 'I am glad to have someone here to talk physics with …': Goudsmit, p. 78.

294 'And finally, suddenly and unexpectedly …': Heisenberg, Elisabeth, p. 105.

294 'I felt like an utterly exhausted swimmer …': Heisenberg, Werner, p. 191.

294 'No, I don't want to leave …' and subsequent quotations: Goudsmit, p. 112.

295 'Sad and ironic.': Goudsmit, p. 113.

296 'Haven't you brought your children with you?': Oleynikov, p. 7.

296 'Demounting and loading of everything …': Oleynikov, p. 7.

CHAPTER 15: TRINITY

301 'I was rarely as depressed as when we left Byrnes' house …': Lanouette, p. 266.

302 'The reputation of the United States …': Rhodes, *The Making of the Atomic Bomb*, p. 640.

304 'The Secretary asked what kind of inspection …' and subsequent quotations: Notes of the Interim Committee Meeting, Thursday 31 May 1945: www.nuclearfiles.org

306 'The Secretary expressed the conclusion, on which there was general agreement …': Notes of the Interim Committee Meeting, Thursday 31 May 1945: www.nuclearfiles.org

307 'All that these advantages can give us …' and subsequent quotations are taken from the Report of the Committee on Political and Social Problems (the Franck Report), 11 June 1945, which can be found at: www.atomicarchive.com

308 'In this note it was necessary …': Compton, p. 236.

309 'In spite of such disastrous damage …': Compton, p. 230.

309 'We didn't know beans about the military situation in Japan …': Bird and Sherwin, p. 300.

309 'We can propose no technical demonstration …' and subsequent quotations: Science Panel's Report to the Interim Committee, 16 June 1945, www.atomicarchive.com

310 'The Committee reaffirmed the position …' and subsequent quotations: Notes of the Interim Committee Meeting, Thursday 21 June 1945, www.nuclearfiles.org

310 'Clearly and unmistakably.': Lanouette, p. 270.

310 'I should like to have the advice of all of you …': Goodchild, *Edward Teller*, p. 103.

311 'The material has not been fully worked over …': Albright and Kunstel, p. 136.

312 'I have been guiding you idiots every step …': Albright and Kunstel, p. 137.

313 'Julius sent me': Rhodes, *Dark Sun*, p. 169.

313 'NKGB USSR received data …': Sudoplatov, p. 475. MLAD is incorrectly identified here as Bruno Pontecorvo.

314 'Making little maps …': Bird and Sherwin, p. 305.

315 You don't worry about it …': Rhodes, *The Making of the Atomic Bomb*, p. 657.

316 'I'll take another bottle of whisky.': Serber, p. 91.

317 'And then without a sound ...': Frisch, p. 164.

318 'There floated through my mind ...' and 'Oppie, now we're all sons of bitches.':
Goodchild, *J. Robert Oppenheimer*, p. 162.

318 'A new weapon of unusual destructive force.': Rhodes, *The Making of the Atomic Bomb*, p. 690.

318 'They're raising the price' and subsequent quotations: Holloway, p. 117.

CHAPTER 16: HYPOCENTRE

319 'I didn't think we were going to use bacteriological weapons ...': Walker, Stephen, p. 56.

320 'Go skimming horribly into the sea ...': Rhodes, *The Making of the Atomic Bomb*, p. 681.

321 'The form of life out here is quickly taking shape ...': Serber, p. 105.

322 'I was one of those who felt ...': quotation from *The White House Years: Mandate for Change: 1953–1956: A Personal Account*, Doubleday, New York, 1963. See www.nuclearfiles.org

323 'His Majesty the Emperor ...': Magic Diplomatic Summary, No. 1205, 13 July 1945, p. 2. National Security Archive, www.gwu.edu/~nsarchiv

323 'With regard to unconditional surrender ...': Magic Diplomatic Summary, No. 1214, 22 July 1945, p. 2. National Security Archive, www.gwu.edu/~nsarchiv

324 'The equivalent of an unconditional surrender.': Rhodes, *The Making of the Atomic Bomb*, p. 684.

325 'This weapon is to be used against Japan ...': quotation from pages from President Truman's diary, 17, 18 and 25 July 1945, www.trumanlibrary.org

326 'We call upon the government of Japan ...': from the Potsdam Declaration, www.atomicarchive.com

327 'Resolutely fight for the successful conclusion of this war.': quotation from press conference statement by Prime Minister Suzuki, 28 July 1945, www.nuclearfiles.org

327 'The war has to be brought speedily to a successful conclusion ...': from A Petition to the President of the United States, 17 July 1945, www.atomicarchive.com

329 'The moment has arrived ...': Walker, Stephen, p. 165.

329 'It is the most destructive weapon ever produced ...': Walker, Stephen, p. 168.

329 'We snickered here and back in the States ...': quotation from Sgt Abe Spitzer Collection, p. 13. Facsimiles of Spitzer's diary can be found at www.mphpa.org

331 'An elongated trash can with fins.': Rhodes, *The Making of the Atomic Bomb*, p. 701.

331 'There will be a short intermission ...': Walker, Stephen, p. 245.

332 'At that moment, a dazzling flash struck my face ...' and subsequent quotations from Shuntaro Hida's personal memoir: 'Under the Mushroom-shaped Cloud in Hiroshima'; www.wcpeace.org

334 'We ran up in a hurry ...': from Shuntaro Hida's personal memoir: 'Under the Mushroom-shaped Cloud in Hiroshima'; www.wcpeace.org

336 'Does not comprise any demand ...': Zenshiro Hoshina, *Secret History of the Greater East Asia War*, Hara-Shobo, Tokyo, 1975, pp. 139–49. This is a report of the Supreme Council meeting, held in an air-raid shelter beneath the Imperial Palace, which began

at 11:30pm on 9 August 1945. An English translation is available from the National Security Archive, www.gwu.edu/~nsarchiv

336 'From the moment of surrender ...': quotation from the Byrnes Note, www.ibiblio.com

336 'Truman said he had given orders to stop atomic bombing ...': Henry Wallace diary entry, Friday, 10 August 1945, p. 2. National Security Archive, www.gwu.edu/~nsarchiv

337 'However, the Imperial Army and Navy ...': Magic Diplomatic Summary, No. 1236, 13 August 1945, p. 3. National Security Archives, www.gwu.edu/~nsarchiv

337 'Tolerating the intolerable.': Hiroshi Shimomura, *Account of the End of the War*, Kamakura Bunko, Tokyo, 1948, pp. 148–52. This is an account of the 'Second Sacred Judgement'. An English translation is available at the National Security Archives, www.gwu.edu/~nsarchiv

337 'The enemy now possesses a new and terrible weapon ...': Imperial Rescript on Surrender, en.wikisource.org

337 'There were an awful lot of guys ...': Serber, p. 115.

338 'I still remember the feeling of unease, indeed nausea ...': Frisch, p. 176.

338 'If atomic bombs are to be added as new weapons ...': Oppenheimer's acceptance speech is reproduced in *The Oppenheimer Years, 1943–45*, Los Alamos National Laboratory, www.lanl.gov/history

CHAPTER 17: OPERATION EPSILON

339 'I wonder whether there are microphones installed here?' and subsequent quotations: Bernstein, *Hitler's Uranium Club*, p. 78.

340 'If it were only to convince ...': Goudsmit, p. 100.

340 'Here was a man ...': Goudsmit, p. 105.

341 'That's impossible ...': Bernstein, *Hitler's Uranium Club*, p. 53.

343 'It's quite possible that they just don't want to say anything' and subsequent quotations: Bernstein, *Hitler's Uranium Club*, p. 81.

344 'I am convinced [the Anglo-Americans] have used these last three months ...' and subsequent quotations: Bernstein, *Hitler's Uranium Club*, p. 102.

344 'But there are many military men in England ...' and subsequent quotations: Bernstein, *Hitler's Uranium Club*, p. 103.

344 'If the Americans have a uranium bomb ...': Bernstein, *Hitler's Uranium Club*, p. 116.

345 'Here is the news ...': BBC Written Archives Centre, reproduced in Bernstein, *Hitler's Uranium Club*, p. 357.

345 'I think it is dreadful of the Americans to have done it ...' and subsequent quotations: Bernstein, *Hitler's Uranium Club*, pp. 117–18.

347 'We wouldn't have had the moral courage ...' and subsequent quotations: Bernstein, *Hitler's Uranium Club*, p. 122.

347 'The point is that the whole structure ...' and subsequent quotations: Bernstein, *Hitler's Uranium Club*, p. 123.

348 'I think it is characteristic that the Germans ...': Bernstein, *Hitler's Uranium Club*, p. 124.

349 'Tell me, Harteck, isn't it a pity that the others have done it?' and subsequent quotations: Bernstein, *Hitler's Uranium Club*, pp. 125–6.

349 'When I was young ...': Bernstein, *Hitler's Uranium Club*, p. 322.

349 'History will record that the Americans and the English made a bomb ...': Bernstein, *Hitler's Uranium Club*, p. 138.

350 'At the beginning of the war a group of research workers ...' and subsequent quotation: Bernstein, *Hitler's Uranium Club*, pp. 147–8.

351 'I should like to consider the U-235 bomb ...': Bernstein, *Hitler's Uranium Club*, p. 169.

352 'They won't let us go back to Germany ...' and subsequent quotation: Bernstein, *Hitler's Uranium Club*, p. 136.

355 'But my speech would be grossly incomplete ...': Bernstein, *Hitler's Uranium Club*, p. 288.

355 'Detained since more than half a year ...': Bernstein, *Hitler's Uranium Club*, p. 300.

356 'The complete suffering of war ...': Cassidy, p. 523.

CHAPTER 18: ДОГНАТЬ И ПЕРЕГНАТЬ!

359 'To avoid the creation of conflicts and misunderstanding ...': Holloway, p. 131.

360 'Hiroshima has shaken the whole world ...': Holloway, p. 132.

360 'I was so stunned that my legs practically gave way ...': Sakharov, p. 92.

360 'The construction of atomic energy facilities ...': Kramer, p. 268.

361 'Beria was harsh and rude to his subordinates ...': Sudoplatov, p. 205.

361 'Beria understood the necessary scope and dynamics ...': Khariton and Smirnov, pp. 20–31.

362 'If a child doesn't cry ...': Holloway, p. 132.

362 'Take measures to organise acquisition ...': Knight, p. 30.

365 'Had been in the hands of the police.': Albright and Kunstel, p. 152.

365 'I have gone this far ...': Rhodes, *Dark Sun*, p. 187.

368 'Those relations may be perhaps irretrievably embittered ...': quotation from Henry L. Stimson, Memorandum on the Effects of Atomic Bomb, 11 September 1945. A copy of this memorandum can be viewed at www.nuclearfiles.org

369 'You don't know southerners ...': Holloway, p. 156.

370 'Given the tension between the Soviet Union and the United States ...': Khariton and Smirnov, pp. 20–31.

371 'The main deficiencies of our present approach ...': Kapitza to Stalin, 25 November 1945, quoted in Kojevnikov, p. 143.

372 'The danger exists that scientific discoveries ...': Holloway, p. 142.

372 'Niels BOHR is famous as a progressive-minded scientist ...': Beria's memorandum to Stalin, dated 28 November 1945, is translated and reproduced in Smirnov, pp. 50–1.

373 'Capable professor of Moscow University ...': Smirnov, p. 56.

373 'I am sure there is no real method of protection ...': Smirnov, p. 59.

374 'Viewing the future development of the work ...': quotation from notes on the discussion between I.V. Kurchatov, lead scientist for the Soviet nuclear effort, and Stalin, 25 January 1946. See the virtual archive at www.wilsoncenter.org

374 'To catch up and to surpass.': Kojevnikov, p. 144.
375 'I have no doubt that if we give our scientists ...': Stalin, election speech, 9 February 1946. A copy of this speech is available at www.coldwarfiles.org

CHAPTER 19: IRON CURTAIN
377 'When I come to think of it ...': Gouzenko, p. 210.
377 'Candidly, everything about this democratic living seemed good ...': Gouzenko, p. 231.
378 'The unbelievable supplies of food ...': Knight, p. 21.
378 'We won't go back, Anna ...' and subsequent quotations: Gouzenko, p. 252.
379 'It's war. It's war. It's Russia.': Knight, p. 33.
379 'It was like a bomb on top of everything ...': Knight, p. 36.
379 'My own feeling is that the individual ...': Knight, p. 37.
380 'Nobody wants to say anything but nice things ...': Gouzenko, p. 312.
380 'Oh, I can't tell you for sure ...': Gouzenko, p. 313.
380 'We can't touch him,': Knight, p. 34.
382 'We have worked out the conditions of a meeting ...': Gouzenko, p. 279.
383 'This clandestine procedure ...': quotation from Nunn May, signed statement taken by Lt. Col. L.J. Burt and Major R.W. Spooner, Intelligence Corps, 15 February 1946. A facsimile copy of this statement is available online in the Security Service file KV 2/2226, UK National Archives, www.nationalarchives.gov.uk
384 'As weakness and the effect of this ...': Knight, p. 78.
384 'The first thing is to define the national problem ...' and subsequent quotations: Rhodes, *Dark Sun*, p. 205.
385 'We recognize that the application of recent scientific discoveries ...': quotation from the declaration on the atomic bomb by President Truman and Prime Ministers Attlee and King. See www.nuclearfiles.org
386 'We have to keep in mind ...': Smirnov, p. 59.
387 'Here's to science and American scientists ...': Holloway, p. 158.
387 'The Commission shall make specific proposals ...': quotation from the Soviet–Anglo-American communiqué on the interim meeting, dated 27 December 1945. See the Avalon Project at Yale Law School, www.yale.edu/lawweb/avalon
388 'He is worth living a lifetime just to know ...': Conant, p. 355.
388 'Everybody genuflected ...': Bird and Sherwin, p. 340.
389 'The device detonated about half a mile in the air ...': Morrison, p. 3.
390 'Unless Russia is faced with an iron fist ...': quotation from draft letter, Harry S. Truman to Secretary of State James Byrnes, 5 January 1946. A facsimile of this letter can be viewed at www.trumanlibrary.org
391 '[Soviet power is] impervious to logic of reason ...': quotation from George Kennan's 'Long Telegram' (Moscow to Washington), 22 February 1946, National Security Archive, www.gwu.edu/~nsarchiv. A facsimile of the telegram can be viewed at www.trumanlibrary.org
391 'From Stettin in the Baltic to Trieste in the Adriatic ...' and subsequent quotations: from Winston Churchill's 'Sinews of Peace' speech, 5 March 1946. This, and many other Churchill speeches, can be viewed at www.churchill.org

393 'Everyone is admiring the Soviet Union ...': Knight, p. 100.
394 'If it means getting any of my late colleagues ...': quotation from Nunn May, signed statement taken by Lt. Col. L.J. Burt and Major R.W. Spooner, Intelligence Corps, 15 February 1946.
394 'The whole affair was extremely painful to me ...': quotation from Nunn May, signed statement taken by Lt. Col. L.J. Burt and Major R.W. Spooner, Intelligence Corps, 20 February 1946. A facsimile copy of this statement is available online in the Security Service file KV 2/2226, UK National Archives, www.nationalarchives.gov.uk
395 'He was just a nice quiet bachelor ...': Moss, p. 114.

CHAPTER 20: CROSSROADS

397 'And as for any post-war problems ...': Rhodes, *The Making of the Atomic Bomb*, p. 530.
398 'Any interest in these industrial and commercial aspects ...': quotation from Quebec agreement, 19 August 1943. See www.atomicarchive.com
398 'Full collaboration between the United States and the British Government ...': Gowing, *Britain and Atomic Energy*, p. 447.
400 'We desire that there should be full and effective co-operation ...': Gowing, *Independence and Deterrence*, p. 76.
400 'The cohesive forces which held men of diverse opinions ...': Brown, p. 311.
401 'Considered it inadvisable ...': Groves, p. 406.
401 'The dissemination of related technical information ...': quotation from the draft McMahon Bill, available at www.rosenbergtrial.org
402 'We need a man who is young, vigorous, not vain ...': Bird and Sherwin, p. 343.
402 'Was the day I gave up hope': Rhodes, *Dark Sun*, p. 240.
403 'We are here to make a choice between the quick and the dead ...' and subsequent quotations: Bernard Baruch, the Baruch plan, presented to the UN Atomic Energy Commission, 14 June 1946. The text can be viewed at www.atomicarchive.com
404 'Quite serious but a somewhat squalid case ...': quotation from the transcript of the shorthand notes taken during the trial of Alan Nunn May, Central Criminal Court, 1 May 1946. A facsimile copy of this transcript is available online in the Security Service file KV 2/2226, UK National Archives, www.nationalarchives.gov.uk
406 'It is likely that a super-bomb can be constructed and will work ...': Rhodes, *Dark Sun*, p. 255.
406 'I still thought it was very optimistic ...': Serber, p. 150.
409 'The damned Air Corps ...': Rhodes, *Dark Sun*, p. 261.
409 'Not so much.': Rhodes, *Dark Sun*, p. 262.
409 'Dressed in all the trappings ...': Graybar, p. 901.
410 'The United States can not hope to win ...': Graybar, p. 897.
410 'At a time when our plans ...': Bird and Sherwin, p. 350.
410 'Common blackmail ...': Rhodes, *Dark Sun*, p. 262.
410 'The term "restricted data" as used in this section ...': quotation from US Atomic Energy Act, 1 August 1946.
411 'The phrase "all data" included every suggestion ...': Morland, p. 1402.
411 'We've *got* to have this ...': DeGroot, p. 352.

412 'Given up all hope that the Russians …': Bird and Sherwin, p. 352.

413 'I have here three affidavits from three scientists …': Chevalier, p. 64.

413 'I had to report that conversation, you know …' and subsequent quotations: Chevalier, p. 70.

414 'A complicated cock and bull story': Bird and Sherwin, p. 359.

414 'Had a rash and is now immune': Bird and Sherwin, p. 365.

415 'I was shocked when I found out …': Rhodes, *Dark Sun*, p. 283.

416 'It is logical that the United States …': quotation from the 'Marshall Plan' speech at Harvard University, 5 June 1947. See www.oecd.org

416 'Further aid to Britain …': Rhodes, *Dark Sun*, p. 300.

CHAPTER 21: ARZAMAS-16

417 'We have to know ten times more than we are doing': Holloway, p. 197.

419 'Finally, after a long search …': Tsukerman and Azarkh, p. 134.

421 'Well, we have reached it': Rhodes, *Dark Sun*, p. 275.

421 'Atomic energy has now been subordinated …': Holloway, p. 182.

421 'Is that all?': Holloway, p. 182.

421 'In the first days of work of the uranium–graphite pile …': Goncharov and Ryabev, p. 89.

423 'Tall, gangling, reserved, obviously intelligent …': Lamphere and Shachtman, p. 84.

423 'Hans BETHE, Niels BOHR, Enrico FERMI …': Venona document images, 2 December 1944, www.nsa.gov/venona

425 'I think the best British heavyweight …': Feklisov, p. 186.

426 'I'm very happy to be with you again …': Feklisov, pp. 187–8.

426 'We had arrived in what was for us a new world …': Tsukerman and Azarkh, p. 49.

427 'It was not merely a regime, it was a way of life …': Rhodes, *Dark Sun*, p. 286.

428 'We worked without heed for ourselves …': Tsukerman and Azarkh, p. 65.

429 'Your No. 5356. Information on LIBERAL's wife …': Venona document images, 27 November 1944, www.nsa.gov/venona

430 'In his shy way he explained …': Lamphere and Shachtman, p. 88.

432 'We still have enough arrogant neighbours.': Holloway, p. 186.

436 'One of the most ruthless efforts …': Rhodes, *Dark Sun*, p. 323.

436 'I don't think we ought to use this thing …': Rhodes, *Dark Sun*, p. 327.

CHAPTER 22: JOE-1

440 'The military services didn't own a single one …': Kohn and Harahan, p. 83.

441 'Darkest night in American military aviation history': Rhodes, *Dark Sun*, p. 341.

443 'Our task would be to investigate …': Sakharov, p. 94.

443 'Possessed by a true war psychology' and 'I understood, of course, the terrifying, inhuman nature …': Sakharov, p. 97.

443 'I envy Andrei Sakharov …': Tsukerman and Azarkh, p. 150.

445 'It's necessary for the project …' and subsequent quotations: Sakharov, p. 105.

445 'To increase its capability to such an extent …': Rosenberg, p. 70.

446 'Incidentally, everybody bemoans the fact …': Rhodes, *Dark Sun*, p. 21.

446 'We practically mapped the place up there …', and subsequent quotations: Kohn and Harahan, p. 86.

447 'I then realized that the combination of the three ideas …': Fuchs' confession, War Office, 27 January 1950, reproduced in Moss, pp. 239–48.

448 'I'd like to help the Soviet Union …' and subsequent quotations: Feklisov, pp. 198–9.

449 'They're supposed to be serious people …': Tsukerman and Azarkh, p. 75.

450 'Nothing will come of it, Igor': Rhodes, *Dark Sun*, p. 366.

450 'An explosion. A bright flash of light …': Tsukerman and Azarkh, p. 77.

451 'We hereby report to you, Comrade Stalin …': Goncharov and Ryabev, p. 91.

452 '[1 group unrecovered] received from REST …': Venona document images, 15 June 1944, www.nsa.gov/venona

452 'On ARNO's last visit to CHARLES' sister …': Venona document images, 16 November 1944, www.nsa.gov/venona

453 'CHARL'Z's information 2/57 on the atomic bomb …': Venona document images, 10 April 1945, www.nsa.gov/venona

454 'We have evidence that within recent weeks …': quotation from Truman statement announcing the Soviet atomic bomb, 23 September 1949, www.atomicarchive.com

454 'Keep your shirt on': Goodchild, *Edward Teller*, p. 141.

454 'It seems to me that the time has now come …': Rhodes, *Dark Sun*, p. 381.

455 'Not very different from what it was …': Bird and Sherwin, p. 419.

456 'If super bombs will work at all …': majority annex, General Advisory Committee's report on building the H-bomb, 30 October 1949, www.atomicarchive.com

456 'Necessarily such a weapon goes far beyond …': minority annex, General Advisory Committee's report on building the H-bomb, 30 October 1949, www.atomicarchive.com

456 'In sum, I believe that the President …': quotation from Lewis Strauss, letter to Truman, 25 November 1949, www.atomicarchive.com

457 'It would be foolhardy altruism for the United States …': Rhodes, *Dark Sun*, p. 406.

457 'Certainly not': Bird and Sherwin, p. 427.

457 'Can the Russians do it?' and subsequent quotations: Bird and Sherwin, p. 428.

457 'It is part of my responsibility as Commander in Chief …': statement by the President on the H-bomb, 31 January 1950, www.atomicarchive.com

458 'Complete with dishevelled appearance …': Lamphere and Shachtman, p. 140.

459 'Were you not in touch with a Soviet official …': Moss, p. 167.

459 'What do you want to know?' and 'I started in 1942 …': Moss, p. 173.

460 'The Soviet government …': Sakharov, p. 99.

EPILOGUE – MUTUAL ASSURED DESTRUCTION

461 'Despite the vision …': J. Robert Oppenheimer, 'Physics in the Contemporary World', second Arthur Dehon Little Memorial Lecture at the Massachusetts Institute of Technology, 25 November 1947. See Oppenheimer, *The Open Mind*, p. 88.

462 'With the discovery of fission …': Rhodes, *The Making of the Atomic Bomb*, p. 751. Rhodes quotes from the galley proofs of C.P. Snow's *The Physicists*, but goes on to observe that the quotation does not appear in the published work.

463 'Our greatest fear finally materialised …': Jean Baggott (now aged 72), communication to the author.

463 'I beseech you, in the bowels of Christ …': Bronowski, p. 374.

464 'It is said that science will dehumanise people ...': Bronowski, p. 374.

464 'It is the tragedy of mankind': Bronowski, p. 370.

465 'May have to sit for a while': Peat, p. 92.

465 'Oh my God, all is lost ...': Peat, p. 92.

466 'I would say definitely ...': Peat, p. 95.

466 'Thorough American.': Peat, p. 95.

466 'US atom scientist's brother ...': *Times Herald*, 12 July 1947.

466 'I will answer, if asked ...': Bird and Sherwin, p. 396.

466 'I think we all have been tremendously impressed ...': Goodchild, *J. Robert Oppenheimer*, p. 192.

467 'I have here in my hand a list of 205 ...': Griffith, p. 49.

467 'Senator McCarthy's crusade ...': Lamphere and Shachtman, p. 142.

468 'I am the man ...': Lamphere and Shachtman, p. 157.

469 'That is where we are just now ...': Rhodes, *Dark Sun*, p. 461.

470 'Simple, great and stupid!': Goodchild, *Edward Teller*, p. 177.

470 'Technically sweet': Rhodes, *Dark Sun*, p. 476.

472 'He congratulates everyone who helped ...': Sakharov, p. 174.

472 'The trouble then is just this ...': J. Robert Oppenheimer, 'Atomic Weapons and American Policy', lecture delivered before the Council on Foreign Relations, New York, 17 February 1953. See Oppenheimer, *The Open Mind*, p. 68.

473 'For my part, I experienced a range of contradictory sentiments ...': Sakharov, p. 193.

475 'Own personal initiative and responsibility': William L. Borden, letter to FBI Director J. Edgar Hoover, 7 November 1953, quoted in full by Borden during his testimony at Oppenheimer's security hearing. See Polenberg, p. 307.

475 'Based upon years of study, of the available classified evidence ...' and subsequent quote: William L. Borden, letter to FBI Director J. Edgar Hoover, 7 November 1953, quoted in full by Borden during his testimony at Oppenheimer's security hearing. See Polenberg, pp. 305–06.

477 'Now let us get back to your interview with Colonel Pash ...' and subsequent quotations: Polenberg, pp. 67–8.

477 'For your information, I might say we have a record of your voice' and 'not one lie ... but a whole fabrication and tissue of lies': Bird and Sherwin, pp. 507 and 509.

477 'You spent the night with her, didn't you?' and subsequent quotations: Polenberg, p. 74.

478 'I would not clear Dr Oppenheimer today ...': Polenberg, p. 81.

478 'Here is a man who is being pilloried ...': Polenberg, p. 204.

478 'I think this board or no board should ever sit on a question ...': Polenberg, p. 207.

478 'Now a question which is a corollary to that ...' and subsequent quotations: Polenberg, p. 253.

479 'I believe ... that Dr Oppenheimer's character ...': Polenberg, p. 264.

479 'I'm sorry' and 'After what you've just said, I don't know what you mean': Bird and Sherwin, p. 534.

479 'Conduct ... sufficiently disturbing as to raise a doubt.': Polenberg, p. 362.

480 'Thinly differentiated from high treason': Moss, p. 202.

480 'Worse than murder': Rhodes, *Dark Sun*, p. 480.
480 'I felt, not satisfaction, but defeat ...': Lamphere and Shachtman, p. 278.
481 'BEK visited Theodore Hall ...': Venona document images, 12 November 1944, www.nsa.gov/venona
483 'I would have done it ...': Albright and Kunstel, p. 240.
483 'But in 1995, when the whole story came out ...': Joan Hall, 'A Memoir of Ted Hall', History Happens website: http://web.archive.org/web/20070607215311/ http://www.historyhappens.net/archival/manproject/joanhalldoc/joanhall.htm
484 'On 14 June a meeting took place ...': Venona document images, 21 June 1943, www.nsa.gov/venona
484 'I later learned that, in the course of tracing the source of leaks ...': Peierls, p. 225.
485 'British security service sources': Farrell, *The Spectator*, 29 May 1999.
486 'It seems paradoxical that the German nuclear physicists ...': Jungk, p. 105.
486 'Science under fascism ...': Goudsmit, p. xxxv.
487 'Apologia for being willing to work ...': Mark Walker, *Nazi Science: Myth, Truth and the German Atomic Bomb*, p. 244.
487 'Why were myths and legends of active resistance ...': Mark Walker, *Nazi Science: Myth, Truth and the German Atomic Bomb*, p. 268.
487 'Eine Legende': Gerald Holton, 'What is *Copenhagen* Trying to Tell Us?', in Dörries, p. 52.
488 'The administration would come back and say ...': Kohn and Harahan, p. 93.

BIBLIOGRAPHY

Albright, Joseph and Kunstel, Marcia, *Bombshell: The Secret Story of America's Unknown Spy Conspiracy*, Random House, New York, 1997

Alvarez, Luis W., *Adventures of a Physicist*, Basic Books, New York, 1987

Arnold, Lorna and Smith, Mark, *Britain, Australia and the Bomb: Nuclear Tests and Their Aftermath*, second edition, Palgrave Macmillan, Basingstoke, 2006 (first published 1987)

Arnold-Forster, Mark, *The World at War*, Fontana/Collins, Glasgow, 1976

Beevor, Antony, *Stalingrad*, Penguin Books, London, 1999

Bernstein, Jeremy, *Quantum Profiles,* Princeton University Press, 1991

Bernstein, Jeremy, *Hitler's Uranium Club: The Secret Recordings at Farm Hall*, second edition, Copernicus Books, New York, 2001

Bernstein, Jeremy, *Nuclear Weapons: What You Need to Know*, Cambridge University Press, 2008

Bethe, Hans A., *The Road From Los Alamos*, American Institute of Physics Press, New York, 1991

Bird, Kai and Sherwin, Martin J., *American Prometheus: the Triumph and Tragedy of J. Robert Oppenheimer*, Atlantic Books, London, 2008

Borowski, Harry R., 'Air Force Atomic Capability from V-J Day to the Berlin Blockade – Potential or Real?', *Military Affairs*, 44, 3, 1980, pp. 105–10

Bronowski, Jacob, *The Ascent of Man*, Book Club Associates, London, 1976

Brown, Andrew, *The Neutron and the Bomb: A Biography of Sir James Chadwick*, Oxford University Press, 1997

Buck, Alice L., *A History of the Atomic Energy Commission*, US Department of Energy, Washington, DC, July 1983

Cairncross, John, *The Enigma Spy*, Century, London, 1997

Cassidy, David C., *Uncertainty: The Life and Science of Werner Heisenberg.* W.H. Freeman, New York, 1992

Chevalier, Haakon, *Oppenheimer: the Story of a Friendship*, Andre Deutsch, London, 1966

Compton, Arthur H., *Atomic Quest*, Oxford University Press, London, 1956

Conant, Jenet, *109 East Palace: Robert Oppenheimer and the Secret City of Los Alamos*, Simon & Schuster, New York, 2005

Cornwell, John, *Hitler's Scientists: Science, War and the Devil's Pact*, Penguin Books, London, 2004

Dawidoff, Nicholas, *The Catcher Was a Spy: The Mysterious Life of Moe Berg*, Vintage Books, New York, 1995

Dawson, Raymond and Rosecrance, Richard, 'Theory and Reality in the Anglo-American Alliance', *World Politics*, 19, 1, 1966, pp. 21–51

DeGroot, Gerard, *The Bomb: A History of Hell on Earth*, Pimlico, London, 2005

Dörries, Matthias (ed.), *Michael Frayn's Copenhagen in Debate*, University of California, 2005

Enz, Charles P., *No Time to be Brief: a Scientific Biography of Wolfgang Pauli*, Oxford University Press, 2002

Farrell, Nicholas, 'Sir Rudolf and Lady Spies', *The Spectator*, 29 May 1999

Feklisov, Alexander, *The Man Behind the Rosenbergs*, Enigma Books, New York, 2004

Fermi, Laura, *Atoms in the Family: My Life with Enrico Fermi*, University of Chicago Press, 1954

Feynman, Richard P., *'Surely You're Joking, Mr Feynman!'*, Unwin, London, 1985

Frank, Charles (Introduction), *Operation Epsilon: The Farm Hall Transcripts*, Institute of Physics Publishing, Bristol, 1993

French, A.P. and Kennedy, P.J. (eds), *Niels Bohr: a Centenary Volume*, Harvard University Press, Cambridge, MA, 1985

Frisch, Otto, *What Little I Remember*, Cambridge University Press, 1979

Gallagher, Thomas, *The Telemark Raid*, Corgi, London, 1976

Gleick, James, *Genius: Richard Feynman and Modern Physics*. Little, Brown & Co., London, 1992

Goodchild, Peter, *J. Robert Oppenheimer: 'Shatterer of Worlds'*, BBC, London, 1980

Goodchild, Peter, *Edward Teller: The Real Dr Strangelove*, Weidenfeld & Nicholson, London, 2004

Goncharov, G.A., 'American and Soviet H-bomb development programmes: historical background', *Physics-Uspekhi*, 39, 10, 1996, pp. 1033–44

Goncharov, G.A. and Ryabev, L.D., 'The development of the first Soviet atomic bomb', *Physics-Uspekhi*, 44, 1, 2001, pp. 71–93

Goudsmit, Samuel A., *Alsos*, American Institute of Physics Press, New York, 1996 (first published 1947)

Gouzenko, Igor, *This Was My Choice*, Eyre & Spottiswoode, London, 1948

Gowing, Margaret, *Britain and Atomic Energy*, Macmillan & Co. Ltd., London, 1964

Gowing, Margaret (assisted by Lorna Arnold), *Independence and Deterrence: Britain and Atomic Energy, 1945–1952*, Volume 1: *Policy Making*, Macmillan, London, 1974

Gowing, Margaret, 'James Chadwick and the Atomic Bomb', *Notes and Records of the Royal Society of London*, 47, 1, 1993, pp. 79–92

Graybar, Lloyd J., 'The 1946 Atomic Bomb Tests: Atomic Diplomacy or Bureaucratic Infighting?', *The Journal of American History*, 72, 4, 1986, pp. 888–907

Gribbin, John, *Q is for Quantum: Particle Physics from A to Z*, Weidenfeld & Nicholson, London, 1998

Griffith, Robert, *The Politics of Fear: Joseph R. McCarthy and the Senate*, University of Massachusetts Press, 1970

Groves, Leslie R., *Now it Can be Told: The Story of the Manhattan Project*, Da Capo Press, New York, 1983 (first published 1962)

Hastings, Max, *Overlord: D-Day and the Battle for Normandy*, Pan Macmillan, London, 1999 (first published 1984)

Haukelid, Knut, *Skis Against the Atom*, North American Heritage Press, 1989 (first published 1954)

Haynes, John Earl and Klehr, Harvey, *Venona: Decoding Soviet Espionage in America*, Yale University Press, 2000

Heilbron, J.L., *The Dilemmas of an Upright Man: Max Planck and the Fortunes of German Science*, Harvard University Press, 1996

Heisenberg, Elisabeth, *Inner Exile: Recollections of a Life with Werner Heisenberg*, Birkhäuser, Boston, 1984

Heisenberg, Werner, *Physics and Beyond: Memories of a Life in Science*, George Allen & Unwin, London, 1971

Herken, Gregg, '"A Most Deadly Illusion": The Atomic Secret and American Nuclear Weapons Policy, 1945–1950', *The Pacific Historical Review*, 49, 1, 1980, pp. 51–76

Herken, Gregg, *Brotherhood of the Bomb*, Henry Holt & Company, New York, 2002

Hewlett, Richard G. and Anderson, Oscar E., Jr, *A History of the United States Atomic Energy Commission*, Volume 1: *The New World, 1939/1946*, Pennsylvania State University Press, 1962

Hida, Shuntaro, 'Under the Mushroom-shaped Cloud in Hiroshima: a Memoir', 2006, available online at www.wcpeace.org

Holloway, David, *Stalin and the Bomb*, Yale University Press, 1994

Howard, Greville (with Peter Browne), 'My Elizabethan Brother, the Earl of Suffolk', *Reader's Digest*, November 1969

Irving, David, *The Virus House: Germany's Atomic Research and Allied Counter-measures*, Focal Point e-book, 2002 (first published in 1968)

Jungk, Robert, *Brighter than a Thousand Suns*, Harcourt Brace & Company, New York, 1958

Karlsch, Rainer and Walker, Mark, 'New Light on Hitler's Bomb', *Physics World*, June 2005, pp. 15–18.

Khariton, Yuli and Smirnov, Yuri, 'The Khariton Version', *Bulletin of the Atomic Scientists*, May 1993, pp. 20–31

Knight, Amy, *How the Cold War Began: The Igor Gouzenko Affair and the Hunt for Soviet Spies*, Carroll & Graf, New York, 2006

Kohn, Richard H. and Harahan, Joseph P., 'US Strategic Air Power 1948–1962: Excerpts from an Interview with Generals Curtis E. LeMay, Leon W. Johnson, David A. Burchinal, and Jack J. Catton', *International Security*, 12, 4, 1988, pp. 78–95

Kojevnikov, Alexei B., *Stalin's Great Science: The Times and Adventures of Soviet Physicists*, Imperial College Press, London, 2004

Kragh, Helge S., *Dirac: A Scientific Biography*, Cambridge University Press, 1990

Kramer, Mark, 'Documenting the Early Soviet Nuclear Weapons Program', *Cold War International History Project Bulletin*, 6/7, Winter 1995, pp. 265–70

Kramish, Arnold, *The Griffin: The Greatest Untold Espionage Story of World War II*, Houghton Mifflin, Boston, 1986

Lamphere, Robert J. and Schachtman, Tom, *The FBI–KGB War*, Berkeley Publishing Group, New York, 1987

Lanouette, William (with Bela Szilard), *Genius in the Shadows*, University of Chicago Press, 1992

Marston, Daniel (ed.), *The Pacific War Companion: From Pearl Harbor to Hiroshima*, Osprey Publishing, Oxford, 2007

Mears, Ray, *The Real Heroes of Telemark*, Hodder & Stoughton, London, 2003

Mehra, Jagdish, *The Beat of a Different Drum: The Life and Science of Richard Feynman*, Oxford University Press, 1994

Morland, Howard, 'Born Secret', *Cardozo Law Review*, 26, 4, 2005, pp. 1401–08

Moorehead, Alan, *The Traitors: The Double Life of Fuchs, Pontecorvo and May*, Hamish Hamilton, London, 1952

Morrison, Philip, 'If the Bomb Gets Out of Hand', in *One World or None*, McGraw-Hill, New York, 1946

Moss, Norman, *Klaus Fuchs: The Man Who Stole the Atom Bomb*, Grafton Books, London, 1989

Oleynikov, Pavel V., 'German Scientists in the Soviet Atomic Project', *The Nonproliferation Review*, 7, 2, 2000, pp. 1–30

Oppenheimer, J. Robert, *The Open Mind*, Simon & Schuster, New York, 1955

Pais, Abraham, *Subtle is the Lord: The Science and the Life of Albert Einstein*, Oxford University Press, 1982

Pais, Abraham, *Niels Bohr's Times, in Physics, Philosophy and Polity*, Clarendon Press, Oxford, 1991

Pais, Abraham, *J. Robert Oppenheimer: A Life*, Oxford University Press, 2006

Pash, Colonel Boris T., *The Alsos Mission*, Award Books, New York, 1969

Peat, F. David, *Infinite Potential: The Life and Times of David Bohm*, Addison-Wesley, Reading, MA, 1997

Peierls, Rudolf, *Bird of Passage: Recollections of a Physicist*, Princeton University Press, 1985

Polenberg, Richard (ed.), *In the Matter of J. Robert Oppenheimer: The Security Clearance Hearing*, Cornell University Press, Ithaca, NY, 2002

Poundstone, William, *Prisoner's Dilemma: John von Neumann, Game Theory and the Puzzle of the Bomb*, Random House, New York, 1992

Powers, Thomas, *Heisenberg's War: The Secret History of the German Bomb*, Da Capo Press, New York, 2000

Preston, Diana, *Before the Fall-out: From Marie Curie to Hiroshima*, Corgi Books, London, 2006

Prins, Gwyn (ed.), *Defended to Death*, Penguin Books, London, 1983

Rhodes, Richard, *The Making of the Atomic Bomb*, Simon & Schuster, New York, 1986

Rhodes, Richard, *Dark Sun: The Making of the Hydrogen Bomb*, Simon & Schuster, New York, 1995

Rose, Paul Lawrence, *Heisenberg and the Nazi Atomic Bomb Project*, University of California Press, 1998

Rosenberg, David Alan, 'American Atomic Strategy and the Hydrogen Bomb Decision', *The Journal of American History*, 66, 1, 1979, pp. 62–87

Sakharov, Andrei, *Memoirs*, Alfred A. Knopf, New York, 1990

Segrè, Emilio, *Enrico Fermi: Physicist*, University of Chicago Press, 1970

Seaborg, Glenn T. (with Eric Seaborg), *Adventures in the Atomic Age*, Farrar, Straus and Giroux, New York, 2001

Serber, Robert (with Robert P. Crease), *Peace and War: Reminiscences of a Life on the Frontiers of Science*, Columbia University Press, New York, 1998

Sherwin, Martin J., 'The Atomic Bomb and the Origins of the Cold War: US Atomic-Energy Policy and Diplomacy, 1941–45', *The American Historical Review*, 78, 4, 1973, pp. 945–68

Sherwin, Martin J., 'Hiroshima as Politics and History', *The Journal of American History*, 82, 3, 1995, pp. 1085–93

Sime, Ruth Lewin, *Lise Meitner: A Life in Physics*, University of California Press, 1997

Smirnov, Yuri, 'The KGB Mission to Niels Bohr: Its Real "Success"', *Cold War International History Project Bulletin*, 4, Fall 1994, pp. 50–1, 54–9

Smith, P.D., *Doomsday Men: The Real Dr Strangelove and the Dream of a Superweapon*, Allen Lane, London, 2007

Snow, C.P., *The Physicists*, Macmillan, London, 1981

Speer, Albert, *Inside the Third Reich*, Phoenix, London, 1995 (first published 1970)

Stalin, Joseph V., 'Speech Delivered by J.V. Stalin at a Meeting of Voters of the Stalin Electoral District, Moscow', 9 February 1946. From the pamphlet collection, *J. Stalin, Speeches Delivered at Meetings of Voters of the Stalin Electoral District*, Moscow, Foreign Languages Publishing House, Moscow, 1950

Sudoplatov, Pavel and Anatoly (with Jerrold L. and Leona P. Schecter), *Special Tasks*, updated edition, Little, Brown & Co., New York, 1995

Szanton, Andrew, *The Recollections of Eugene P. Wigner*, Basic Books, Cambridge, MA, 2003

Tsukerman, Veniamin and Azarkh, Zinaida, *Arzamas-16*, Bramcote Press, Nottingham, 1999

Walker, Mark, *German National Socialism and the Quest for Nuclear Power, 1939–1949*, Cambridge University Press, 1989

Walker, Mark, *Nazi Science: Myth, Truth and the German Atomic Bomb*, Perseus Publishing, Cambridge, MA, 1995

Walker, Mark, 'Otto Hahn: Responsibility and Repression', *Physics in Perspective*, 8, 2006, pp. 116–63

Walker, Mark, 'Nuclear Weapons and Reactor Research at the Kaiser Wilhelm Institute for Physics', in Heim, Susanne, Sachse, Carola and Walker, Mark (eds), *The Kaiser Wilhelm Society under National Socialism*, Cambridge University Press, in press

Walker, Stephen, *Shockwave: The Countdown to Hiroshima*, John Murray, London, 2006

Waltham, Chris, 'An Early History of Heavy Water', arXiv: physics/0206076v1, June 2002 (available from http://arxiv.org)

West, Nigel, *Mortal Crimes*, Enigma Books, New York, 2004

Wheeler, John Archibald (with Kenneth Ford), *Geons, Black Holes and Quantum Foam*, W.W. Norton & Company, New York, 2000

INDEX